PRINCIPLES AND APPLICATIONS OF SOIL GEOGRAPHY

Edited by

E. M. Bridges and D. A. Davidson

Longman London and New York

Longman Group Limited
Longman House
Burnt Mill, Harlow, Essex, UK

Published in the United States of America
by Longman Inc., New York

First published 1982

British Library Cataloguing in Publication Data

Principles and applications of soil geography.
1. Soils
I. Bridges, Edwin Michael
II. Davidson, Donald Allen
631.4 S591 80–41509

ISBN 0-582-30014-2

Printed in Singapore by Kyodo Shing Loong Printing Industries Pte. Ltd.

CONTENTS

PREFACE

A textbook, published recently in the United States of America, contains the comment that 'these are exciting times in pedology'; the editors of this book share this enthusiastic viewpoint. A great deal of activity is taking place in the scientific study of the soil and as a result, many papers and books are being written upon the subject.

As university teachers, the editors appreciate the need for a university or college textbook for undergraduates which deals with the underlying principles of soil geography and which introduces the practical applications of the subject. This book attempts an answer to the question: why an interest in soils? It explains man's involvement with soils and stresses his complete dependence on them for food supplies. The text describes how soil maps are made, how information from soil surveys is stored and manipulated and explains how such data may be used.

The authors of the chapters of this book can claim collectively about 200 years' experience in soil studies and can modestly be said to have a considerable expertise in the subject. The editors were conscious of the need to provide a coherent treatment of the subject and they assume the responsibility for the overall structure of the book, but the detailed form of separate chapters has been modified by contributors in conjunction with the editors to give the necessary coherence. The editors are extremely grateful to the contributors for their forebearance in the lengthy process of arriving at the present format and it is our hope that the result is a satisfactory account of current principles and applications of soil geography.

In the preparation of the separate chapters, several people have kindly read and commented upon draft manuscript material for individual contributors. The editors are pleased to acknowledge with gratitude the valuable assistance of B. W. Avery, E. Grant, R. Hartnup, D. Mackney, A. McBratney, J. M. Ragg, A. P. A. Vink and P. S. Wright. The editors also wish to acknowledge the assistance given by their wives, and particularly to Mrs Glenys Bridges, for compilation of the index. Sources of illustrative material are acknowledged elsewhere but initial drafting has been done by the cartographic staff at the University College of Swansea and the final diagrams were prepared by the publisher's staff. In conclusion, the editors wish to record their thanks to the staff of Longman for their help and encouragement during the preparation of this book.

E. M. Bridges University College of Swansea
D. A. Davidson University of Strathclyde
June 1980

CONTRIBUTORS

Dr D. A. Davidson, Senior Lecturer in Geography at the University of Strathclyde, holds the B.Sc. degree from the University of Aberdeen and the Ph.D. degree of the University of Sheffield. He was formerly a lecturer at St David's University College, Lampeter and has also lectured at Carleton University, Ottawa. He has studied relationships between soil and archaeology in Scotland and Greece and is author of *Soils and Land Use Planning* (Longman) and *Science for Physical Geographers* (Arnold). He is a member of the British Society of Soil Science, the Institute of British Geographers and the Institution of Environmental Science.

Dr E. M. Bridges, Senior Lecturer in the Department of Geography at the University College of Swansea, holds B.Sc. and M.Sc. degrees from the University of Sheffield and the Ph.D. degree of the University of Wales. Formerly a member of the Soil Survey of England and Wales, he has worked on surveys in the United Kingdom and Bahrain and has taught in the Universities of New England, Australia, the University of Khartoum and the University of the West Indies, Trinidad. He is author of *The Soils and Land Use of the District North of Derby* (Harpenden), *World Soils* (Cambridge University Press) and is a member of the British Society of Soil Science, the Institute of British Geographers and is a Fellow of the Royal Geographical Society.

Mr B. Clayden, Principal Scientific Officer with the Soil Survey of England and Wales at its Headquarters at Rothamsted Experimental Station, holds the degree of B.Sc. from the University of Sheffield. He worked on soil surveys in Somerset, Devon and Iraq before serving as Regional Officer for Wales, and now has a national responsibility for soil classification and correlation. He is author of *Soils of the middle Teign valley district of Devon* and *Soils of the Exeter district*, and joint author of *Soils in Dyfed I* and *The classification of some British soils according to the comprehensive system of the United States*, all published by the Soil Survey. He is a member of the British Society of Soil Science and the Quaternary Research Association.

Dr C. C. Rudeforth, Principal Scientific Officer with the Soil Survey of England and Wales, holds the degrees of B.Sc., M.Sc. (with distinction) and Ph.D. of the University of London. He is the Regional Officer for Wales in the Soil Survey of England and Wales and has worked on soil surveys in the north of England and Iraq. He is the author of *Soils of North Cardiganshire* (Harpenden), and is joint author of *Hydrological Properties of Soils in the River Dee Catchment* (Harpenden) and *Soils, Land Classification and Land Use of West and Central Pembrokeshire* (Harpenden). He is a member of the British Society of Soil Science.

Dr R. J. Huggett, Lecturer in the Department of Geography in the University of Manchester, holds the degrees of B.Sc. and Ph.D. of the University of London. After holding a school teaching post for a short period in Hertfordshire, he took up his present position. He is the author of *Systems Analysis in Geography* (Oxford) and a joint author of *Modelling Geography: a mathematical approach*. Dr Huggett is a member of the Mervyn Peake Society and of the Institute of British Geographers.

Mr M. G. Jarvis, Principal Scientific Officer with the Soil Survey of England and Wales holds the degree of B.A. from the University of Oxford. He is the Regional Officer for the Soil Survey of England and Wales in the south-east of England and has a particular interest in the application of soil surveys. He is the author of *Soils of the Wantage and Abingdon district* (Harpenden), joint author of the *Soils of Berkshire* and joint editor of *Soil Survey Applications* (Harpenden). He is a member of the British Society of Soil Science.

Professor H. D. Foth, Professor of Soil Science at Michigan State University, holds the degrees of B.S. and M.S. from the University of Wisconsin and the Ph.D. from Iowa State University. He is the author of *Fundamentals of Soil Science* and *Soil Geography and Land Use*, both published by Wiley. Professor Foth has received numerous teaching awards including the Agronomic Education Award, the Oahus Award and the Ensminger Distinguished Teacher Award. He is a fellow of the American Association for Advancement of Science, member of the Soil Science Society of America, the American Society of Agronomy, a Teacher Fellow of the National Association of Colleges and Teachers of Agriculture and is a member of the Association of American Geographers.

INTRODUCTION

The aim of soil geography is to record and explain the development and distribution of soils on the earth's surface. It is a branch of learning which lies between soil science and geography and is of particular importance to both subjects. One view of geography is that it focusses attention on the interaction of man and the physical environment. Nowhere is this more obvious or so vital as it is with the man–soil relationship. Soils are a major resource of all countries, and as the basis of agriculture they play an important role in the sustenance of mankind. The biology, physics and chemistry of soils, all of which converge in a study of soil fertility, are the concern of soil science. These studies merge to form soil geography.

Mankind's relationship with the soil is long and complex, reaching back to prehistoric times. For primitive cultures, the soil was simply an insignificant part of the total environment, but with the development of agriculture came the necessity for man to seek fertile soils and the technology with which to manage them. Except for a few very favoured locations, fertility of soils proved to be transitory, and agricultural systems evolved which attempted to prolong or enhance the productive capacity of the soil. By the nineteenth century, man's understanding of soils in temperate regions of the world was such that a sustained production of crops could take place, but in tropical regions much still remains to be learnt. At the present time, with virtually all the best soils already under arable cultivation, the need for a knowledge of the distribution of soils and their properties has been brought sharply into focus as further increases in agricultural productivity are sought from finite soil resources.

The professional skills of a soil surveyor enable him to record both vertical and horizontal variation of soils and their characteristics on a two-dimensional map. Soil maps are produced at many different scales for a wide variety of purposes and different users' requirements. It is important that the limitations and possibilities of the different scales are fully appreciated by soil map users. The information contained in the many soil maps and reports which are available constitute the material of soil geography. The arrangement of this material in meaningful categories draws the soil geographer into problems of soil classification, data handling and conceptual modelling, all of which are discussed in Chapters 1–5 of this book.

The last three chapters move away from the basic principles of soil survey and soil geography to discuss some of the applied aspects of the subject. These are discussed conveniently in two chapters, respectively concerned with agricultural and non-agricultural uses of soil survey data. Land classification, which follows from a knowledge of the soils, has become an important tool in farm management and land-use planning generally. It enables the best economic use to be made of the available soil resources. A dispassionate assessment of the facts of soil geography can often provide information which can help resolve disputes between those who wish to extend urbanization and those who wish to retain rural uses of land.

On a global scale, soil geography has recently received a great impetus with the production of the FAO/UNESCO Soil Map of the World. Previously, soil maps of the world had been compiled on an empirical basis, but the vast increase in knowledge in recent years has enabled soil scientists to produce this new map based on pedological criteria. This has enabled a more accurate estimate to be made of the extent of different soils throughout the world. In turn, this allows a more realistic approach to the determination of total world agricultural productivity as there is an established link between soil morphology and crop yields. With millions of people in developing countries under-nourished and a continually increasing world population, every effort must be made to obtain the maximum productivity from the world's soils. This must happen without causing the degradation or destruction of the soil mantle which is so essential for life on earth.

Soil geography has been a traditional but minor part of the courses offered in most university departments of soil science or geography. With the growth in knowledge during the last few years the stature of soil geography has increased. It is now widely acknowledged that it has a contribution to make to the world outside the academic and research institutions which have nurtured its development. Soil studies are one means of integrating large parts of physical geography in a way which is of immediate relevance to mankind and in this way gives coherence to geography as a whole. Interpretations of soil survey data have expanded the need for soil–geographical studies from a rather narrow agricultural base to be of wider applicability as forecast by Bartelli *et al.* (1966) and Simonson (1974).

ACKNOWLEDGEMENTS

The editors and publishers wish to acknowledge with gratitude the following journals and individuals who have kindly allowed diagrams and tables which first appeared elsewhere to be reproduced within these pages:

American Society of Photogrammetry for our Fig. 4.7; Catena for our Fig. 5.11; C.S.I.R.O. for our Fig. 2.4; Department of Regional Economic Expansion, Toronto, Canada for our Fig. 7.1; Earth Surface Processes for our Fig. 5.12; Elsivier Scientific Publications for our Fig. 4.10, 4.16, 4.17, 5.6, 7.6, 7.7, 7.8, 7.9, 7.10; F.A.O. for our Figs. 2.3, 6.12, 6.13; Geographical Association for our Fig. 2.1; Geographical Journal for our Figs. 2.2, 6.1 and 6.11. H.M.S.O., MOD and Institute of Civil Engineers for our Fig. 7.3; The International Development Research Centre, Ottawa for our Fig. 4.8; The International Institute for Land Reclamation and Improvement for our Fig. 6.9; The Journal of Soil Science for our Figs. 1.3, 4.4, 4.9, 4.13, 4.14, 4.15, and 5.4; Dr. W. Junk for our Fig. 1.4; The Land Resources Division of the Department of Overseas Surveys for our Fig. 2.5; Michigan Department of State Highways for our Figs. 7.4 and 7.5; The Ministry of Agriculture for our Figs. 6.4 and 6.5; Dr. F. C. Peacock for our Fig. 5.8; D. Reidel Publishing Company for our Fig. 4.12; Rothamsted Experimental Station and the Soil Survey of England and Wales for our Figs. 2.6, 2.7, 4.3, 6.2, 6.3, 6.6, 6.7, 6.8, and 7.2; Dr. D. S. Sasser for our Fig. 5.9; The Soil Science Society of America for our Figs. 4.5 and 4.11; The State Agricultural University, Wageningen for our Fig. 6.10; Dr. A. Warren for our Fig. 5.5; The Williams Wilkins Company, Baltimore, for our Fig. 5.7; The United States Department of Agriculture for our Figs. 8.1, 8.2, 8.3, and 8.4.

Agriculture Canada for our Table 3.8 from Table 8, publication 1646 *The Canadian System of Soil Classification* by the Canada Soil Survey Committee, by permission of the Minister of Supply & Services, Canada; George Allen & Unwin (Publishers) Ltd for our Table 3.11 from *The Soils of Europe* 1953 by Kubien, published by Thomas Murby; the author, Kurt W. Bauer for extracts from 'The Use of Soils Data in Regional Planning' as published in *Geoderma 10*; Canada Land Inventory for 'Capability Classes' from the *Land Capability Classification for Forestry, Report No 4* by R. J. McCormack; Commission de Pedologie et Cartographie, Ecole National Superior for our Table 3.12; Food & Agriculture Organisation of the United Nations for our Tables 3.9, 8.1, 8.4 and quotations from *A Framework for Land Evaluation* Soils Bulletin No 34; Forestry Commission for our Tables 7.1 and 7.2 from *Forest Record 69* by D. G. Pyatt; *Geoderma* and the authors for our Table 5.1 by A. J. Conacher and J. B. Dalrymple 1977, 5.2 and 5.3 by D. H. Yaalon 1975 and 7.10 by G. J.W. Westerveld and J. A. van den Hurk; The Geographical Society of New South Wales and the author Dr. Conacher for our Table 5.4 from *Australian Geographer* 12(3) 1973; International Institute for Land Reclamation and Improvement for our Table 2.3 adopted from Vink 1963, *ILRI Publication No. 10*; Michigan Department of Transportation for our Table 7.8 from *Field Manual of Soil Engineering* 1970; Office of Science & Technology Policy for our Table 8.2 adapted from *The World Food Problem*; Oxford University Press for our Tables 3.13 by Avery 1973 and 4.4 after Rudeforth 1975 from *Journal of Soil Science* (c) Oxford University Press; Relim Technical Publishers for our Table 3.14 by Stace et al, 1968; Scripta Publishing Co for our Table 3.10 by Rozov & Ivanova from *Soviet Soil Science* 1967 No 2 pp 147–156. Reprinted by permission; *Soil Science Society of America Journal* for our Table 5.5 by Smeck and Runge from Volume 35, 1971; Soil Survey of England & Wales for our Tables 2.4 by J. K. Coulter, 6.2 and 6.3 by D. Mackney and 6.4 by L. J. Hooper from *Technical Monograph No 4* 1974 edited by D. Mackney, our Table 2.5 compiled from *Field Handbook* (Hodgson 1974), our Table 6.1 by J. S. Bibby and D. Mackney compiled from *Technical Monograph No 1* and our Table 7.3 by Farquheson et al from *Soil Survey Special Survey No 11* 1978; The Williams Wilkins Co. and the authors Thorp and Smith for our Tables 1.1 from *Soil Science 99* 1965 and 3.2 from *Soil Science 67* 1949.

We are unable to trace the copyright owner of our Table 6.5 from Veldkamp and would appreciate any information which would enable us to do so.

1

SOILS AND MAN IN THE PAST

D. A. Davidson

Ever since man has made land-use decisions, it has been to his advantage to assess the suitability of land for particular purposes. Indeed, his existence on earth is only possible through husbandry of plants and animals which in turn depend upon the physical resources. Modern agricultural systems of course have to operate within economic, social and political situations as well as environmental ones, but there is no doubt that the physical nature of land imposes varying constraints on agricultural development in different parts of the world. Such limitations become more apparent as population increases and as energy becomes progressively more expensive and difficult to obtain. The basic physical resources for agriculture are climate and soils; the former provides the necessary inputs of energy and moisture, whilst the latter supplies the medium in which crops grow in terms of physical support, moisture and nutrients. The studies of climate and soils are fundamental topics, as man is so dependent upon these resources.

Soil geography is interpreted as the study of the nature, formation and distribution of soils, and how soil characteristics and man's activities are interrelated. According to FitzPatrick (1971) 'pedology is the study of soils as naturally occurring phenomena, taking into account their composition, distribution and method of formation'. Thus pedology constitutes a substantial part of soil geography, but with the important addition that soil geography also examines the ecological relationship between man and soils. It is sometimes suggested that soils and man can be separated to allow the study of soils as naturally occurring phenomena. In practice this is extremely difficult since there are probably very few areas in the habitable world where man either in the past or present has not influenced soils. Thus, a very appropriate way to begin a book on soil geography is to examine the ways by which man through time has influenced soils. Such a focus underlines the symbiotic relationship between man and soils as well as stressing that many soil attributes are only explicable with reference to man's activities in the past. In order to examine how man has influenced soil conditions, an outline is given from prehistoric times until the present of man's effects on and use of soils.

1.1 PRE-NEOLITHIC COMMUNITIES AND SOILS

From our advanced technological stance, it is easy to forget the extent to which human prehistory is dominated by hunting-gathering societies of the Palaeolithic and Mesolithic periods. For this long period of the order of 3.5 million years, humans existed in nomadic or semi-nomadic bands in environments ranging from African savannas and open woodlands, to temperate forests and forest-tundra. In such societies, man evolved and adapted to the particular challenges and opportunities of his environment. At least in the early Palaeolithic, these small groups of widely dispersed people had virtually no effect on the natural environment (Butzer 1964). Little is known about the deliberate use of fire by man in Palaeolithic times though it is known that Pekin man tended fire in his caves. Man may have used fire to make clearings so that hunting was more successful. Fire also might have been employed to encourage the growth of certain food plants or to drive game. However, such suggestions must remain in the realm of speculation. Although the potential effect of fire must be recognized even in the Palaeolithic cultures (Stewart 1956), it can be suggested that Palaeolithic man had made at most only a highly localized impact on vegetation and thus soils through his use of fire.

At the economic level, the Mesolithic period was very similar to the preceding Palaeolithic; the distinguishing characteristic was the advanced development of a stone-working technology, expressed in the small highly worked flint implements called microliths. The gradual increase in population, combined with a better technology, led man to make the first clearly discernible impact upon the landscape. In Britain, researches by Dimbleby (1961, 1962), Simmons (1969a, b) and Smith (1970) indicate that Mesolithic man had a significant localized effect on the vegetation of such areas as the North York Moors and Dartmoor. Jones (1976) has observed inwash stripes of mineral material in a Mesolithic context in a peat-filled glacial drainage channel in north Yorkshire. Such sedimentation is correlated with adjacent forest clearance and resultant erosion. Mellars (1975) has argued that a deliberate policy of forest clearance by Mesolithic communities would have had clear benefits through improving the general health and reproductive rate of animals and by controlling the distribution of herds at different times of the year.

The possible consequences of deforestation in terms of soils are summarized in Fig. 1.1. It should be emphasized that the eventual outcome depends very much on the nature of the parent materials, the climatic regime, the new vegetation type, the slope of the land and the time period available for forest regeneration. In uplands, or on lower areas of glacial sands and gravels, the processes of soil degradation fol-

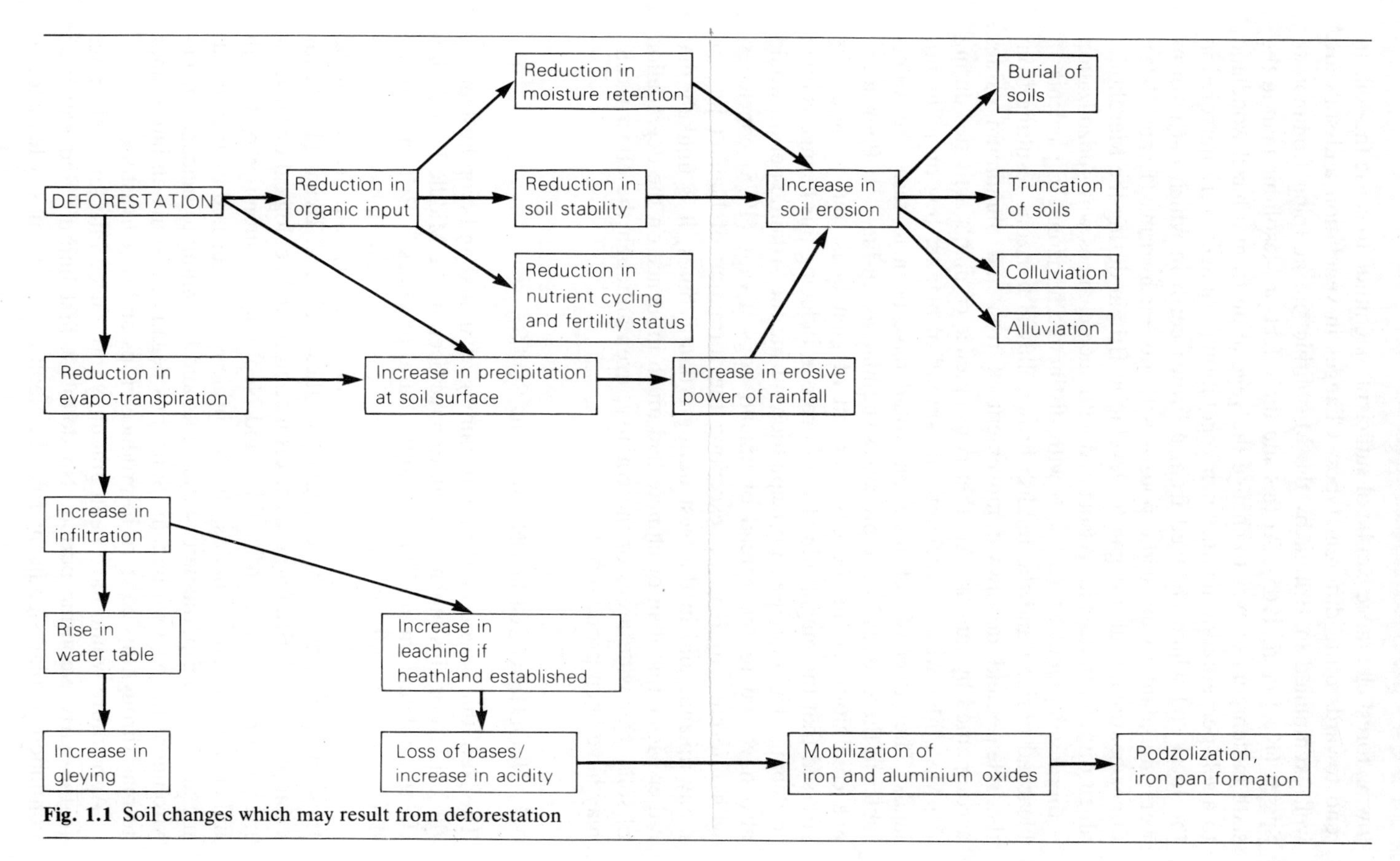

Fig. 1.1 Soil changes which may result from deforestation

lowing forest clearance can be of sufficient magnitude to cause the soils to tend towards quite different types. Changes in vegetation and soils are well exemplified in Britain by the Mesolithic site at Iping Common in Sussex (Keef *et al.* 1965). At this site the effect of Mesolithic man in the sixth millennium BC was to change the vegetation from a hazel woodland to a self-perpetuating heath. Such vegetational changes were mirrored in the local soil which changed from a brown earth in which earthworms were abundant to an acidic podzol with no earthworms (Evans 1975). Other heathlands also began to develop in Britain during the Mesolithic, albeit on a limited scale. Associated with such changes was podzolization – the development of acid soils with distinctive horizons, in particular a bleached horizon and the reddish brown illuvial horizons resulting from the release and downward movement of iron and aluminium oxides accompanied by humus. Another consequence of man's activities during the Mesolithic was to accelerate the spread of blanket bog on many uplands. The effect of deforestation would have been to raise the water-tables in these areas by the processes summarized in Fig. 1.1. Poor drainage conditions led to the development of peat bogs, but it should be stressed that the magnitude of man's role in inducing this change is open to debate. The increase in precipitation during the Atlantic period would also have led to the spread of upland bogs. Evans (1975) concludes, with reference to Britain, 'excepting the formation of blanket bog on many uplands and in the west during Atlantic times, it is unlikely that either Mesolithic man or climate had much influence on the degradation of soils over wide areas of the country, even although the effect locally may have been pronounced'.

1.2 NEOLITHIC COMMUNITIES AND SOILS

The Neolithic period marks the beginning of animal and crop husbandry, the earliest origins of which seem to have been in the Middle East during the 9th millennium BC. The first agricultural practices in Britain occurred during the 4th millennium BC. It is very easy to overemphasize the contrast between a Mesolithic and a Neolithic economy, but archaeological thinking now stresses the continuity of these cultures. Evans (1975), for example, sees in Britain no clear distinction in archaeological or environmental terms between Mesolithic and Neolithic communities of the 4th millennium BC. Nevertheless, the Neolithic is distinguished by man building at least semi-permanent settlements and beginning to manage the environment for his own benefit. Soils thus began to assume an importance to man through his concern to produce crops and rear livestock.

Agriculture would have been gradually introduced into forest clearings in the early Neolithic period. No doubt at first hunting and gathering continued to play significant roles in Neolithic economies. Clearance of trees may have been on a selective basis whereby trees were ring-barked and allowed to die slowly (Walker 1966). The intervening spaces were gradually opened up by cattle and particularly by pigs (Bradley 1978). Alternatively, complete clearance could have been achieved by fire or by cutting down all the trees. The change in vegetation, the application of wood ash, the effect of livestock and to a limited extent, the use of simple ards as ploughs necessarily led to changes in soil-forming processes (Fig. 1.2), but it would be wrong to consider that all such processes led to soil degradation. Many present-day soils support intensive agriculture without deterioration and the farmer's aim is to improve and maintain soil fertility. In the Neolithic era, a certain amount of mixing of the soil would have resulted from the use of the ard. These benefits, of course, are short-lived in a slash-and-burn agricultural system, with yields decreasing very quickly. Coles (1976) notes that clearances in Russia and Canada during historical times yielded uneconomic returns after the first year of cropping. He relates the rapid exhaustion to the thin layer of soil available to crops. The advantages of the wood ash quickly became dissipated and weeds assumed dominance. By the time the land was abandoned, it would have been in very poor condition in terms of fertility. Weedy land must have posed very serious problems since stone ards were unable to cut through such secondary vegetation.

In Britain the transition from the Mesolithic to the Neolithic period coincides with the change from Pollen Zone VIIa (the Atlantic) to Zone VIIb (the Sub Boreal) (Fig. 1.3). The distinguishing characteristic in the pollen record is the decline in elm pollen between these zones. There has been much discussion on the cause of this decline; some workers stress the role of climatic change given the apparent synchroneity of the decline over the British Isles whilst others emphasize the role of man in reducing elms through his selective gathering of leaf fodder for livestock. A review of such themes is provided by Bradley (1978), who also explores a hypothesis fusing both ideas, *viz.* climatic deterioration combined with man's rising numbers forced him to adopt a more settled and intensive form of land use. The implication is that the elm decline does not mark the beginning of the Neolithic clearances, but rather the beginning of a period of intensified agricultural activity. Thus small agricultural clearances must have been created during the 4th millennium BC in the Atlantic period, but they were of sufficiently small extent not to have been reflected in the pollen record.

Neolithic man certainly began to develop an awareness of soil character. There is a close correspondence between the spread of Neolithic settlements across central Europe and loessic soils, whilst a British example is the association on Anglesey of Neolithic sites and well-drained soils derived from glacial deposits (Grimes 1945). An insight into the mechanisms whereby pioneering agricultural communities selected particular

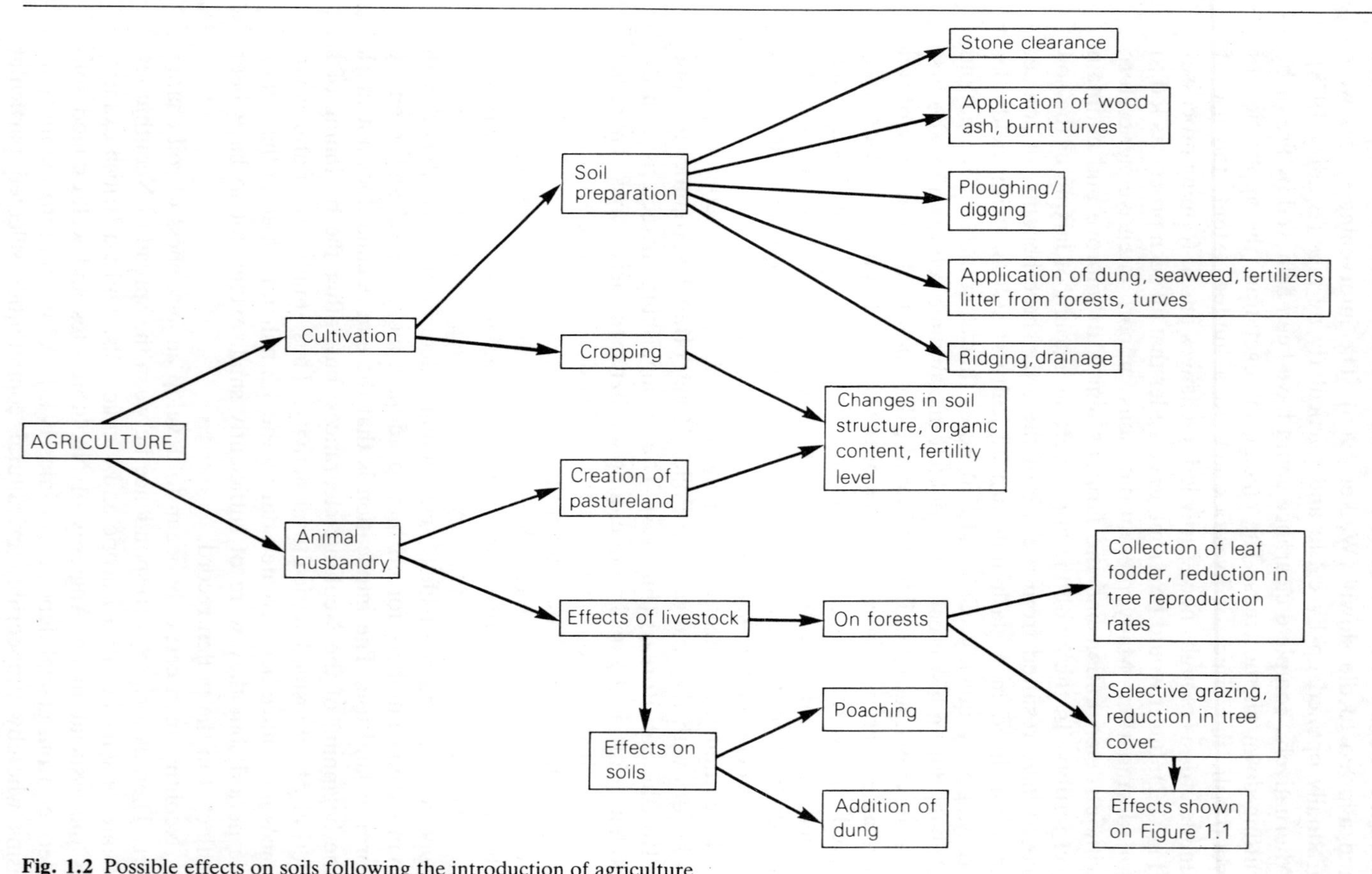

Fig. 1.2 Possible effects on soils following the introduction of agriculture

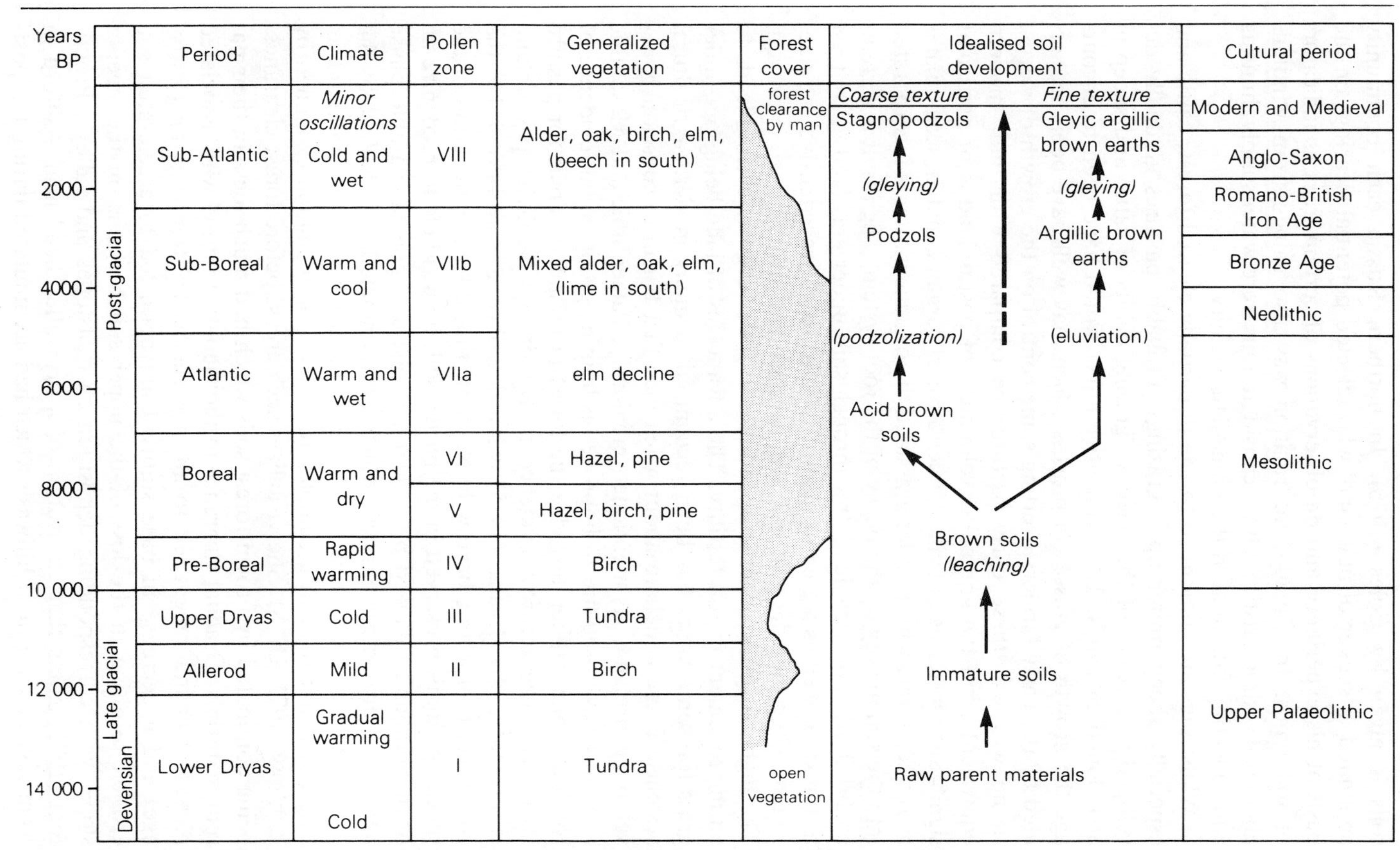

Fig. 1.3 Relationship between climate, pollen zones, vegetation and soils in the post-glacial period (after Bridges 1978)

soils is given by Coles (1976). In northern Russia each community appointed a 'seeker of new land' who selected potential clearings on the basis of his experience and on observations of certain trees as indicators of soil types. In Canada, information was also gained from windfalls about soil characteristics. Other considerations were water supply, hunting and collecting potential in the immediate vicinity.

Whilst pigs are content to forage in partially cleared forest, cattle and especially sheep prefer open grazings. Probably pastures gradually developed as a result of the selective grazing habits of cattle and sheep in abandoned clearings. The cumulative effect would have been to encourage the growth of grasses, a process which may well have been accelerated by the early farmers exerting some control on the growth of weeds. In no way can these early pastures be compared with their modern equivalents, but the gradual development of pastureland is of particular significance in terms of soils. Under a forest, organic matter accumulates as surface litter, whilst under grassland the organic matter is incorporated into the soil mainly by the decay of the root system. The result is a greater distribution of organic matter through soils under grassland in contrast to forest, as well as a greater quantity. Grass roots are extensive and fine, so that when they die soil drainage, aeration and structure are improved. The higher organic content in addition to faster organic cycling leads to an improvement in soil fertility. The advantages of land being put under grass for some years has been recognized in many modern agricultural systems. There is debate about when man first began to consciously develop pastureland. As mentioned previously, Palaeolithic and Mesolithic man may have used fire to improve his hunting grounds, but evidence of Neolithic man deliberately clearing forest in temperate areas for pasture is limited. The clearest evidence in Britain for Neolithic pastureland is presented by Evans (1975) who describes a soil buried by the construction of a long barrow in Neolithic times. By identification of the species of snails preserved in the buried soil, he was able to recognize an open woodland followed by a clearance phase, a grassland/arable phase and then a dry grassland phase which was terminated at *c.* 2800 BC by the construction of the barrow. An intricate land use history for chalklands in England is mirrored in a complex history of soil evolution examined by Limbrey (1975: ch. 8). She visualizes such areas, before forest clearance, being dominated by brown forest soils which had much loess in their parent materials. Gradual clearance in the Neolithic period with associated phases of abandonment led to an increase in clay translocation (lessivage) and a decrease in base status. Cultivation led to accelerated soil erosion resulting in the loss of the upper soil horizons on the steeper slopes with corresponding deposition on footslopes and valleys. Proudfoot (1971) notes that excavation of a dry valley head near Ashford in Kent revealed 1.5 m of hillwash which had accumulated during the postglacial period. He attributed half of this accumulation to the results of woodland clearance in early Iron Age times. A suggestion made by Limbrey (1975) is that the gradual truncation of chalkland soils resulted in the formation of shallow rendzina soils. Truncation caused illuvial horizons to appear at the surface and the effect of ploughing was to incorporate calcareous material from the C horizon into the new surface horizons. According to Limbrey, then, one possibility is that gradual clearance of the forest on chalklands induced soils to change from brown forest soils to sols lessivés to calcareous rendzinas. These rendzinas may well have become pastureland which was subsequently difficult to clear since grass roots are resistant to burning. Any clearance would only have been possible by breaking up the turf with a stone axe prior to scratch ploughing with an ard.

Three aspects of this interpretation are open to debate. First, the development of an argillic horizon may be favoured by a forest cover as well as by a climate with a distinct dry season. Thus the proposal for an increase in clay translocation following forest clearance can be questioned. Second, there is debate as to whether the chalklands were forested. Barker and Webley (1978) argue, for example, that the Wessex uplands have had an open grassland environment since early Neolithic times and they reject the hypothesis of gradual forest clearance with shifting cultivation. In the discussion of this paper, Evans states that there is clear evidence reflected in the types of preserved snails at archaeological sites to indicate woodland on the chalklands of England. Third, Barker and Webley (1978) suggest that the chalklands have been more stable in a geomorphological and pedological sense than previously accepted and thus the rendzina soils, for example, are interpreted as the natural types to occur on the grassland downs. The alternative view is that chalk slopes have been subjected to massive erosion, an outlook supported by Dimbleby (1976) and exemplified by a detailed stratigraphic and palaeopedological investigation by Valentine and Dalrymple (1976) in Buckinghamshire. They identified a sequence of deposits in a scarp foot valley which they related to man's clearance and cultivation of the land beginning in late Neolithic or early Bronze Age times.

It is clear that in the Neolithic period man began to make rational decisions about soils in terms of where he settled and at the same time he also became a significant force in soil development. His influence was not limited to the lowland areas of the British Isles. Pennington (1975) has assembled much palynological data from north-west England to show that Neolithic man had a significant effect on vegetation and soils both in lowland and upland localities. She dates the elm decline to *c.* 3100 BC, a time of increasing surface runoff and rising water-tables. In upland areas forest soils gave way to peat and soil instability spread on lowland areas with the expansion of cultivation.

1.3 BRONZE AND IRON AGE COMMUNITIES AND SOILS

In the British Isles the Bronze Age began *c.* 1800 BC and perhaps the most significant change was an increase in population, reflected in the spread of the 'Beaker' culture. The rate of deforestation was correspondingly increased and the resultant soils changes were induced (Figs 1.1 and 1.2). During Bronze Age times man in Britain considerably extended his influence. For example, Mesolithic man had achieved local clearances of the forest on the North York Moors, but it was not until the Bronze Age that the main phase of clearance took place. According to Dimbleby (1962), the clearance was primarily for pasture, a view challenged by Fleming (1971), who argues that cereal cultivation was the objective. The ultimate effect in terms of soils was the production of impoverished podzolic soils under a heathland vegetation. Evidence for other Bronze Age clearances is summarized by Evans (1975). For example, clearances on the Cumberland lowland and on the Somerset Levels seem to have been for cereals. Pollen analysis from sites on Bloak Moss in Ayrshire indicates a pattern of shifting cultivation during the Bronze Age with clearances primarily for pasture (Turner 1970, 1975).

The use of metal might suggest a sudden increase in efficient tools designed to cut down trees and work the resultant timber. However, it was only towards the end of the Bronze Age that the bronze axe in its socketed form was perfected. Such a development, combined with a range of wood working tools, speeded up the rate of forest clearance. The smelting of tin and copper to make bronze required fuel, another factor which accelerated forest clearance in particular areas.

A feature of many upland areas of Britain is the frequent occurrence of cairns, many dating from the Bronze Age. Some evidence has been found for the rough alignment of cairns or for their occurrence along small banks, suggesting that they mark the edge of plots (Feachem 1973). In terms of soils, the significant points are the practice of stone clearance and the fixing of defined areas for agriculture. Celtic fields, well preserved on the chalklands, date from the middle of the Bronze Age though many were formed during the 1st millennium BC. No doubt a form of shifting cultivation continued to be practised, but on a more organized spatial pattern. Fields would have been gradually improved through stone clearance, through the use of the ox-drawn plough and some fields near settlement sites may have been kept in longer use through the practice of folding cattle on them, an early version of in-bye land. The evolution of such an organized agricultural system led to permanent settlement sites. It was then in the farmers' interests to learn the practical skills of soil management and conservation so that some fields in proximity to their settlements were always in a state fit for cultivation or livestock.

The last major period of European prehistory is the Iron Age, which had its origins in Anatolia and in the area to the north of the Caucasus where the Cimmerians lived. The first Celts made their appearance in Britain in the seventh century BC, but the arrival of Celts did not accelerate until around the fifth century BC. There is some evidence of climatic deterioration towards the end of the Bronze Age, which may have further accelerated the abandonment of fields in upland areas. On low lying areas, a deterioration in drainage conditions was expressed in the extension of bogs and marshes. The Iron Age saw an influx of new peoples, tools and crops which resulted in a renewed period of forest clearance and resultant soil modification. The use of iron axes was integral to such a process. A hardy form of wheat called spelt was introduced, the seed being sown in the autumn. Other new crops were the Celtic bean, rye and clubwheat. The effect of a broader farming base was to lead to greater food security as well as a better spread of produce over the year which in turn allowed the population to increase. Fields became more extensive and some form of crop rotation would have been practised. These small Celtic fields were ploughed by an iron-shod ard; mould-boards were not introduced until Roman times. The effect of ploughing resulted in the lowering of the soil surface on the upper side of sloping fields and the corresponding accumulation on the lower side – the formation of lynchets. Bridges (1978a) notes that virtually all known Celtic fields in England and Wales occur on rendzinas, brown calcareous earths and argillic brown earths, though he notes that such spatial correlation is illusory since Celtic fields were formerly much more widespread.

The distribution of Celtic fields in Britain demonstrates that by late prehistoric times man had made a significant impact on forest clearance and thus on soils. Information is also available from Europe indicating that accelerated erosion was occurring from late Bronze Age times. Butzer (1974) summarizes the evidence which indicates that soils developed on loess were being eroded in central Europe at this time. The result in certain instances was the burial of younger Bronze Age valley-floor sites by as much as 1–2 m of loessic soilwash. In detail it is difficult to establish causes for such erosion, but deforestation and increased runoff, cultivation and subsequent abandonment are likely factors. In Britain, soil changes in late prehistory are most clear on uplands such as Wales and North York Moors with the onset of podzolization, and on the chalklands with the production of rendzinas. However, it must be borne in mind that there were many areas of Britain where man had made little impact before the arrival of the Romans. In Scotland the central valley was not extensively cleared until Medieval times. Attention must turn to the Middle East and the Mediterranean basin for evidence of man's more extensive and irreversible influence on soils during the first two millennia BC.

1.4 SOILS AND CIVILIZATIONS IN THE MIDDLE EAST AND THE MEDITERRANEAN BASIN

It is no coincidence that early civilizations flourished in similar environments. The Nile valley, Mesopotamia and the Indus basin all had dependable water supplies which permitted irrigation, relatively flat land which was not liable to serious erosion, and fertile alluvial soils which were periodically recharged with new alluvium and nutrients. The Egyptians were farming the Nile valley on an extensive basis by 3500 BC and by 3000 BC centralized government at Memphis had been formed. Around 2500 BC pyramids were being built, a clear indication of a well-organized and richly endowed society. A similar pattern is evident in Mesopotamia where the Sumerians founded city states *c.* 3500 BC. By 3000 BC many irrigation schemes were in operation. The irrigation canals required more frequent clearance than in the Nile valley since the Mesopotamian rivers had higher sediment loads. The first written records were provided by Hammurabi, who described an intricate and widespread irrigation network in the eighteenth century BC. The history of Mesopotamia is dominated by a sequence of invasions culminating in Alexander the Great's conquest. The break-up of his empire (323 BC) can be taken as marking the beginning of the end for Mesopotamian civilization. Irrigation canals gradually became silted and abandoned and had to be replaced by new systems in subsequent periods. The head of the Persian Gulf has receded 200 km through sedimentation since the time of the Sumerians. Such sedimentation has resulted directly from deforestation and overgrazing on the surrounding hills, especially in the headwater regions in Turkey.

The Indus civilization in Pakistan has many similarities to those in Eygpt and Mesopotamia. Irrigation of alluvial soils was again the basis of the economy. Increased sedimentation through deforestation of the foothills of the Hindu Kush and Himalayas as well as the spread of saline soils contributed to the decline of the civilization, in this case at *c.* 1700 BC.

It is often forgotten that the introduction of irrigation also requires good drainage conditions. If such conditions are not present, calcium and magnesium precipitate as carbonates whilst sodium ions are left to be adsorbed by soil colloids. The effect of the sodium is to destroy soil structure and make it almost impermeable as well as saline. The salinity of ground water also increases, so that salinization spreads to areas where the capilliary fringe above the water table comes near to the surface. The evidence for salinization in Mesopotamia has been investigated by Jacobsen and Adams (1958), who identify three major occurrences; a serious spell from 2400 BC to 1700 BC in southern Iraq, a less serious period from 1300 BC to 900 BC in central Iraq, and after AD 1200 in the area to the east of Baghdad. For the first period, these authors illustrate how crop choice became limited as salinization spread, with the eventual abandonment of wheat and, concurrent with a shift to barley, was a decline in yields.

However, it is important to note that salinization is not explicable solely in terms of badly designed or maintained irrigation systems. Hardan (1970) demonstrates that salinization was present in Mesopotamia before or at the beginning of irrigation and that intensive irrigation may have accelerated natural processes of salinization. According to Yahia (1971), the saline groundwater is the main source of the soil salinity. The origin of this saline groundwater is due to salt exchange with the soil as well as the groundwater gaining salt from bedrock. Increase in salinization can also follow from clearance of natural vegetation in arid and semi-arid areas. The level of the water table in the soil is increased, causing the accumulation of salts at or near the soil surface, producing bare salt scalds in areas which were previously agriculturally productive. The spread in recent times of salt scalds in non-irrigated areas of Western Australia has been investigated by Conacher (1975), who notes that between 1955 and 1970 the area of salt affected soils had increased by 122 per cent. He argues that the main cause of salinization is the increase in throughflow rather than the rise in water-table.

The rise of civilizations in eastern Mediterranean was based on rain-fed agriculture, in contrast to irrigation in the Middle East. On Crete there developed the focus of the Minoan civilization whose influence extended throughout the Aegean. Their civilization came to an abrupt end around 1500 BC, in part due to the ascendancy of the Mycenaens as well as to earthquakes and other effects of the cataclysmic eruption of the Santorini volcano. Part of the explanation may have been the deteriorating soil conditions. All the archaeological and palaeoenvironmental evidence suggests that Greece was well forested in Bronze Age times and the forests grew on much deeper soils than exist today. Gradual forest clearance and selective grazing by animals resulted in accelerated soil erosion which was under way in some areas by the end of the 2nd millennium BC (Davidson 1980). Since then a virtual transformation of soil types and their distribution has occurred in Greece.

The Phoenicians lived on the coastal strip of Lebanon during the 1st and 2nd millennia BC and they gradually deforested their lands, in the process of obtaining their famous cedars. Soil erosion soon became widespread and the Phoenicians resorted to terracing to try to preserve some soil on hillslopes. The extent of environmental change is well expressed in the remains of Antioch, once a capital of the Seleucid Empire in ancient Syria. In the second century AD this city had a population of half a million and now is buried by as much as 9 m of sediment derived from eroded hillslopes (Carter and Dale 1974).

The Classical period in Greece and Italy saw the extension of soil erosion and writers began to make comment about the serious situation.

Plato provides a vivid description of Attica (Greece) in the fourth century BC:

> *. . . what now remains of the once rich land is like the skeleton of a sick man, all the fat and soft earth having wasted away, only the bare framework is left. Formerly, many of the present mountains were arable hills, the present marshes were plains full of rich soil; hills were covered with forests, and produced boundless pasturage that now produce only food for bees. Moreover, the land was enriched by yearly rains, which were not lost, as now, by flowing from the bare land into the sea; the soil was deep, it received the water, storing it up in the retentive loamy soil. . . .*
>
> (*Critias, quoted from Carter and Dale 1974*)

The widespread occurrence of soil erosion in the Mediterranean basin is well established, but little is known in any detail about pedogenetic history. Research by Spaargaren (1979) has produced some interesting results on this matter. He investigated the weathering and soil-forming processes in a limestone area near Pastena in south-central Italy. He was able to calculate that a regolith of 1 m thickness on limestone required at least 500 000 years for development and thus instances of regolith of 4 m which he observed required a substantial part of the Quaternary period for their evolution. He noted that many of the soils in his study area were severely eroded. He correlated the most recent major phase of erosion with human activity, the most rapid change taking place during the fourth century BC. The result was the generation of what Spaargaren calls anthropogenic colluvium which occurs to a depth of *c.* 6 m in the centre of a closed basin. This permitted him to calculate that *c.* 1 m of soil must have been eroded from the surrounding slopes to produce such a depth of colluvium. Thus a considerable period of landscape stability must have existed prior to the onset of this last major erosional/depositional phase.

Alluviation also resulted from hillslope erosion, a process which was dominant in the Mediterranean basin between Classical and Medieval times. Vita-Finzi (1969) calls these alluvial deposits 'Younger Fill'. Hill-slope erosion meant the degradation of the land resource base in such localities, but the corresponding deposition on lower slopes, in alluvial fans or in valley floor alluviam has proved an asset to present day agriculture. These flat or very gently sloping areas offer scope for intensive agriculture through the application of irrigation. In many valleys in Greece, for example, a sharp discontinuity can be observed between the intensively cultivated alluvial soils on the valley floors and the unproductive thin stony soils on the valley side slopes. The stripping of soil from the hill-slopes and the corresponding valley floor deposition can be related to a large extent to man's influences.

1.5 SOILS AND MAN IN WESTERN EUROPE FROM ROMAN TIMES UNTIL THE EIGHTEENTH CENTURY

The preceding section stressed the magnitude of soil change which had been initiated in Mediterranean lands by Classical times. In western Europe man had begun to induce significant soil changes prior to the Roman conquest – the Neolithic changes on the chalklands of England and the Bronze Age pressures on the uplands. Hoskins (1955) estimates the population of Britain at more than 400 000 on the eve of the Roman conquest. The growing Iron Age population combined with the introduction of the heavy wheeled plough led to colonization of more difficult terrain. The extent of such deforestation is illustrated by the work of Turner (1979) in north-east England. On the basis of pollen analysis she concludes that much of the area was deforested and used for farming during late pre-Roman Iron Age times. The pollen record indicates that such clearances for agriculture were maintained during the Roman period and lasted until the sixth century. The extent of such forest clearance is also reflected in the distribution of Roman villas. These villas were the centres of large agricultural estates; new field systems were introduced and the major objective was the production of cereals in lowland areas. Such agricultural extensions introduced the possible soil changes indicated on Figs 1.1 and 1.2. The Romans were also the first to introduce large drainage schemes in the Fens (e.g. Car Dyke). The native population in Britain seems to have been able to maintain the agricultural systems for up to two centuries following the departure of the Romans, but gradually the landscape in many areas must have reverted to a semi-natural state.

The next period in European history is distinguished by the movement of peoples, implying the development of population pressure or political instability or some combination of these factors. In Britain we think of the arrival of the Angles, Jutes and Saxons with later Scandinavian settlers in the north and west, and the movements culminating in the Norman conquest. Credit is usually given to the Anglo-Saxons for the initiation of open field agricultural systems. The effect of ridge and furrow was to produce marked microrelief designed to improve soil drainage. Little is known in detail about these early open field systems, but it can be assumed that strips were cultivated for cereals, with fields being alternated between crops and fallow. Soil conditions influenced the Anglo-Saxon colonization pattern, a theme investigated by Wooldridge (1948) who stresses the importance of 'loam terrains' in south-east England. The use of a plough fitted with a coulter to cut the soil as well as a mouldboard to turn it permitted heavier soils to be cultivated. Such technological and agrarian advances were limited to lowland Britain; in the uplands lighter ploughs were used when relief and soil conditions permitted, otherwise a footplough called a caschrom was used.

The Domesday book presents a picture of the English landscape twenty years after the Norman conquest. Hoskins (1955) estimates the total population as *c.* 1.25 million with about 10 per cent resident in 'boroughs'. Despite the spread of agriculture during earlier centuries, there were still extensive areas of woodland. For example, Sussex was only densely settled in its southern part in 1086 whilst another third was part of a natural forest which extended into Surrey, Kent and Hampshire (Darby 1948). East Anglia was the most densely occupied part of England at this time. Moorland, marshland and fenland were also being reclaimed on a limited scale. Continual increases in demand for timber, primarily for construction and charcoal, further accelerated woodland clearance and the spread of feudal agricultural in lowland England. In western Europe as a whole, the eleventh to thirteenth centuries witnessed the doubling or even quadrupling of the cultivated area (Carter and Dale 1974). Such agricultural expansion provided the economic foundation for the development of Medieval towns, feudal manors and monasteries. The net result was that by the middle of the fourteenth century much of the cultivable area of western Europe was in agricultural use.

This major phase of colonization came to an abrupt end in the fourteenth century. Part of the problem was that there had been over-expansion on to marginal areas, such as the sandy Breckland soils in England. This was the time when many English villages were abandoned – over 2000 deserted villages are known. No doubt economic decline was also involved, but an important factor was bubonic plague, also called Black Death, which reached England in 1348. Between one third and one half of the population died and such a disaster had a marked effect on soils. The pressure on the land was suddenly reduced, leading to land reversion and there was also a switch in emphasis from arable to pasture. Jacks (1954) describes how the feudal open field system led in many instances to soil deterioration with the loss of crumb structure and the depletion of plant nutrients. The problem was exacerbated by the rigid system of cropping on a communal system. The Black Death led to the breakdown of the feudal system which coincided with the rise in demand for wool. Enclosures were made for sheep and the effect was to give soils a rest. The effects of putting land down to grass are all beneficial in terms of improving soil structure, organic content and stability.

In western Europe, the Dutch and Flemish were the pioneers in introducing crop rotations. From the fourtheenth century onwards these farmers were growing such crops as wheat, turnips and clover in rotation. The rise of owner-occupiers in these countries meant that it was in the interests of the farmers to try to conserve and improve the fertility status of their soil. Thus liming and manuring of fields became more common.

Associated with such agricultural improvements was land reclamation in the coastal zone of the Netherlands. Dike building began on a significant scale *c.* AD 1200 and gradually increased to reach a maximum during the period AD 1600–1625 when 32 000 ha were reclaimed (de Bakker 1979). The creation of polders is probably the best known example of major land reclamation.

In continental Europe as in Britain there is a dearth of detailed research on the relationships between land use history, geomorphological processes and soil evolution. A notable exception is a collaborative research project in the Luxembourg Ardennes. Imeson and Jungerius (1974), on the basis of monitoring present-day erosional processes in wooded localities, conclude that erosion rates are low and they suggest that such a situation would have prevailed during much of the Holocene as long as the forest cover was maintained. They ascribe such stability to high porosity and stability of the surface soil. Forest clearance for agriculture resulted in accelerated soil erosion though this was not on a catastrophic scale. Kwaad and Müchler (1975, 1977) identify a more detailed pattern of soil evolution in the Luxembourg Ardennes. They propose a period from before AD 1400 until AD 1800 of severe erosion resultant upon forest clearance for agriculture. Buckwheat was introduced in the area *c.* AD 1400 and it is a crop which requires deep tillage. Riezebos and Slotboom (1978) associate the introduction of this crop with the fallow land which is integral to the 'three field system' in accelerating soil erosion. Reafforestation from *c.* AD 1800 markedly reduced the erosion rate from the previous period. Thus the research in the Luxembourg Ardennes neatly demonstrates the intimate association between land use history and soil evolution.

Plaggen soils provide the clearest illustration of man's effect on soil formation and they occur extensively in northwestern Europe (Fig. 1.4). These are man-made soils on the areas underlain by Pleistocene sands which supported heathland vegetation. Sheep and cattle were grazed on these heathlands and taken every night into a sheepfold or cowshed where heather sods were used as bedding material. The sods, called plaggen in Dutch and German, were cut using special scythe and pitchfork from the top organic dominated layer of podzols. This dung-impregnated bedding was added as manure to arable land, a process described by de Bakker (1979). The effect of adding this material was to fertilize the land as well as to gradually raise the surface. In the Netherlands there are many instances of plaggen soils with a depth of *c.* 1 m, which would have taken of the order of 750 years to develop. Black plaggen soils resulted from the use of heather turves whilst grass turves produced brown plaggen soils. Man's actions over the centuries in these sandy districts of north-west Europe has thus been to produce not only a man-made soil, but also to create local relief of the order of 1 m. These slightly raised

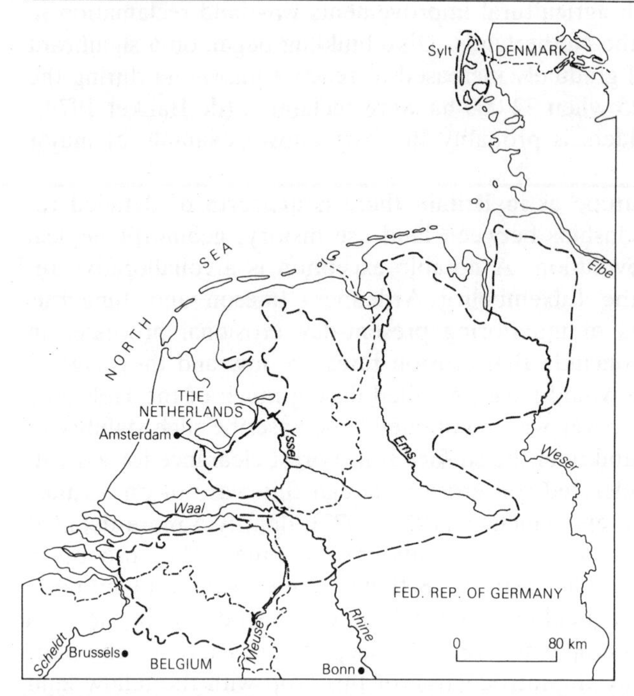

Fig. 1.4 Distribution of Plaggen soils in northwest Europe (after de Bakker 1979)

plaggen soils are a distinctive feature of the Pleistocene sandy districts in the Netherlands. On occasion plaggen soils have been buried by wind-blown sand and these inland dunes have only been stabilized in recent decades by afforestation schemes. This history of land use is well reflected in the sequence of soils, from the plaggen soils through the unstable period of moving sand, followed by the development of a surface micro-podzol under the recent coniferous forest.

A different type of plaggen soil has been described and investigated in Ireland by Conry (1971). According to this study, Irish plaggen soils resulted from the addition of large quantities of calcareous sea sand, which besides adding carbonate to soils, also yielded phosphorus, magnesium and sodium. In detail, the sand was applied either alone, or mixed with stable dung or seaweed or peat mould. Conry (1971) describes a plaggen soil 85 cm thick. These soils are restricted to coastal localities and are important in a present-day agricultural and horticultural context since they are more fertile than surrounding non-plaggen soils. In Scotland plaggen soils of sufficient extent to be mapped have been reported for Orkney (Macaulay Institute for Soil Research 1975–76).

In England and Wales Inclosure Acts between the fourteenth and nineteenth centuries reduced the extent of common land and gave greater ownership to individuals. As already noted, the effect of increased grassland was to improve soil conditions and individual farmers and land owners began to invest more time and resources in land improvement. Bridges (1978b) quotes a remarkably varied range of soil additives from hogs' hair to sugar bakers' scum! But the long standing practice of marling was probably the most beneficial, especially on light textured soils. The introduction of enclosures paved the way for the agricultural revolution. Jacks (1954) summarized the major events of agricultural history of importance to soils before this revolution as: (1) open-field farming based on the feudal manor; (2) the Black Death; (3) the rise of the wool trade; and (4) the Tudor enclosures.

1.6 THE EFFECT OF MODERN AGRICULTURE ON BRITISH SOILS

Before AD 1700 about one half of the English arable area was still cultivated on the openfield system with fragmented strip holdings and common rights to pastureland and waste (East 1948). During the eighteenth century the pace of enclosures accelerated, with final extensive enclosures taking place in the first half of the nineteenth century. Enclosure is a term which applies not only to the rearrangement of common fields, but also to the reclamation of heathland, fenland or woodland. Such enclosures resulted from the efforts and imagination of individuals. For example, the famous Coke of Holkham was responsible for transforming heathlands on his estate in north Norfolk to cornfields. Marling of the sandy soils played an important role in this reclamation.

The enclosure of common fields and land reclamation during the eighteenth century was part of a broader agricultural change – often referred to as the agricultural revolution, though the process was more of a gradual evolutionary nature. Population was increasing, creating a greater demand for food and thus increasing agricultural prices. Important innovations were also being made including new agricultural machinery, new crops, better breeds of livestock, new schemes of crop rotation and new methods of cultivation. This upsurge in new techniques was fostered by individuals such as Jethro Tull, Lord Townshend, Bakewell of Dishley, Coke of Holkham and Arthur Young. Agricultural societies and shows helped to disseminate the new techniques and encouraged a spirit of enterprise.

Jacks (1954) describes an individual on his enclosed pastures as a hoarder of soil fertility whilst a communal farmer on his open fields was a consumer of soil fertility. Thus a rotation of pastoral and arable farming was a means of spreading the benefits from the latter to the former. The introduction of clover and new and improved grasses increased hay production, with the clovers also raising the nitrogen levels in soils. Turnips solved the problem of winter feed for cattle and sheep. From a soil standpoint, a more significant innovation was the seed drill, invented by Jethro Tull. The consequence was that seeds were planted in the soil in rows rather than broadcast over the surface. Weeds became less of a problem and with drill-sown turnips, cultivation between rows was possible.

The widespread use of lime was also integral to the agricultural revolution. The early practice was to roast limestone in kilns and the resultant calcium oxide or hydroxide (quicklime) was spread on the fields. Several clear advantages are consequential upon the application of lime. Adequate supplies of calcium become available to crops and the presence of calcium reduces soil acidity which improves the availability of other nutrients. The presence of calcium also improves nitrogen fixation and other bacterial activities, reduces the potential for soil toxicity problems and aids soil structure by flocculating clays. Many derelict kilns are evident throughout Britain, indicating that the application of ground limestone has superseded quicklime. The application of calcareous sand or marl has similar merits.

The agricultural improvement movement came later to Scotland and displaced a different type of communal agriculture. The pre-agricultural revolution landscape in Scotland was dominated by the infield–outfield system centred on 'ferm touns'. The infield was kept constantly in cereals (barley and oats) with the part under barley receiving all the dung. The outfield was subdivided into the folds and the faughs. The folds were further subdivided into ten and each year one of these subdivisions was brought into tillage after the cattle had been folded on it the previous year. Each fold was cropped for four or five years. The faughs never received any manure, but parts were ploughed at intervals of four or five years and then cropped for three or four years. In effect the old Scottish agricultural system was a type of shifting cultivation linked with black cattle husbandry. Improvements began to be made in the later part of the eighteenth century. The writings and efforts of Sir John Sinclair did much to encourage the introduction of new techniques in Scotland. He provided much detailed advice on such topics as stone clearance, land drainage and ploughing techniques. He was also responsible for organizing the survey of all parishes throughout Scotland – the result is known as the Old Statistical Account and these 21 volumes describe the state of the Scottish landscape in the last decade of the eighteenth century. A graphic account of the conflict between the old and new types of agriculture is provided by the writer for the parish of Birse in Deeside, north-east Scotland.

'It must be confessed that agriculture in this country is rather in an imperfect state; many of the people continue rivetted in their prejudices against the modern improvements in husbandry. All, however, must allow,

that these improvements have commenced in this parish, and that some of the tenants have done more in the way of farming to purpose, than many of their neighbours in several places around. Several have upon their possession a small limekiln; they purchase the limestone at some of the quarries in the parish, and burn it with peat, mixed often with wood, or even with some coals from Aberdeen. The lime they lay upon the ground has much effect. . . . . Some are going on with spirit, inclosing and clearing their ground of stones. Others are throwing every impediment in the way of the improver, by trampling down the fences, and by not only neglecting to remove the stones from the fields, but even by alleging, that the stones are beneficial to the soil, and tend to nourish the crop.'

(*from Sinclair, pp. 109–110, 1793*)

By the time of the New Statistical Account for Birse in 1845, the new agriculture was well established.

The agricultural soils in Scotland have been gradually improved and transformed since the early nineteenth century. The magnitude of stone clearance is expressed in broad consumption dykes (walls) in north-east Scotland. The installation of stone or tile drains very much improved drainage conditions. Heavy applications of dung and deeper ploughing over the decades also helped to thicken the topsoil. Thus it is impossible to interpret present-day soils in such areas as north-east Scotland without close reference to the enormous efforts at soil improvement by generations of farmers.

Technological and agricultural innovation has continued up to the present day with the consequence of increasing farmers' influence on the nature of their soils. This is a theme pursued in Chapter 6. Industrial and urban changes over the last 200 years have also had a significant impact on soils. Extractive industries have caused the loss of soils through the excavation of quarries for rock or sand and gravel, and through the spreading of mining waste on the surface. Less obvious is soil pollution caused by the mining of metals. In Britain this topic has been investigated by Davies (1976), who found, for example, high levels of mercury and other metals in floodplain deposits which had been polluted by mine waste. Pollutants can also be added to soils by air-borne mechanisms. In another study Davies (1978) relates lead and other metals in soils to emission from vehicles, factories, mining and coal ash from chimneys. The spread of urban areas means the transformation of land use and thus soil type. The usual practice of property developers is to strip the topsoil from an area prior to building, and soil, possibly from another area, is added when construction is complete. In new areas of residential development, subsequent cultivation of gardens further causes modification of soil properties from the original state. To a large extent such garden soils can be considered man-made and the same interpretation has to be applied to soils created in city redevelopment schemes where urban landscaping is introduced.

1.7 THE COLONIZATION OF NEW LANDS AND ASSOCIATED SOIL CHANGES

Attention has been focussed in the preceding sections on areas which have been increasingly utilized since prehistoric times. The impact of man on soils became significant when he began to practise agriculture. In continents such as North America and Australasia extensive cultivation of land was introduced by European settlers during the colonial period from the early seventeenth century onwards. It is obviously important to assess man's impact on the soils of these new lands, but it is also of interest to note that the agricultural potential of these areas was a major incentive to potential settlers. It was very much to the benefit of development companies to portray their land as offering high yields. Johnson (1979) has examined the images of soil fertility created in promotional literature for New Zealand in the period 1839–55. She found that the expectations of emigrants were largely based on the images presented by the pamphlet writers and promoters of the New Zealand Company. Assessments of soil quality in the first instance were made either on the luxuriance of the vegetation or on the colour and texture of the soil. In practice the image of soil quality was often very different from the reality of the situation. However, human reaction to soil conditions is dependent upon the perceived nature of soils and this still applies in a present-day context even though objectively based soil survey data may be available.

Man's impact on the landscape and soils of the New World has been compressed into a much narrower timespan than in the old settled areas of Europe and the Middle East. The gradual European colonization of North America led to extensive deforestation for agriculture as well as the cultivation of former grasslands in the Great Plains and Prairies. Such land use changes and the introduction of agriculture led to the modification and often the transformation of soil properties. The acceleration of soil erosion was the most serious result, expressed most clearly in the huge dust storms of the American mid-west from 1934 to 1938 and in the 1950s. Although some attempts were made at soil conservation in the eighteenth and nineteenth centuries in the USA, serious consideration was not given to this problem until the major efforts of Hugh Bennett over the first four decades of the twentieth century. He has been described as soil conservation's 'fiery apostle' since he was constantly trying to get recognition of the problem. He estimated that in the 1930s nearly

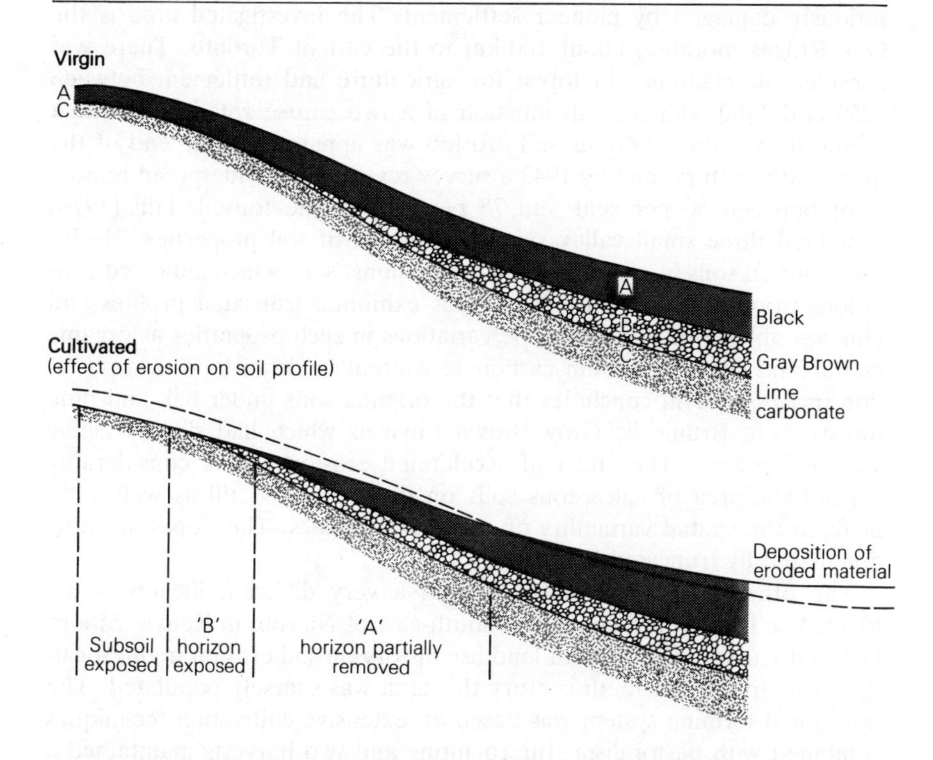

Fig. 1.5 Effects of soil erosion in Manitoba (after Ollier 1969 from Ellis 1938)

20 million hectares of cropland had been more or less ruined in continental USA with another 20 million hectares reduced to a marginal state of productivity. The consequences of soil erosion need no emphasis – the loss of topsoil, the burial of fields by loose windblown silt, the sedimentation of drainage ditches and the increased sediment load of rivers and resultant reservoir siltation. These physical problems were aggravated by the economic and social ills of the depression years. The effects of erosion on a soil sequence in Manitoba are illustrated in Fig. 1.5. The exposure of subsoil and a B horizon in the upper part of the slope is matched by the burial of the original A horizon on the lower slope. To illustrate in greater detail man's effects on soils associated with recent colonization, examples from Canada and Kenya are selected.

Hill (1976) provides a detailed study which examines the effects of accelerated erosion and resultant deposition on soils in an upland area seriously damaged by pioneer settlement. The investigated area is the Oak Ridges moraine, about 100 km to the east of Toronto. There was considerable clearance of forest for agriculture and settlement between 1850 and 1860 with the introduction of a two course rotation of wheat followed by fallow. Serious soil erosion was apparent by the end of the nineteenth century, and by 1942 a survey revealed the widespread removal of between 50 per cent and 75 per cent of the topsoil. Hill (1976) examined three small valley systems in terms of soil properties. He located buried soils in all valleys and depressions, soils which indicated conditions prior to settlement. Slope soils exhibited truncated profiles and Hill was able to relate downslope variations in such properties as organic matter and silt and calcium carbonate content to downslope transportation processes. Hill concludes that the original soils under oak and pine forests were Brunisolic Gray Brown Luvisols which had slightly acidic surface horizons. The effect of accelerated erosion was to considerably expand the area of calcareous soils on the calcareous till as well as to increase the spatial variability of soil characteristics. The slopes are now dominated by truncated Luvisols.

The other example is selected from a very different locality – the Machakos Hills to the immediate south-east of Nairobi in Kenya. Moore (1979) discusses the effects of land use history on soil erosion in this locality. Prior to the twentieth century this area was sparsely populated. The traditional farming system was based on extensive cultivation techniques combined with pastoralism. Interplanting and two harvests maintained a cover to soils for much of the year. When the soil was exhausted through cultivating cassava and sweet potatoes, land was left to revert to bushland. According to Moore (1979) 'the mobility and flexibility to the farming system was well adjusted to the soil and rainfall limitations'. Population rapidly increased after 1900 following colonial administration by the British. The more demanding crop of maize was introduced and the Europeans also used pure stands rather than interplanting techniques. They also created native reserves which soon suffered from overpopulation. The loss of certain grazing rights with consequential overstocking compounded with an increase in native arable areas and shortening of fallow periods led to soil deterioration and erosion which had become very evident in the 1930s. Eroded areas soon had subsoils exposed at the surface with the consequences of lower nutrient status and lower infiltration capacity. Moore (1979) summarizes the various attempts which were made in the 1930s to introduce grazing rotations, more extensive grasslands and terracing, and lower stocking rates, but such attempts seem to have had little success.

The Canadian and Kenyan examples illustrate how a critical threshold in ecological balance can be quickly broken when new land use practices and population pressures are introduced. In both instances such situations were achieved after about 30 years, and these areas are still suffering from the consequences.

CONCLUSION

The traditional approach to soil formation is to consider the interplay of five factors, *viz.* parent material, climate, organisms, topography and time. The effect of man is included in the organism factor. In many ways this is an unsatisfactory approach since it tends to underemphasize the role of man in influencing soils. An alternative strategy is to consider how man influences the five factors, an approach adopted by Bidwell and Hole (1965), who examine the beneficial and detrimental effects of man on each of these soil forming factors (Table 1.1). Yaalon and Yaron (1966) propose that 'the vast activities of man as a soil modifier have not been as yet fully incorporated into the body of pedological investigations or into its general conceptual framework'. In similar vein Bridges (1978) notes for British soils that '. . . the impact of man has been very much underestimated in past pedological studies'. Yaalon and Yaron (1966) propose the study of man-made changes within the structure of a process-response model. They call man-induced processes in soils 'metapedogenesis', and they illustrate how a wide range of soils can be modified by metapedogenic processes. For example, the action of deforestation and ploughing in temperate areas can change podzols to acid brown forest soils. It is also possible for brown forest soils to be transformed into rendzinas or podzols depending upon local circumstances.

There is still a strong tradition in many soil surveys that the prime concern is with identifying and mapping soils as naturally occurring phenomena. As Yaalon and Yaron (1966) state, '. . . the genetic processes altered by man's intervention are looked upon merely as deviations from the normal genetical soil type, which do not concern the pedologist desirous of producing a soil map'. Some idea of the extent to which the effects of metapedogenesis are taken into account can be obtained from examining the classification schemes used by soil surveys, a topic discussed in Chapter 3. The classification system of the Soil Survey of England and Wales has a major group labelled man-made soils which is subdivided into man-made humic soils and disturbed soils (Avery 1973). Thus man's potential effect of creating a topsoil (more than 40 cm thick) or causing major topsoil disturbance is appreciated. The American system of soil classification incorporates man's influence on soils through defining two types of diagnostic surface horizons (epipedons) and one diagnostic subsurface horizon (Soil Survey Staff 1975). An anthropic epipedon is a dark surface horizon, high in phosphate and bases, and is formed by long cultivation during which large amounts of organic matter and fertilizer are added. The other surface horizon is the plaggen epipedon, a man-made

Table 1.1 Suggested effects of the influence of man on five classic (though arbitrary) factors of soil formation (Bidwell and Hole 1965)

	Beneficial effects*	Detrimental effects*
1. Parent Material	(*a*) Adding mineral fertilizers; (*b*) accumulating shells and bones; (*c*) accumulating ash locally; (*d*) removing excessive amounts of substances such as salts; (*e*) marling; (*f*) warping	(*a*) Removing through harvest more plant and animal nutrients than are replaced; (*b*) adding materials in amounts toxic to plants or animals; (*c*) altering soil constituents in a way to depress plant growth
2. Topography	(*a*) Checking erosion through surface roughening, land forming, and structure building; (*b*) raising land level by accumulation of material; (*c*) land levelling	(*a*) Causing subsidence by drainage of wetlands and by mining; (*b*) accelerating erosion; (*c*) excavating
3. Climate	(*a*) Adding water by irrigation; (*b*) rain-making by 'seeding' clouds; (*c*) release of CO_2 to atmosphere by industrial man, with possible warming trend in climate; (*d*) heating air near the ground; (*e*) subsurface warming of soil, electrically, or by piped heat; (*f*) changing color of surface of soil to change albedo; (*g*) removing water by drainage; (*h*) diverting winds	(*a*) Subjecting soil to excessive insolation, to extended frost action, to exposure to wind, to compaction; (*b*) altering aspect by land forming; (*c*) creating smog; (*d*) clearing and burning off organic cover
4. Organisms	(*a*) Introducing and controlling populations of plants and animals; (*b*) adding organic matter (including 'nightsoil') to soil directly or indirectly through organisms; (*c*) loosening soil by plowing to admit more oxygen; (*d*) fallowing; (*e*) removing pathogenic organisms, as by controlled burning	(*a*) Removing plants and animals; (*b*) reducing organic matter content of soil through burning, plowing, overgrazing, harvesting, accelerating oxidation, leaching; (*c*) adding or fostering pathogenic organisms; (*d*) adding radioctive substances
5. Time	(*a*) Rejuvenating the soil through additions of fresh parent material or through exposure of local parent material by soil erosion; (*b*) reclaiming land from under water	(*a*) Degrading the soil by accelerated removal of nutrients from soil and vegetative cover; (*b*) burying soil under solid fill or water

* The terms 'beneficial' and 'detrimental' imply a value judgement, and the table is admittedly oversimplified and patently biased, but as a device to stimulate discussion, the presentation of such a table is felt to be justifiable.

layer more than 50 cm in depth which is produced by cutting sods for bedding material subsequently added to the soil. The formation of plaggen soils is described in section 1.5. The only diagnostic subsurface horizon resulting from man's influence is the agric horizon. It forms immediately below the ploughed layer and contains accumulated clay and organic matter. Such illuvial processes are accelerated by the improvement in soil structure and drainage following cultivation of the upper layer. In the American *Soil Taxonomy*, the presence of diagnostic horizons is used to classify soils. The details of this procedure are considered in Chapter 3, but the relevant point in the present context is that the American soil classification system does incorporate man's influence on soils to a limited extent. Without doubt, soil surveyors will have to increasingly develop data collection and soil classification techniques which accommodate the magnitude of man's influence on soils.

2 TECHNIQUES OF MODERN SOIL SURVEY

E. M. Bridges

INTRODUCTION

The task of the soil surveyor may be succinctly described in five words: 'to make a soil map'. This simple phrase glosses over a large number of difficult problems concerning theoretical concepts of soil and practical difficulties of recording a three-dimensional phenomenon upon a map which is two-dimensional. For the present, soil will be considered as a three-dimensional natural body developed at the surface of the earth. It is the unconsolidated mineral or organic material at the earth's surface capable of supporting plant growth. Soils have an internal organization which includes a characteristic profile in depth, composed of recognizable horizons, produced from the parent material by the action of the soil-forming processes.

Variation in the impact of soil-forming processes results in soils with different assemblages of horizons. These enable the soil surveyor to develop a grouping of soil profiles for mapping units as is described in this chapter. The different profile characteristics also enable soils to be arranged into related classes for the purposes of classification.

Once the soils of an area have been mapped, the soil surveyor is, or should be, in a position to make precise statements about the soil mapping units. From his knowledge of the soils, their properties and distribution, the soil surveyor is able to give factual information which will enable decisions to be made regarding the use of land whether for horticulture, agriculture, forestry or indeed any other use. It is this theme which underlies the material presented in this book, with its practical importance and its relevance to studies of physical and human geography.

2.1 SOIL MAPPING UNITS

The basic unit employed in modern soil studies is the soil profile. This comprises a vertical section through the horizons in which features of pedogenesis can be observed. The soil profile includes those horizons from the parent material which have been altered by the processes of soil formation. They are layers of variable thickness which are approximately parallel to the surface of the soil. Obviously, the features useful to the soil scientist are those which are least susceptible to change. However, the soil profile is not an immutable feature like a geological rock succession but is a dynamic body produced by the interaction of different factors in an open system. It is either in a state of dynamic equilibrium or is a slowly evolving entity (Ch. 5).

Surface horizons with an intimate mixture of organic matter and mineral material are referred to as A horizons, and immediate subsurface horizons which have lost material by leaching and eluviation are indicated by the letter E. Deeper horizons which have gained material from the upper or eluvial horizons are designated B horizons and the parent material, the letter C, if unconsolidated and R if solid rock. Subscripts placed after these major soil horizon designations indicate specific characteristics (Table 2.1). Lists of these horizon symbols are given in most handbooks of field soil description. The A and B horizons, most intensively exploited by the plant roots, are sometimes referred to as the solum. In certain cases, the presence of specific characteristics in the soil profile to an arbitary depth is regarded as an important criterion (Ch. 4).

As a soil profile is only two-dimensional, it cannot be used conceptually to make a soil map because it is necessary to have a three-dimensional unit for this purpose, as well as for a general understanding of the spatial characteristics of soils. Such a unit is the pedon; alternative names suggested include the soil area or pedounit (Fig. 2.1). The pedon is described as the smallest volume which may be called a soil, and it is defined as having lateral dimensions large enough to encompass the natural variation of the horizons present. The area of a pedon ranges from 1 to 10 m^2 depending upon this variability. The three-dimensional character and limited variability of the pedon gives a more acceptable basic unit of study than the soil profile. Theoretically, by grouping together closely

Table 2.1 Soil horizon nomenclature

Litter layers and organic horizons	Mineral horizons
L Fresh litter, original plant structures little altered.	**A** Mineral soil horizon formed at or near the surface, characterized by incorporation of humified organic matter intimately associated with the mineral fraction. Incorporation of organic matter is presumed to result from biological activity or artificial mixing during tillage. Ah: uncultivated A horizon; Ap: cultivated A horizon; Ag: gleyed A horizon

Table 2.1 (cont.)

Litter layers and organic horizons	Mineral horizons
F Partly decomposed or comminuted litter remaining from earlier years; some plant structures remain visible.	**E** Subsurface mineral horizon underlying the A horizon that is lighter in colour and contains less organic matter, sesquioxides of iron and/or clay than the horizon beneath. Ea: bleached horizon of podzol soils; Eb: lighter coloured horizon of brown soils; Eg: gleyed E horizon
H Well decomposed litter in which original plant structures cannot be seen; may be mixed with mineral matter.	**B** Mineral soil horizon differentiated from adjacent horizons by colour and structure. It usually underlies an A or E horizon and is characterized by illuvial concentration of silicate clay, iron, aluminium or humus. Other forms of B horizon result from alteration of the parent material by removal of carbonates; formation, liberation or residual accumulation of silicate clays or oxides. Bfe: sharply defined iron pan; Bs: ochreous coloured, sesquioxide enriched B horizon; Bt: clay enriched B horizon Bw: B horizon showing evidence of alteration by weathering, leaching or reorganization *in situ*. Bg: gleyed B horizon Bh: B horizon containing translocated organic matter associated with aluminium or iron and aluminium.
O Peaty horizons accumulated under wet conditions and saturated for at least 30 consecutive days in most years or have been artificially drained. Of: fibrous peat; Om: semi-fibrous peat; Oh: uncultivated amorphous organic material; Op: cultivated surface of amorphous organic material; Omh and Ohh: Om or Oh horizon containing black glossy coats on structure faces or channels	**C** Unconsolidated or weakly consolidated mineral horizon which retains evidence of rock structure and lacks pedological properties. The C horizon may possess accumulations of $CaCO_3$ or more soluble salts, it may have dense, brittle properties and it may be modified by gleying. Cu: unconsolidated materials without gleying; Cr: weakly consolidated rocks (chalk, shale, glacial till); Ck: containing secondary $CaCO_3$ at least 1 per cent; Cg: gleyed C horizon; CG: intensely gleyed C horizon; Cm: cemented material; Cx: compact dense fragipan
	R Hard or very hard bedrock

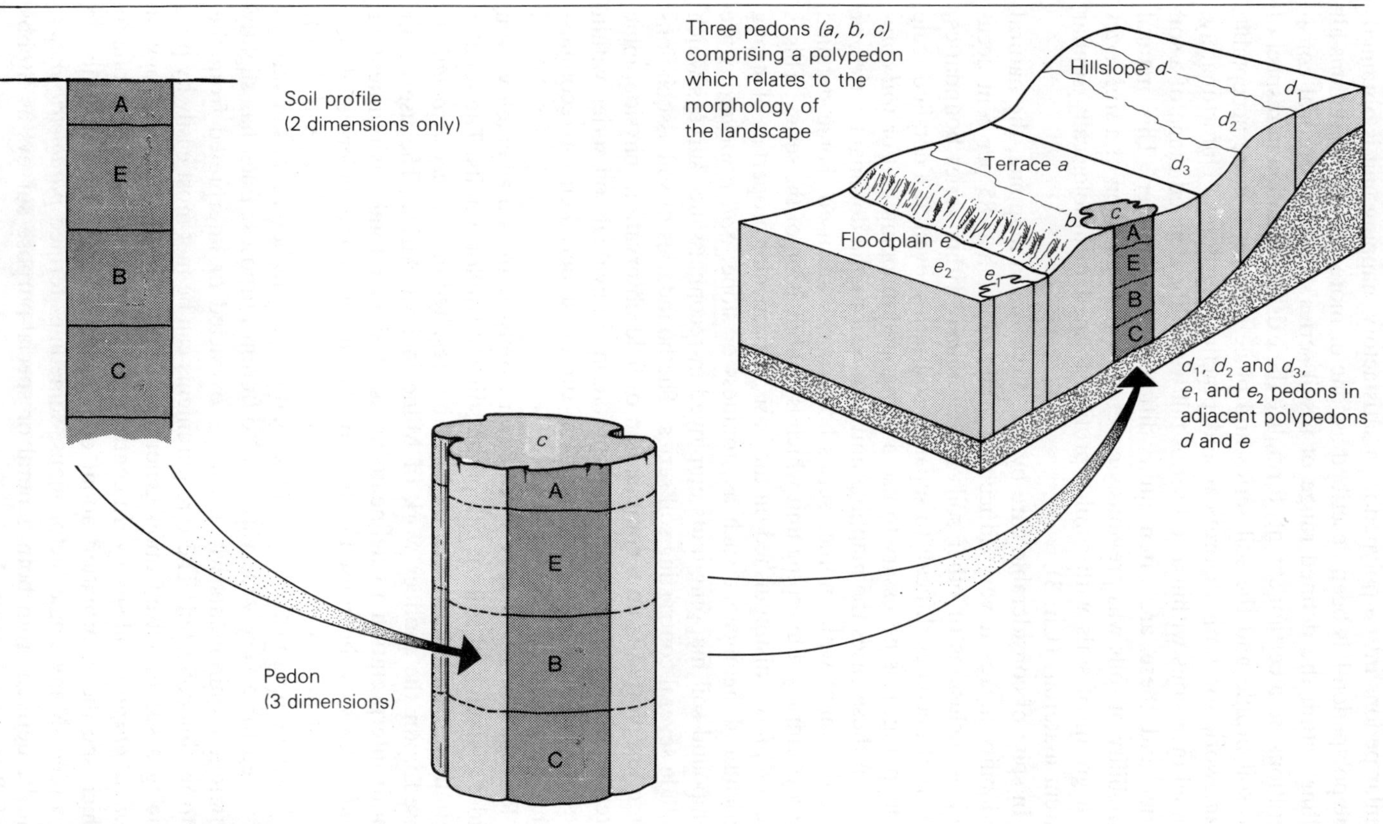

Fig. 2.1 The soil profile, pedon and polypedon (Bridges 1978)

similar pedons into a polypedon, a satisfactory mapping unit is obtained. The polypedon has been described as 'one or more continuous pedons all falling within the defined range of a soil series'. Such theoretical considerations, to a certain extent, formalize the existing situation in respect of the soil profile and the soil series and place them in a more acceptable framework for both theoretical and practical purposes. Unfortunately, not all inclusions within a soil series fall within the defined range of a soil series and these are often quite different in character. This natural variability of soils was previously covered by the definition of a *soil series* as a group of soils with similar profiles formed in lithologically similar parent materials (Ch. 3).

In spite of considerable care by soil surveyors in mapping, the natural variability makes it very difficult to attain more than 85 per cent accuracy as inclusions of other soils are contained within series boundaries. Where it becomes difficult to separate soil series, even at a detailed scale of mapping, it is necessary to use a compound mapping unit or *soil complex*. In these cases the mapping unit is known by the dominant series, or if co-dominant with another series both names are used. Some detailed surveys employ a mapping unit which is a subdivision of the series. This is the *soil phase*, distinguished on the basis of particular properties of strong agricultural significance such as stoniness or slope. *Soil associations* are compound soil mapping units employed in reconnaissance surveys which include several contrasting soil series. The boundaries of soil associations are often plotted from a combination of field observation, physiographic interpretation and air photo interpretation. Individual soil series within the associations are not delineated in surveys undertaken for reconnaissance purposes.

Reconnaissance surveys which cover extensive areas may employ even wider groupings of soils or composite soil and landform units. The regular variation of soils in relation to slopes has resulted in *catenary associations* based upon the pioneer work of Milne in East Africa. The use of air photo interpretation to delineate areas of similar total environment or *land systems* has been used with particular success in the under-developed parts of northern Australia and Papua-New Guinea (e.g. Story *et al.* 1976).

It will be demonstrated in the second part of this book that soil surveys are used for a very wide variety of different purposes. Each has slightly different requirements which can be extracted or interpreted from the information collected. These requirements can be met most easily by producing a survey based upon general soil properties and morphology, a *general purpose* soil survey. In contrast, a *special-purpose* survey is one in which specific information about one, or a few soil properties only, is recorded. Many single factor maps, sometimes termed *parametric maps*, can be derived from both general or special-purpose surveys as Stobbs (1970) has demonstrated.

2.2 SCALE AND PURPOSE OF SOIL SURVEY

One of the most critical features of a map is its scale, for this determines what can be shown, especially in terms of the smallest area capable of being represented. On the map, this is limited by practical considerations to an area of about 0.25 cm^2. Thus any soil area which occupies less than 0.25 cm^2 when represented upon the map must be combined with adjacent soil mapping units. Soil maps are published at all scales from those in atlases which cover large areas at small scale to large scale maps of small areas such as an individual farm, development project or nature reserve. The information shown, and particularly its detail, varies with the scale and with the methods used to compile the map (Fig. 2.2). The mapping scales and different types of soil survey are considered in the following paragraphs. (See also Soil Survey Staff 1951, Stobbs 1970, Young 1973, 1976 and Western 1978.)

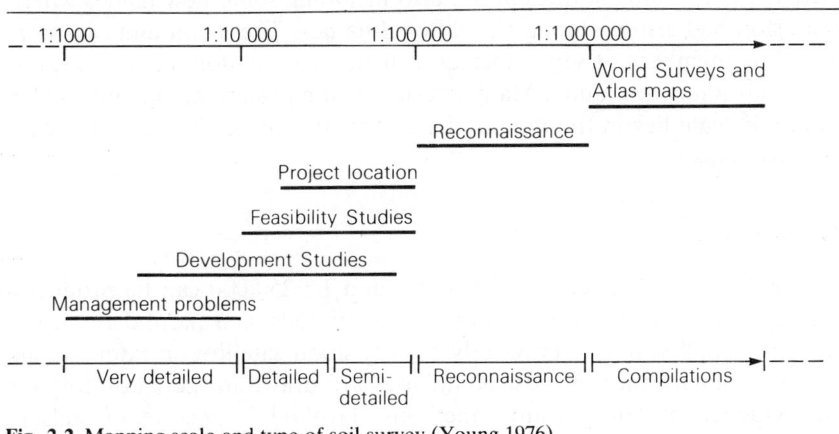

Fig. 2.2 Mapping scale and type of soil survey (Young 1976)

Compilations

As the name suggests, these maps are compilations derived from existing soil surveys together with intuitive extrapolation about the soils in areas where no soil surveys have taken place. In the past small-scale maps of continental-sized areas were produced by indirect means, using the factors of soil formation as a guide to the soils which might be present. Thus many soil maps were but intepretations of climate or geology maps. At present the greater availability of knowledge about soils enables a more soundly based approach to be made with road-traverses linking information already available from previous surveys. Inevitably, the mapping units employed in these surveys are broad, normally the Soil Orders or the Great Soil Groups of the World, and a very general picture of soil

Table 2.2 Type of survey, scale and characteristic mapping units (After Stobbs 1970; and Young 1973)

Type of survey	Mapping scale	Mapping units
Very detailed	larger than 1 : 10 000	Soil series and phases
Detailed	1 : 10 000 to 1 : 50 000	Soil series, phases and complexes
Semi-detailed	1 : 50 000 to 1 : 100 000	Soil series and complexes
Reconnaissance	1 : 100 000 to 1 : 250 000	Soil associations
Exploratory	1 : 250 000 to 1 : 1 000 000	Soil orders, great soil groups or land systems
Compilations	smaller than 1 : 1 000 000	Soil orders

distribution is given. The FAO/UNESCO Soil Map of the World at 1 : 5 000 000 is a good example of this type of soil map (FAO/UNESCO 1974) (Fig. 2.3). It has 106 'soil units' in 26 groups bearing familiar names such as podzol and chernozem but also including some new names where confusion had arisen amongst traditional names. These soil units are portrayed as members of soil associations including one dominant and up to three subsidiary soil units. Many national soil maps are compilations also and their scale lies in the region of 1 : 1 000 000 or smaller as in the case of atlas maps.

Exploratory surveys

Surveys at scales between 1 : 1 000 000 and 1 : 250 000 can be produced for large areas as general inventory maps of soils as a natural resource. At these small scales, surveys rely heavily upon satellite or extensive air photo reconnaissance as most boundaries are drawn in the laboratory on the evidence of interpretative methods. Ground inspection of soils is limited to approximately 1 per km^2 and takes the form of checking areas defined upon air photographs to see if the boundaries are valid and that the soils belong to particular great soil groups or soil orders. Maps produced at these scales are often the integrated environmental type in which the soil is only one of several environmental features examined in the delineation of composite land systems (Fig. 2.4). In addition to the general inventory purposes already mentioned, these maps may be used for large-scale comparative exercises and for identifying sites for future work of a more intensive nature.

Reconnaissance surveys

Surveys of soils at scales between 1 : 250 000 and 1 : 100 000 fulfil a similar purpose to those described as exploratory surveys and are often commissioned for specific purposes rather than as a general inventory survey

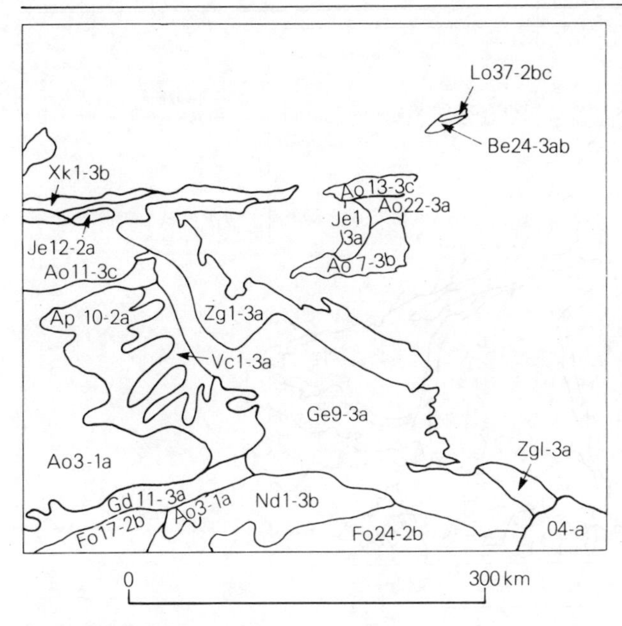

Legend for soil groups shown:

Ao3–1a Ferratic Arerosols
Ao7-3b Orthic Acrisols with Orthic Ferralsols
Ao13-3c Orthic Acrisols with Lithosols
A011-3c Orthic Acrisols with Nitosols and Lithosols
A022-3a Orthic Acrisols with Plinthic Ferralsols
Ap10-2a Plinthic Acrisols with Regosols
Be24-3ab Eutric Cambisols
Fo17-2b Orthic Ferralsols with Acrisols, Lithosols
Fo24-2b and Dystric Regosols
Gd11-3a Dystric Gleysols with Plinthic Acrisols and Planosols
Ge9-3a Eutric Gleysols with Eutric Fluvisols and Dystric Gleysols
Je1-3a Eutric Fluvisols with Calcaric Fluvisols
Je12-2a Eutric Fluvisols with Eutric Gleysols and Mollic Gleysols
Lo37-2bc Orthic Luvisols
Nd1-3b Dystric Nitosols
04-a Histosols with Thionic Fluvisols, Mollic and Eutric Gleysols
Vc1-3a Chromic Vertisols
Zg1-3a Gleyic Solonchaks with Plinthic Gleysols
Xk1-3b Calcic Xerosols and Lithosols

Fig. 2.3 Extract from FAO/UNESCO Soil Map of the World 1 : 5 000 000 Volume South America

(Fig. 2.5). Their major purpose is to identify possible areas for further intensive soil survey work as might be done before locating new irrigation schemes. Air photographic interpretation is the basis of boundary location but the soil observations become slightly more numerous and the

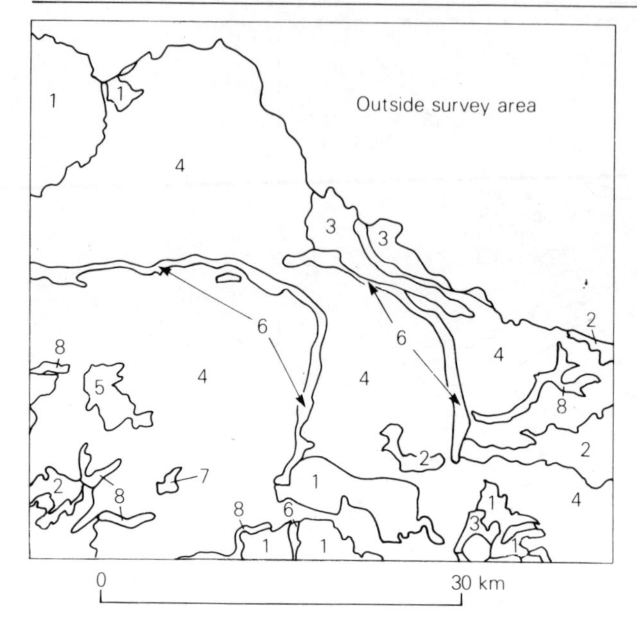

Legend for soils and land systems shown:

1. Skeletal soils on Buldiva land system
2. Sandy soils on Queue land system
3. Coarse sandy soils on Cully land system
4. Gravelly, lateritic soils on Knifehandle and Kysto land system
5. Skeletal soils on the Nova land system
6. Sandy alluvial soils on Effington land system
7. Krasnozems on Viney land system
8. Organic silty brown earths on Argoolook land system

Fig. 2.4 Extract from CSIRO 1 : 500 000 Soil map of the Alligator Rivers area Northern Territory, Australia (Story *et al.* 1976)

mapping units comprise individual great groups or associations of great groups.

Semi-detailed surveys

In surveys at scales between 1 : 100 000 and 1 : 50 000 considerably more detail can be shown (Fig. 2.6). Air photographic interpretation may have its part to play in boundary location at these scales but ground observation becomes increasingly important as the scale becomes larger. Thus maps compiled at 1 : 100 000 would have soil associations as their mapping units, but maps at scales larger than 1 : 50 000 would have soil series as the major mapping unit. In both cases, the associations or series may be given a physiographic grouping for purposes of discussion.

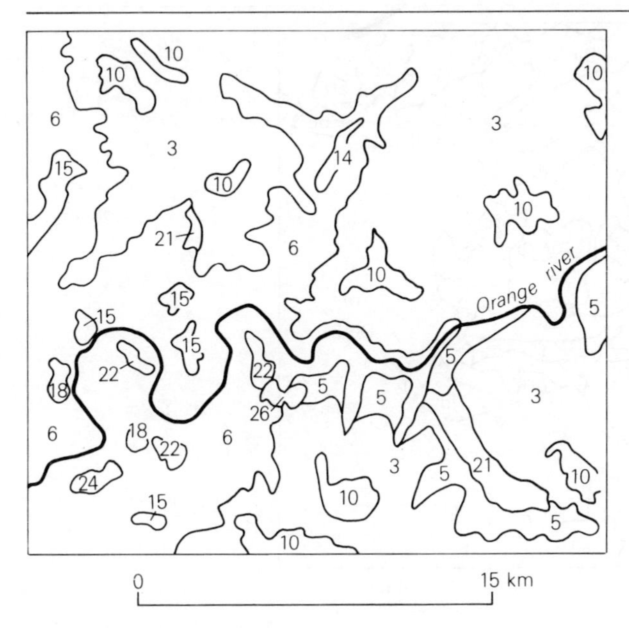

Legend for soil mapping units shown:

3. Lithosols on lavas/Calcimorphic soils
5. Lithosols on lava rich in ferromagnesian minerals/ Eutrophic brown earths
6. Lithosols on sedimentary rocks/sedimentary rock debris
10. Calcimorphic soils/Lithosols on lava
14. Vertisols of topographic depressions
15. Claypan soils
18. Claypan soils/Fersiallitic soils
21. Eutrophic brown soils/Vertisols
22. Fersiallitic soils
24. Fersiallitic soils/Claypan soils

Fig. 2.5 Extract from Land Resources Division 1 : 250 000 Soil map of Lesotho (Carrol and Bascomb 1967)

Soil surveys published at scales between 1 : 100 000 and 1 : 50 000 are used for general resource appraisal purposes at regional level and to establish areas for detailed investigations. With greater detail than is possible to show on reconnaissance surveys, these semi-detailed surveys provide a basis for developing alternative strategies for land use, settlement or agricultural development. Accordingly, FAO has referred to surveys at these scales as '*pre-investment surveys*'.

Detailed surveys

At scales between 1 : 50 000 and 1 : 10 000 it becomes possible for the

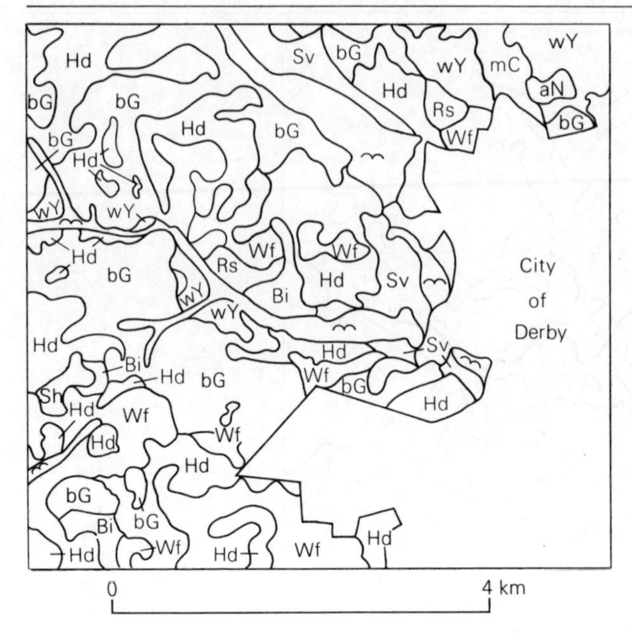

Legend for soil mapping units
aN Alton series
Bi Brockhurst series
bG Bromsgrove series
Hd Hodnet series
mC Mercaston series
Na Newport series
Rs Risley series
Sh Salop series
wY Windley series
Sv Sutton complex
⌒ Soils on alluvium

Fig. 2.6 Extract from Soil Survey of England and Wales 1 : 63 360 Sheet 125 Derby (Bridges 1966)

cartographer to indicate field boundaries upon the topographical map. Consequently in a soil survey produced at these scales, soils can be related directly to the parcels of land which enclose them. At 1 : 25 000 the smallest area which can be conveniently shown true to scale is 0.25 ha and a soil boundary line on the map represents a zone 5 m wide on the ground. When the scale is increased to 1 : 10 000 the smallest area which can be conveniently shown is 0.625 ha and at this scale a boundary line on the map represents a zone only 2 m wide on the ground. The soil surveyor can indicate with considerable accuracy the location of soils on the landscape and can show the intricacy of their boundaries on the map. Air photography interpretation is used less at these detailed scales of mapping, but it may be of value in certain circumstances. It has been found useful in areas of very low relief such as the fenlands of eastern England where old river courses (roddens) can be plotted from air photographs but are very difficult to see on the ground. In arid climates, the distribution of areas of soil affected by salt accumulation can often be seen clearly on air photographs.

Publication of soil maps at scales of 1 : 25 000 or 1 : 10 000 is very useful in areas where soil-related problems are known to exist. Agricultural advisers can see at a glance if a particular field is likely to have a soil with a specific problem, and as the soil map units are at series or even phase level, extrapolation of information from one area to another is possible. Maps at these scales are valuable for advisory personnel when giving technical advice as well as forming the basis of economic forecasts for the viability of development schemes. The soil detail known at these scales is such that it should be possible to indicate whether or not a development scheme is feasible. Hence the alternative name '*feasibility surveys*' is sometimes given. In this category, there is clearly some overlap between surveys which are for feasibility and those which take place following a decision to proceed with a scheme; the surveys are then known as *development surveys*.

Very detailed surveys

Surveys published at scales of 1 : 10 000 and larger are usually concerned with the precise location of high-cost projects or management problems of specialized crop production. Surveys in this category would have specific objectives and the data to be collected would be contractually agreed. Boundary lines between different soil series, or even phases of soil series, occupy zones less than 2 m wide and so the soil surveyor can show the soil boundaries with exactitude (Fig. 2.7).

In order to map soils in great detail, surveyors have found it necessary to use the grid mapping technique described later, inspecting the soil at regular predetermined intervals according to a sampling plan which gives all soils an equal chance of being represented. Although air photo interpretation and free survey methods can be used in combination with the grid survey method, boundaries are located by interpolation between field observations.

Surveys at scales of less than 1 : 10 000 are used only for very detailed projects and research investigations. An example would be the fields of an experimental farm where experiments into crop management and fertilizer requirements are to be set up using 'benchmark' soils from which the results could be extrapolated by agricultural advisers to the soils of surrounding areas.

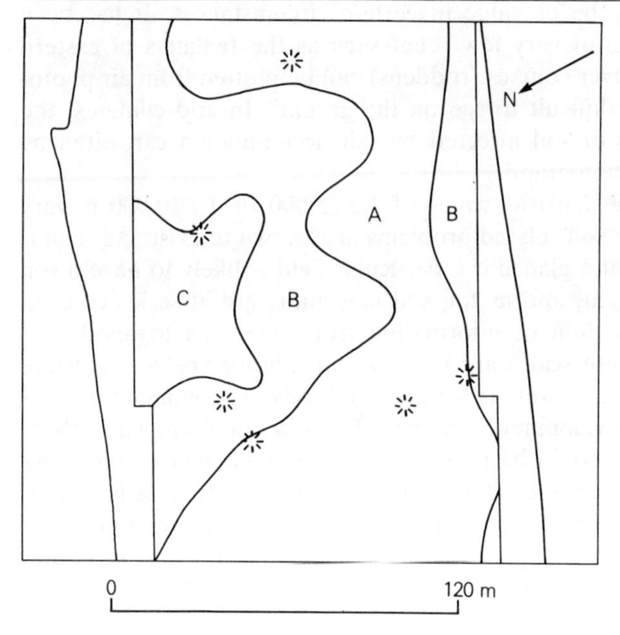

Legend for soil mapping units:
Batcombe series:
Phase A: silty clay loam or clay loam with yellow-red clay at <30 cm depth
Phase B: silt loam with brown silty or finer subsurface horizon over yellow-red clay at depths between 30 and 60 cm.
Hook series:
C: brown friable silt loam over silty clay loam subsoil of 60 cm overlying clay with flints.
Construction of the map of the whole field entailed 190 auger holes and 5 soil pits. ⁂ marl pits or 'dells'

Fig. 2.7 Part of soil map of Broadbalk Field approximately 1 : 2000 Rothamsted Experimental Station Report for 1969 (Avery and Bullock 1969)

2.3 SCALE AND NUMBERS OF OBSERVATIONS

As the scale of the map becomes larger, it becomes possible to make and record more observations and to show more detail of the soil pattern. Inspection of the soil itself, as well as observations of landscape characters which reflect soil conditions, are used by the soil surveyor in the compilation of a soil map. The skilled surveyor integrates in his mind the effects of the soil-forming factors and processes with the physical landscape and hypothesizes the distribution of soils. While he is in the field, therefore, the soil surveyor continually makes observations. At certain intervals the surveyor will make a direct inspection of the soil, using either a spade to dig a pit or an auger to extract a sample from lower horizons to ascertain the soil profile. As it is clearly impossible for the surveyor to inspect the soil at all points of the landscape, the process of soil surveying is one of sampling, followed by the determination of boundaries to obtain the most acceptable pattern of soil distribution at the scale of the map.

All soil maps must be based ultimately upon field observations, but those which are produced at small scales for atlases and generalized appreciation of soil distributions are mainly schematic, being derived from empirical data supported by information from existing surveys. In such cases, the ground information is minimal and the mapping units are the broadest in the classification system such as the soil orders of the American *Soil Taxonomy* or the major groups of the classification employed in England and Wales.

In reconnaissance surveys at scales between 1 : 250 000 and 1 : 1 000 000 air photo interpretation is a most helpful aid in rapidly determining soil boundaries, especially in under-developed areas where the topography, geology and vegetation often reflect soil conditions. This can economise greatly on time, expense and the number of ground observations which have to be made. Maps such as these would be produced with an average of one observation per km^2 with the use of air photo interpretation but at least double this for surveys not using it. Air photo interpretation can still be of great assistance in semi-detailed and detailed surveys but within the range of these scales direct inspection of soil characteristics becomes increasingly important. Figures quoted by some authorities recommend between 12 and 25 observations per km^2 for survey without air photo interpretation and only between one and three with it for maps of a detailed or semi-detailed nature.

In surveys for 1 : 10 000 maps in the Netherlands, Steur *et al.* (1961) describe a density of observations lying between 400 and 900 per km^2. In the author's experience working in rolling country of the English Midlands approximately 1 km^2 could be mapped per day on the basis of between 30 and 50 soil inspections for eventual publication at 1 : 63 360. This is confirmed by Beckett (1978), who quotes similar figures for surveyors in different parts of Britain adding that on average surveyors produce about 3–4 km of boundary per day working at that scale. In very detailed surveys the emphasis rests heavily upon direct soil inspections, and although air photo interpretation can still be helpful, there is less economy in time and numbers of observations made.

One of the major constraints of soil survey is that the surveyor must walk across the landscape, making his observations and inspections as he goes. Not that this time walking between sites is wasted, for the experienced soil surveyor will be using his expertise to interpret the soil pattern from the features of the landforms and vegetation present. He does not rely entirely upon soil-based inspections and interpolation of boundaries to make his map. The problem of numbers of observations needed to produce a map of a particular scale can be approached from the opposite point of view. At each scale there is a minimum-sized area which can be depicted upon the finished map. This minimum map area is about 0.25 cm^2 (whatever the scale) and it is the minimum area that the surveyor can use because if he identifies a smaller area on the ground, it will have to be incorporated into an adjacent mapping unit as an impurity. Thus for any mapping unit to be shown it must have at least one observation in the field to prove its presence which suggests there is a minimum of four observations or inspections per cm^2 of the published map. The various sizes are indicated in the Table 2.3 (Vink 1963).

Table 2.3 Scale, area of 1 cm^2 on the map and minimum number of observations per km^2

Scale of map	Size of 1 cm^2 area on the map	Minimum number of observations per km^2
1 : 2 500	0.000 625 km^2	8000
1 : 10 000	0.01 km^2 (1 ha)	400
1 : 25 000	0.062 5 km^2	64
1 : 50 000	0.25 km^2	16
1 : 100 000	1 km^2	4
1 : 250 000	6.25 km^2	0.7
1 : 500 000	25 km^2	0.15
1 : 1 000 000	100 km^2	0.05

Soil variability and soil maps

There have been few studies of soil variability, a surprising fact as there is no practical way of making a soil map which would ensure completely uniform mapping units. The natural variability of soils is often commented upon but little has been written of its importance and implications when soil maps are used for practical purposes. In the process of mapping, surveyors subjectively decide on the variation allowed in the mapping unit, but rarely is the extent of this variation adequately defined. Soil surveyors are very aware of the problem of soil variability and their aim is to minimize it within their mapping units. In statistical terms surveyors try to keep the within class variance as low as possible. This variability is often relatively local in the sense that at least half the variance within a field may be present within an area of 1 m^2 and this appears to remain true with increasing size of field (Webster and Beckett 1968; Beckett and Webster 1971).

A problem which emerges from a brief investigation of the literature concerning soil variability is that no two soil scientists appear to have approached the problem with the same sampling design. Agricultural advisers would normally sample on a 'whole field' basis except where substantially different soils were obvious. An accepted procedure is to sample with an auger along a zig-zag line across the whole field to give a bulk soil sample for analysis. Ball and Williams (1968) found a high degree of local variability in uncultivated brown soils from Snowdonia and emphasized the changes in the range between 1 cm and 1 m. Their samples were taken at 27 m intervals along a transect parallel to the slope. At each sampling position samples were taken in a circular pattern 1 m in diameter to demonstrate seasonal changes, replicate samples were taken on a circular pattern 0.75 m in diameter and finally bulk samples were taken from the centre of the circles. Cipra *et al.* (1972) attempted to measure the variability using a sampling pattern at distances of 3 m, 9 m, and 8–145 km. Such multiple-sampling investigations have drawn attention to the problem at a research level. Definition of this variability can be given in statistical terms and Nortcliff (1978) has used a hierarchical sampling design based on the major parent materials present in a study of soil variability in Norfolk. This design enabled him to investigate the implications of variability at different scales of mapping. Other workers have used a simple grid form of sampling and this is supported by Campbell (1979), who suggests that if samples are collected on a regular spatial interval, place to place variability can be described by autocorrelation. This is defined as a measure of the degree of association between neighbouring measurements of a spatial distribution. For further discussion see Chapter 4.

At the practical level, the soil scientist is interested in whether sampling on the basis of soil series, or other mapping units gives a more accurate and useful result compared with the whole field method which may include one, two or more mapping units. Coulter (1974) presents results from four sites where samples from series and from whole fields are compared. Table 2.4 shows that in most cases the coefficients of variation are much reduced by series sampling within soil series boundaries. However, the variations measured in these studies are of soil chemical characteristics such as pH, $CaCO_3$ content, exchangeable cations and organic carbon and not the morphological criteria used by soil surveyors to map soil distributions.

At present it is not known if the variability of soils can be related to any other easily determined or observed soil property. If this were possible it would help overcome one of the limitations of soil maps highlighted by Avery (1970): 'lateral variability is the major factor limiting the predictive value of soil maps'.

2.4 PROCEDURES FOR SOIL SURVEY

Weston (1978) has stated that a soil survey is bound by normal contrac-

Table 2.4 Coefficients of variation for soil sampled by series and by field (Coulter, 1974)

Field measurement	Site A		Site B		Site C		Site D	
	CV%	$(CV)_w$%	CV%	$(CV)_w$%	CV%	$(CV)_w$%	CV%	$(CV)_w$%
2 m clay	6.83	25.00	3.85	21.48	4.13	27.59	3.91	9.96
0.3 m clay	19.43	66.89	2.94	26.80	7.94	31.31	5.70	22.30
$CaCO_3$	12.76	106.6	49.15	120.4	26.35	59.15	34.99	33.28
Exchangeable K	7.58	226.22	19.26	23.39	11.11	27.34	12.27	14.63
$CaCl_2$ soluble P	17.88	16.55	53.74	57.10	54.19	55.60	10.63	24.73
$NaHCO_3$ soluble P	55.23	53.46	18.32	27.90	26.27	32.88	7.51	13.36
Per cent K in 2 m clay	8.74	40.79	4.80	5.67	2.70	6.49	5.97	6.49
Organic carbon	7.28	21.58	6.08	8.26	5.61	13.48	4.62	13.52

CV% Coefficient of variation when sampled by soil series.
$(CV)_w$% Coefficient of variation when sampled on a 'whole field' basis.

tual obligations between the client who requires the survey and the surveyor or his employer. Therefore, before any surveying begins negotiations must take place regarding the scope of the work, thc plan of operation and date of completion, the cost and method of payment, the obligations of employer and employed, the type of personnel to be used and any permitted variations from the contract. Assuming that investigations indicate the feasibility of the operation is possible and that contracts have been agreed, it is possible for the survey to begin.

The first step is to appoint a leader or party chief who will be responsible for supervising the survey and the maintenance of its technical standards. This person will normally be a graduate soil scientist with appropriate experience and he will subsequently be responsible for recruiting his team of professional assistants. Once these essential preliminary steps have been taken, work on the survey itself can begin. The work falls into three stages: preliminary work, ground survey and preparation of maps and report.

Preliminary work

During negotiations for the contract it will have been necessary to study the proposed area of survey so that estimates of time, personnel required and material costs can be made. At an early stage of the preliminary work this background knowledge must be passed to the survey team so that all the relevant information is known to the individual surveyors. All relevant maps, and geographical, geological, climatological and biological accounts of the proposed survey area should be examined. The surveyors should obtain a clear knowledge of the road and track network and topography of the area as well as the hazards likely to be met. The effects of the incidence of rainfall upon the rivers and the trafficability of the landscape could be severely limiting at certain times of the year.

It may be necessary, in the absence of topographical map coverage, to commission a full coverage of air photographs. For reconnaissance purposes photographs at a scale of 1 : 70 000 are considered to be satisfactory, but for detailed work cover at a scale of 1 : 10 000 is preferable. As photo scale increases, the number of prints to cover a specific area increases greatly, resulting in problems of cost and handling. It is helpful to have separate field and laboratory sets of the photographs, especially if the mapping has to be done using the photographs as a base map. In addition to the separate air photographs it is extremely helpful to have a controlled mosaic of the survey area constructed from the air photographs. If photographic cover is not available, additional time must be allowed to enable the photographs to be taken and processed. Only when the photographs have been supplied can the work of interpretation of the soil features begin.

Transport and equipment has to be organized. If vehicles are not readily available at the location of the survey they will have to be purchased and adapted if necessary for soil survey use before being freighted to the survey area. For soil survey work in the developing countries it is safer to assume that no supplies of equipment or spares will be available, so it is necessary to take everything likely to be required. Even where spares and equipment can be obtained, frustrating delays may occur which could unnecessarily prolong the period of field work.

During the initial stages of setting up a base near the survey area, local counterpart staff can be used to recruit drivers and labourers and any other ancillary staff. Every opportunity should be taken to teach local officials how and why certain things are done; in this way the future continuation of any agricultural improvements can be safeguarded. It is also desirable from the outset that all opportunities are taken to acquaint local people with the nature and purpose of the work. Dissemination of this knowledge will prepare for the fieldwork and save explanations later. It must be appreciated that not everyone sees the implementation of modern agriculture as desirable and some surveyors have found difficulties with worried victims of change. Counterpart officials who can explain to people in their own language are invaluable for, inevitably, a crowd of interested spectators soon gathers when a pit is being dug.

It is the responsibility of the party chief to ascertain by reconnaissance how great is the range of soils within the project area and to compile provisional mapping units so that the field survey can begin. In doing this crucial piece of work the party chief must take care the mapping units are suitable for the scale of the survey and for the variability of the soil pattern to be encountered by the assistant surveyors in their day-to-day work. This operation normally relies very much upon experience, but Beckett and Bie (1975) have shown that the intricacy, or length of boundary, which will be encountered in the survey can be estimated by the number of intersections made by soil boundaries with a random array of sampling lines. It is theoretically possible to make a statistical estimate of the likely total length of boundary, number of soil mapping units and the amount of time necessary to survey the whole area before routine survey begins.

The party chief will have to visit each significant physiographic region of the area to be mapped and will examine all the parent materials present and the range of soils developed upon them. Advantage will be taken of any excavations which already exist such as quarries, ditches and road cuttings. Failing any chance exposures, small inspection pits have to be made with a post-hole spade or auger samples are taken. Essentially the party chief does in a practical manner what the statistical approach attempts to do and his samples include examples of all the soils the assistant surveyors are likely to meet.

Methods of survey

The routine part of any soil survey consists of an examination of the soil profile at intervals over the area to be mapped. Examination may take place from chance exposures, but in general the surveyor has to rely upon his own means to reveal the soil profile. The most convenient and rapid method of looking at the soil profile is by the use of an auger. There are several patterns but they fall into two broad groups: screw augers and bucket augers. Both types break the soil and so cannot show structure, but texture and colours can be easily identified. The screw auger is useful for rapid inspections down to about 1 m and has been extensively used in British surveys. Where soils are deeper, or it is desired to know about the deeper subsoil conditions, as in irrigation studies, the bucket or 'Jarrett' type of auger has been found preferable. Although small inspection holes are often dug in the reconnaissance phase of a survey, complete soil pits are time-consuming and expensive, so this method is reserved normally for special sampling sites and other points of particular interest. A tractor-mounted excavator if available can save a great deal of time and energy, assuming trafficability of soils is good.

In carrying out the ground survey, the surveyor adopts one of two basic methods: grid survey or free survey. Both methods may be augmented by the use of air photographic interpretation.

Grid survey. The point has been made earlier that it is impossible to look at the soil profile everywhere, consequently soil maps are based on a series of observations scattered across the countryside. Where these observations are made on the intersections of a grid or at fixed intervals along a survey line the term 'grid survey' is used to describe the procedure. Such a survey gives all soils present in the area a statistically good chance of being represented on the final map and is a useful basis if further statistical work is envisaged.

Grid survey is an approach which can be used in landscapes covered with thick forest where visibility is severely limited. Under such conditions position location is difficult by other means and it would be difficult to use the morphological features of the landscape as a guide to the position of soil boundaries. Instead boundaries have to be interpolated between points of contrasting soil mapping units. At the opposite extreme, grid survey can be used where there is very little relief or other geographical features to guide the surveyor in his positioning of soil observations and boundaries. Although grid surveys have been used for reconnaissance or semi-detailed work in difficult country, they are usually employed where a survey of detailed or very detailed nature is required. Grid survey is most suitable for scales of soil mapping greater than 1 : 10 000 and is used for all intensive soil surveys.

Examination of the soil at fixed points throughout the survey area eliminates the subjective element of interpretation and soil boundaries are interpolated between differing observations. Such a method requires less skill and judgement as all the surveyor has to do is record the soil features at predetermined intervals. It is possible to employ less experienced staff using this approach of soil mapping. However, one major disadvantage of the method is that it can be inflexible in areas where the soil pattern is simple and can lead to wasted effort. It can also result in inspections taking place too close to hedgerows or roads or in one case (theoretically) in the river! (Fig. 2.8). Equally, if the grid is made finer to deal with an intricate pattern of soils, the number of observations increases enormously, and once again effort may be wasted.

Free survey. At semi-detailed and reconnaissance scales of soil mapping the most economical method of constructing a soil map is by 'free survey'. As the surveyor walks across country he positions his soil observations according to his interpretation of the landscape and his soil boundaries according to morphological breaks and changes of slope. By integrating in his mind, the implications of different facets of the landscape, combined with the natural vegetation patterns he can see, the surveyor deduces where soil boundaries occur and positions his auger borings to prove or disprove his hypothesis (Fig. 2.9). Although this is an effective and relatively rapid method of soil surveying, it can lead to a concentration of observations around the edge of soil mapping units within which 'impurities' of other soils could occur.

The great advantage of the method is that the surveyor is free to vary

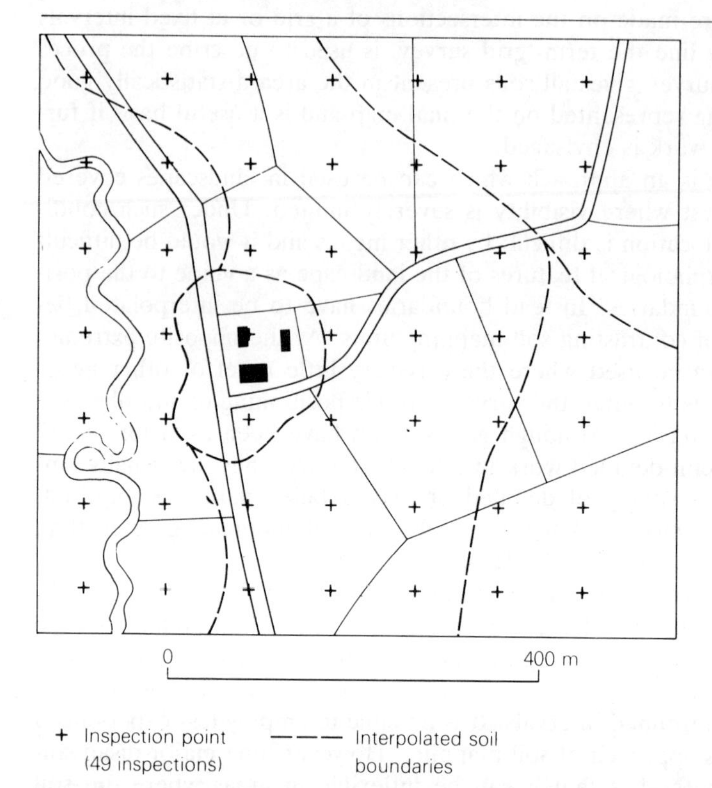

Fig. 2.8 Grid survey. Sites for soil inspections are controlled by the grid.

the intensity of his observations according to the intricacy of the soil pattern. This results in a greater accuracy when soil patterns are complex and does not waste time and energy when conditions are uniform. The use of the free survey method necessitates a good base map or air photograph upon which the surveyor can work without any problems of location so that observations and boundaries are correctly placed. In densely settled areas like Britain, the surveyor can position himself accurately by the field pattern but this could be difficult where fewer landscape features occur or where visibility is restricted by dense vegetation.

A very laborious method of surveying a soil boundary is to continually cross and re-cross it, extending it on the map as it is followed on the ground. This is not a method to be recommended as it is laborious and little knowledge is gained about the soils within the boundary. Each time a soil observation or inspection is made it is advisable to record as full details as possible. The amount of extra time involved is not great and if some reinterpretation is necessary the information is invaluable. If only a mapping symbol is recorded, and this is sometimes done when the

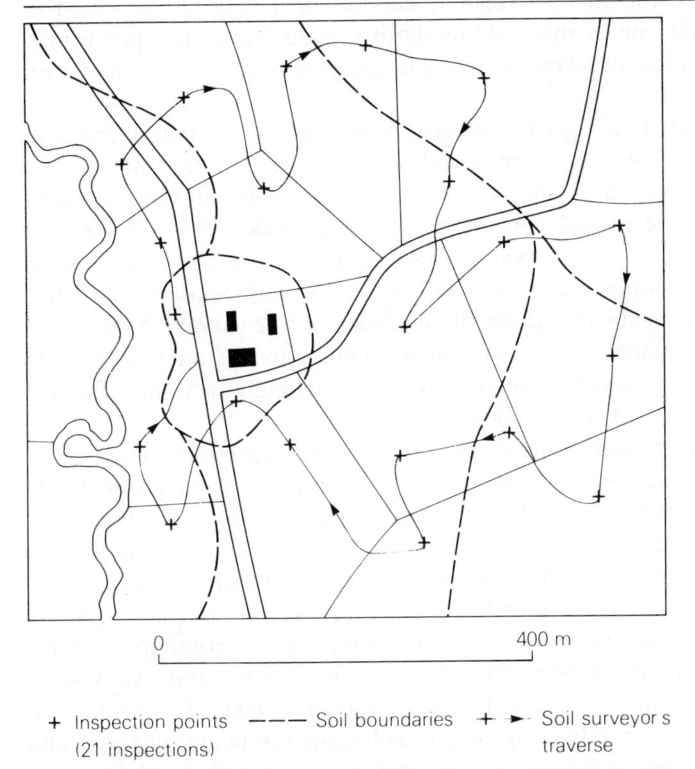

Fig. 2.9 Free survey. Sites where soil inspections took place through the surveyor's experience.

legend is finalized, reinterpretation is impossible and re-survey has to take place.

Experience has shown the free survey is best adopted for the production of maps between the scales of 1 : 100 000 and 1 : 25 000. Free survey has the additional advantage for the geographer that the resulting soil mapping units can usually be portrayed in a physiographic subdivision of the area. Thus the soil pattern can be seen as fitting logically into the landscape as part of a natural environment in which the soils are an integral part. It follows also that the land capability or suitability ratings suggested for the area can be viewed as an extension of the natural environment, requiring a particular combination of management practices.

Combination of mapping methods

The mapping method most suited to the aims of the survey and the terrain of the area in question must be decided upon by the party chief at the beginning of the survey. He may choose, or it may be stipulated in the contract that one of the basic methods is used, but in practice it may be most economical in terms of time and manpower to use a combination of methods.

In very detailed surveys the extent of the survey is limited and grid mapping is most commonly employed so the scope for other methods is restricted. However, a combination of grid inspections and sampling with free survey can be usefully employed at survey scales of 1 : 25 000 and smaller. The free survey provides a rapid ground coverage and the grid inspections and samples can be used to give an independent and complementary statistical support for the methods of free survey. As the map scale becomes smaller, the scope for a combination of air photo interpretation and free survey method increases, reaching a peak of efficiency in the range 1 : 25 000 to 1 : 100 000.

The nature of reconnaissance surveys with their extensive coverage of large areas at small scale ensures that a combination of air photo interpretation with limited ground checks is used to map the soils. It is most important in the use of air photo interpretation that the landscape can be readily subdivided into recurring landscape units significant in terms of soil patterns. This type of mapping is often referred to as 'physiographic survey' and soil associations are mapped within the physiographic regions identified on the air photographs. Both reconnaissance and exploratory surveys may rely on the 'external' factors, such as climate, for positioning soil boundaries but random checks are subsequently made by traversing the area in vehicles or landing at scattered locations with a helicopter to ascertain the soils present and to check the validity of boundaries.

In virtually all soil surveys, the published map is produced at a smaller scale than that at which the field work took place. For example, the field work for the 1 : 63 360 and 1 : 25 000 published maps of the Soil Survey of England and Wales was done using the 1 : 10 560/1 : 10 000 maps of the Ordnance Survey as base maps. The use of a larger scale map for field work has the advantage that the surveyor has more space in which to write his observations and when reduced to the smaller published scale, minor inconsistencies are arbitrarily removed by the cartographic process of reduction. A rule of thumb suggests that the field maps should be approximately at a scale twice the size of the published map.

Concluding phases of the survey

As the mapping programme nears its end, some of the survey team's effort will be diverted from routine mapping to taking representative samples. This is particularly true for a survey based on free survey methods, but in the case of a grid survey full descriptions and samples are systematically taken from pre-selected points on the grid as the survey proceeds. Full soil descriptions and samples are usually taken from pits specially sited to provide an example of the mapping unit for the report and for laboratory analysis. In general soil surveys, samples are taken from the horizons of the soil profile to provide additional information about soil genesis, but in fertility surveys samples may be taken from pre-selected depths regardless of the horizonation of the profile. These samples are simple bulk samples taken from the profile and placed in a suitable bag, labelled carefully and sent to the laboratory. In some cases, oriented samples are required for thin-section analysis; these are taken in 'Kubiena boxes' specially designed for taking a small sample from the profile face, the orientation of which is carefully recorded. If material is required for public display, a complete soil monolith may be taken from a soil pit while it is open.

In surveys for irrigation schemes, the infiltration capacity of the topsoil must be measured and the hydraulic conductivity of the lower horizons. If these are saturated, then a measurement can be made on the hydraulic conductivity by the 'borehole' method in which water is extracted from the hole and the time taken to re-fill is used to calculate the rate of water movement through the soil material. Infiltration capacity and hydraulic conductivity are also used in drainage design (see Ch. 6). The amounts of salts and their composition are extremely important in arid regions where most irrigation schemes occur. Samples must be taken back to the laboratory for electrical conductivity measurements so that the surveyors can be informed of the salt content of the soils they are mapping. It is helpful if samples of groundwater and the water available for irrigation can be measured for soluble salt contents.

It will usually be the responsibility of a senior person, or the party chief, to institute checks to see if the mapping has been done to an acceptable standard. Any uncompleted boundary lines must be completed and some of the early stages of mapping may have to be revised slightly in the light of subsequent experience. Ideally, once the fieldwork has ended there should be no need to return to the area for further checking. This is vital in any survey of an inaccessible or far distant part of the world, but even in areas where return is easy, it would add to the time and expense of completing the work.

As a guide to profile description, most surveyors acknowledge the influence of the *Soil Survey Manual* (Soil Survey Staff 1951; hereafter abbreviated to SSS), but with the passing of the years most countries have produced field handbooks with their own modifications of this invaluable book. A new version of the United States Soil Survey Manual is imminently expected. The contents of these field handbooks act as a framework within which soil description can be successfully and uniformly carried out by many different people. Many of these handbooks have been described and summarized by Hodgson (1978) and it is unnecessary

Table 2.5 Major headings for soil profile description (after Hodgson, 1976)

General information	Site description	Profile description
Profile number	Elevation	Recognition of horizons – see Table 1.1
Grid Reference	Relief, slope and aspect	Depth of horizons and thickness
Described by	Soil erosion and deposition	Colour according to Munsell Charts
Date	Flooding	Organic matter status
Weather conditions	Rock outcrops	Particle-size Class
Locality	Land use and vegetation	Stoniness
	Soil surface, from and condition, stoniness	Soil water state
		Soil Structure
		Consistence
		Roots and other soil flora
		Fauna
		Carbonates
		Features of pedogenetic origin
		Boundary to next horizon
		Soil reaction

Profile descriptions may be written out in full (Fig. 4.14) or may be encoded on standard soil description cards (Fig. 4.13). Alternatively, a description may be dictated into a portable tape-recorder and transcribed later in the laboratory. The full range of the terminology is given in the Soil Survey Field Handbook (Hodgson 1976).

to go into detail here, other than to present the main headings in Table 2.5. The development of soil information systems, in which information is stored in computers, requires information to be uniformly handled and presented in a form which can be handled by the data manipulating system (Chapter 4). With the appropriate recall program, the complete profile can be obtained on a print-out from the data store (Webster *et al.* 1976) (see Fig. 4.13).

2.5 LABORATORY SUPPORT

Modern soil survey has become increasingly aware of the need for clear criteria in the mapping and characterisation of soils. In the past laboratory data have been like the icing on the cake, a luxury which often played very little part in the survey as a whole. At the present time, there is a trend to require soils to meet certain criteria before they can be classified, which necessitates laboratory facilities. The supportive role of the laboratory in soil survey work can be discussed under three headings: physical, chemical and microscopic criteria.

Physical criteria

Facilities for particle-size analysis are essential for checking the field surveyors manual estimation of soil texture and for accurate determination of the particle size of samples from representative profiles. Before a B horizon can be called an argillic horizon, it must generally have at least 1.2 times as much clay as some horizon above. If the soil has less than 15 per cent clay then the argillic horizon must contain 3 per cent more clay and if the eluvial layer has more than 40 per cent clay, then the B horizon must have more than 8 per cent more clay to qualify as argillic. Criteria such as these are impossible to fulfil without laboratory support.

The physical properties of infiltration and conductivity have already been mentioned in connection with the field mapping programme, but characteristics such as bulk density, liquid and plastic limits and compression tests are widely used by many soil scientists. Moisture holding and release properties of soils are of vital significance to growing crops and a start has been made in England and Wales to quantify these properties for major soil series of wide geographical extent (Hall *et al.* 1977).

Chemical criteria

One of the first questions experts ask about a soil usually concerns its pH value. This is a simple test which reveals a great deal to the soil scientist about the capacity of the soil for plant nutrition and growth; it also appears frequently as a criterion for soil classification. Together with the electrical conductivity (EC), these are the most significant soil characteristics required by almost all soil surveys. Some classifications throw weight on the adsorbed cations in soils, but as these may have been changed by management techniques, their usefulness as a criterion is least in areas with a long settled history. However, a subdivision into soils with high and low base status has proved useful in the past. At the present time the percentage base saturation of less than 35 is used to separate the Ultisols from the Alfisols in the Soil Taxonomy of America.

The presence of calcium carbonate (free lime) in the soil is another characteristic which is important in pedological work as well as in the more practical aspects of soil science. Most crop plants are reasonably tolerant of the presence of lime in soils and even gypsum, but only a very restricted group can grow in soils rich in sodium salts. In arid and semi-arid parts of the world, salinization and alkalization limit the productivity of many thousands of hectares of otherwise cultivatable, productive land. Unfortunately, the area of salted land is increasing through poor and inefficient management. Laboratory tests for salts can be made quickly by the electrical conductivity of the saturated paste, or longer and more complex methods involving 'wet' chemistry can be used to determine individual salts present.

Modern soil classification systems require a knowledge of the iron content of soil horizons before the different varieties of podzols and podzolic soils can be correctly placed in the right taxonomic group. Several extractants are used separately or in sequence to obtain figures for the forms of iron involved in pedogenesis. Readily movable iron, 'free' iron and organically bound iron as well as residual forms which are difficult to move and only play a small part in soil formation, can be determined.

Organic matter is a characteristic feature of soils and its presence greatly influences the chemical, physical and biological properties. Organic matter can be roughly measured by 'loss on ignition' after the organic constituents have been burnt but an alternative approach is to determine the organic carbon content which can be converted by a conventional factor into a figure for the organic matter. The relationship of carbon to nitrogen is of interest to pedologists and nitrogen is one of the major plant nutrients. The plant nutrients and trace elements are more the province of the agronomist than the soil geographer, but as the different soil mapping units have markedly differing responses and powers of nutrient supply, they become of interest to the soil surveyor, especially when he is asked to compile derivative maps of land capability (see Ch. 6).

Microscopic criteria

Since its introduction by Kubiena 40 years ago, the technique of soil microscopy has become widely used by soil scientists throughout the world. It has been used to support soil survey work as well as being a study in its own right. The study of sand mineralogy has for many years been a technique whereby the parent material of the soil was investigated and rates of weathering have been calculated using index minerals. Both qualitative and quantitative studies are used, ranging from the low-magnification examination of humus forms to high-magnification studies of soil fabrics including the oriented clay domains using the petrological microscope or even a scanning electron microscope.

Oriented samples brought in from the field are air-dried and impregnated with synthetic resins and the resulting rigid block of soil can be cut into thin slices and mounted on to a microscope slide. After further grinding and polishing a thin-section is produced which can be examined under the petrological microscope. Characteristic fabric patterns can be identified from particular horizons and in the case of the argillic horizon a minimum of 1 per cent of clay skins should be visible in the section. In the case of brown podzolic soils a pellety fabric is characteristic which helps to distinguish these podzolic soils from the closely related brown soils which lack this property.

Microscopy is also used in the study of soil fauna. In addition to the well-known earthworms, termites and ants, many other small creatures, only visible with the aid of a microscope, live in the soil. Their role is concerned with the breakdown of organic matter and their life-histories and feeding habits are not known in great detail.

2.6 THE SOIL SURVEY MAP AND REPORT

The result of a soil survey is a map upon which is shown the distribution of the soil mapping units. Once the field surveyors have compiled a 'fair' copy of their map it is passed to the cartographers who have the responsibility of preparing it for publication. High standards of skill and accuracy are required to transfer the boundaries accurately to scribed plates. One is needed for each colour used, and there may be many combinations of colour necessary to make clearly distinguishable shades on the finished map.

Although a map is the logical outcome of a soil survey, much of the information gained is presented in a written report which accompanies the map and which explains in detail the mapping units, their properties and relationships. According to the Soil Survey Manual, every soil report should contain: an explanation of how to use the soil map and report; a general description of the area; descriptions of the individual mapping units shown on the map, supplemented with tables showing their characteristics and their relationships; predictions of the yields of common crops under specifically defined sets of management practices for all the soils mapped; explanations of the management problems of each soil with special emphasis on how the characteristics of the soil influence the problems and their solutions (Soil Survey Staff 1951). It is not always possible to meet all these requirements, but the soil survey report should go as far as possible to present such information.

The form of the report varies in detail and according to the purpose it is intended to serve. The memoirs and records of the Soil Survey of England and Wales are written with the qualified agricultural advisory officer in mind; many reports are specifically for experts who have to make a policy decision before money is allocated and spent, other reports are aimed specially at the farmer working the land. A format which is commonly used is as follows:

Table of contents. This should serve as an outline to the user, giving the names of the mapping units employed on the soil map.

General description of the area. This presents the physiography and drainage, the climate, water supply, natural vegetation, settlement and population, communications and industries and markets. This section is of particular interest to geographers whose methodology is used to present the information. The environmental conditions of the area should be illustrated with maps, diagrams depicting how climate, geology, geomorphology, vegetation and drainage affect the development of soils.

Methods of survey. For many users of the report it will be necessary to explain how soils are mapped and classified. The field procedures should be outlined briefly and the arrangement of the mapping units according to the classification used should be demonstrated.

Detailed description of the mapping units. This depends very much on the material to be presented. Authors of reports discuss the soils they have mapped either according to soil orders, major groups, etc., or they use a physiographic framework within which the soils of each physiographic area are presented. For each mapping unit there should be presented the environmental conditions in which it is found, its extent and importance as well as a discussion of the profile utilizing features such as colour, texture, structure, consistence, soil-water state, soil organic matter status, roots and soil fauna, presence or absence of carbonates and other features of pedogenetic origin. Although repetitive, this section represents the heart of the soil survey report and every effort should be made to present the information clearly and concisely. Tables should be employed to portray the extent of the different soil mapping units and their location in the landscape can be depicted on isometric block diagrams. Representative soil profiles should be given indicating the ideal mapping unit and any variations from it.

Land use and capability. Except where a soil survey is being done for resource assessment purposes only, it may be assumed that changes in land use could follow as a result of the increased knowledge available. Many authors give the history of land use preceding the time of the survey. This gives a basis from which any proposed changes may be evaluated. Where possible, the yields of major crops on the soil mapping units should be given; this is a strong point of many American surveys and is normally lacking in European surveys. From the basic soil survey it is possible to produce derivative maps of land use capability which can be used for farm planning and for the presentation of soil information to a wider audience including planners who do not necessarily have a detailed knowledge of soil science. Land use capability maps can also be made more precise in the form of maps for the suitability of soils for specific crops or for specific methods of cultivation (see Ch. 6).

Analytical results. Many of the findings of the laboratory will be brought into the presentation of the soils in earlier chapters, and some authors feel it desirable to give the analytical figures with the representative profile descriptions. If this is done there is much tabular material scattered throughout the text and other people feel it more concise to present the analytical data all together preceded by a short account of the methods used to obtain them.

CONCLUSION

A great deal of hard work and scientific expertise is put into a soil survey and the map and report form the culmination of the project. The map and report are a vital part of the exercise, for, without communication to others, the knowledge gained is of limited value. Many survey organizations find that the production of reports is a time-consuming business and that map production is ahead of the report. Consequently, they have found that an extended legend is extremely useful because it enables the map to be used more fully without the report. It is also possible to use a map in the field on its own, although it is much more difficult to cross-reference to a book at the same time; used together, however, the map and report can be the source of much factual information of concern to those who work the land and those who have to make decisions concerning its use. Soil surveys are the basis of soil geography, but at present only about one-fifth of the soils of the world have been surveyed. Excluding arid and permafrost areas, systematic soil surveys in which the boundaries are based on physiographic data and the composition of the mapping units based on field studies, cover approximately 11 per cent of Africa, 23 per cent of Asia, 15 per cent of Australia, 80 per cent of Europe, 46 per cent of North and Central America and 15 per cent of South America (Dudal 1978). 'Other surveys', which interpret general information on landforms, geology, climate and vegetation in terms of soils, increase our knowledge of soil geography, but much basic work remains to be done upon the remaining four-fifths of the world's soils.

3

SOIL CLASSIFICATION

B. Clayden

INTRODUCTION

Soil has been defined in an earlier chapter as the unconsolidated mineral or organic material at the earth's surface capable of supporting plant growth. This broad definition includes raw soils like those of dune sands, screes and coastal marshes that are too young or unstable to have been significantly affected by soil-forming processes. The soil mantle, broken only by water, barren rock or ice, is impossible to comprehend in its entirety except in very general terms. Soil classification involves the identification of individuals within this continuum and the formation of classes with common characteristics. Its essential purpose is to organize existing knowledge so that the properties and relationships of different kinds of soils can be systematically recalled and communicated (Cline 1949).

Many geographers and other environmental scientists find classification a particularly unpalatable part of soil studies and the reasons for this are not hard to find. It is evident that there is no single universally accepted system applicable to the soils of the world and that most countries with soil survey organizations have independent national systems which have been developed to suit particular needs. The number of different systems can be taken as an indication of the inherent difficulty of applying classification procedures to a universe which is initially ill-defined and continuously variable in space and time. Nevertheless it reflects badly on international communication even within Western Europe or between the USA and Canada. It can almost be seen as a matter for despair when in the British Isles the three separate soil survey organizations of England and Wales, Scotland and Ireland use different schemes to define map units. A further explanation for the multiplicity of systems may be that in most countries soil classifications have developed from the needs of soil survey organizations financed by government departments of agriculture who are rightfully concerned that the soils of their particular country are classified in a helpful and meaningful way in relation to the land capability of particular environments. Thus the Dutch system (p. 87) is very specifically framed to cater for the soil landscapes and land-use history of the Netherlands, even though Dutch pedologists are far from parochial and have made considerable contributions to the development of more comprehensive systems. Knowledge of the soils of the world is still very uneven and perhaps a satisfactory international system can only emerge after the development and testing of many more national or regional schemes.

Apart from the number of schemes extant, a glance at the literature will indicate that particular systems are regularly modified, a process of evolution testifying both to the youthfulness of the science and the strength of research. A classification system reflects the existing knowledge and concepts concerning the population being classified (Cline 1949). It must be modified as knowledge grows and new concepts develop.

Faced with a formidable array of different systems a considerable effort is then required to come to terms with the multilingual nomenclature. This problem is exacerbated by the different meanings attached to similar names for soil classes, particularly in the case of old favourites like brown earths, podzols and gley soils (Tavernier and Smith 1957; Wilde 1953).

Before the student can tackle the subject at all he must be clear about preliminary concepts and aims. First, it must be recognized that soils are not found as discrete individuals like plants and animals, so that the unit of classification must be defined. Second, any classification must have a purpose and with soils the aims can vary widely. For the most part, general systems developed for the definition of map units in soil surveys designed to aid land use will be discussed, but others have been constructed with less practical aims and more emphasis on soil formation and evolution to summarize relationships for scientific purposes. However, the reader should not be unaware of other ways in which soils can be grouped to serve special purposes. For example, a mole catcher might be interested in a scheme that simply distinguished soils with or without earthworms, whereas one of use in locating sites for temporary airfields might involve features like soil wetness, bearing strength, rockiness and the nature of the soil surface.

The unit of classification is generally taken to be the soil profile with certain lateral and vertical dimensions. In England and Wales for example the unit is taken to have a volume of about 1–2 m^3 'that in practice can be adequately described and sampled as a single entity' (Avery 1980, and see Hodgson 1976). The notion that soil classification should be based on soil profiles has been challenged since such units are out of keeping with the common conception of the 'natural individual' (Knox 1965). It can be reasonably claimed that a natural soil individual is a three-dimensional body, considerably larger than the profile pit, bounded horizontally by lines of inflexion where there is a maximum rate of change of many soil properties with distance. Such soil bodies comprise some of the more homogeneous units shown on large scale soil maps which are classified in terms of the soil profile at isolated points, rather than by the characteristics of the body as a whole.

The Soil Conservation Service of the US Department of Agriculture attempts to overcome this anomaly by defining the *pedon* as a unit of sampling and the *polypedon* as a unit of classification (SSS 1975). The pedon is similar in concept to the soil profile with a vague lower limit where the solum passes to little altered material. Its horizontal area can extend from 1 to 10 m^2 to take account of intermittent or cyclic horizons though, in most cases, where horizons are continuous and of roughly uniform thickness, the pedon has an area of 1 m^2. The polypedon 'consists of contiguous similar pedons that are bounded on all sides by "not soil" or by pedons of unlike character'.

It may be questioned if it is feasible to classify soils using 'individuals' considerably larger than the soil profile or pedon. Clearly this is the intention in the new approach adopted in the USA, though classes of the system are represented by pedons rather than polypedons (SSS 1975). Most other widely used schemes define classes by criteria derived from similarly detailed morphological, chemical and physical studies of particular soil profiles. Thus, for the purpose of this chapter, the unit of classification will be taken as the soil profile or pedon since, as Knox (1965) concludes, 'Future developments may lead to a soil classification system based on some kind of soil landscape unit, but at present the difficulties seem overwhelming.' This topic is developed in another chapter which deals with soil classification in relation to soil mapping.

Soils can be classified in many different ways depending on the particular attributes chosen as a basis for classification and the attributes selected depend on the purpose (Gilmour 1937). The theoretical and practical purposes have been discussed at length by Cline (1949 and 1963). De Bakker (1970) in an extensive literature review notes that a definite statement about purpose is often lacking but divides classification schemes as follows.

1. Theoretical or scientific. Classifications of this kind place emphasis on origin and the relationships between classes. Such systems which make use of as many known properties of the soils as possible without a single or applied objective are also known as 'natural classifications' (Cline 1963; Kubiena 1958; Gilmour 1951; Muir 1969). However, despite the persistence of this term, all soil classifications are contrived by man and are in that sense 'artificial' (Leeper 1956).

2. Practical. These are designed for application to agriculture or other technical use of soils, and are mainly devised to serve the needs of soil survey. Thus the system of Avery (1980) is specifically designed as a basis for soil mapping in England and Wales. Other specific purposes may be served by a *special* classification (Turrill 1952; Mulcahy and Humphries 1967) which is based on few relevant attributes. However, every gradation is found between the different types.

The impact of the theory of evolution on biological taxonomy has exercised a strong influence on methods of soil classification, despite the fact that soils are not developed from a common ancestor but from a great variety of different parent materials. Speculations about 'soil genesis' – meaning mode of development – are involved in most general or natural soil classification schemes which involve assessment of the degree and kind of alteration of the parent material. Thus, unlike biological system that emphasise inherited characters, morphogenetic soil classification focuses attention on characteristics acquired by soil-forming (pedogenic) processes. With extreme emphasis on genesis the scheme can no longer be considered a 'natural classification' but rather a special classification made for the particular purpose of studying soil development.

Most systems of soil classification are *hierarchical*, with a framework resembling a family tree, in which successively lower categories have more classes than the one above. At each categorical level properties are chosen to define mutually exclusive classes and this involves ranking the criteria to define successively lower categories. Hierarchies assist memory and the construction of keys for identification as in biological systems but have limitations when applied to the continuous universe of soils (Avery 1968; Webster 1968 and 1977b).

The hierarchical arrangement is well suited to the classification of populations of higher plants and animals in which natural groups can be recognized empirically. Such populations have inherent nested clustering in that the individuals forming a group have many shared attributes not found in individuals of other groups. Existing knowledge suggests little or no such clustering in the whole universe of soil profiles. Thus a hierarchical system inevitably separates very similar soils into different classes because of the arbitrary criteria used for class definition. It will be seen in the description of particular hierarchical systems below that two soils separated at the highest categoric level can resemble each other in every respect except the characteristic used for differentiation.

The most promising alternative is a *co-ordinate system* (Avery 1968; Webster 1968) in which differentiating criteria are not necessarily ranked as in hierarchies. It consists of two or more special classifications, each based on one set of attributes, that overlap to give a reference framework for creating classes. Schemes of this type used in Russia, East Germany and Belgium are summarized below. It is possible to conceive complex schemes with separate axes for each attribute measured and the class boundaries identified within a multi-dimensional graph as in numerical methods of classification (p. 94). In principle, co-ordinate classifications should have advantages over hierarchical systems for soil, but their application has been limited. This can be explained by analogy with the problem of 'uncontrolled soil legends' (SSS 1951) in that the number of classes can become unwieldy when defined by the possible combinations available in a co-ordinate system. Most modern systems of classification prefer to define classes by 'diagnostic horizons', involving a number of different kinds of attribute, in order to create a manageable number of divisions.

In a chapter of this length it is possible only to summarize the development of ideas on soil classification and to outline the basis of a selection of internationally recognized modern systems. No attempt is made to cover the multitude of schemes proposed in scientific journals or texts that have been little used (see De Bakker 1970). Application of any particular scheme necessitates careful reference to the original text, particularly since the criteria used for differentiation are often detailed and because classes within a particular category can be 'keyed out' in a particular order. It is most profitable to understand first the way in which familiar soils are classified using a modern soil survey report in conjunction with the national classification system. There has been an unfortunate tendency in the teaching of soil geography to expect students to be familiar with the global sweep of soils before they know the difference between the A and B horizons of their local soils (Taylor 1960). A reasonably thorough understanding of a particular national system will assist greatly when progressing to more comprehensive schemes or those designed for contrasting environments.

3.1 THE FORMATIVE PHASE

Man's awareness of soil and its variability developed in very early times particularly with the emergence of civilizations based on agriculture as described in Chapter 1. Interesting early attempts to describe British soils are reviewed by Muir (1961), but it was not until the nineteenth century that serious attempts at soil classification began with those of Thaer (1853) and Fallou (1862) in Germany. These schemes together with that of Richthofen (1886) are outlined by Buol *et al.* (1980) under what they call the 'early technical period' of soil classification. It coincided with a time of rapid growth in geological knowledge, and attempts to classify soil, regarded then as an unorganized mixture of disintegrated rock and decaying organic matter, relied largely on the nature of the underlying rocks. In Britain this approach continued in soil surveys until about 1920. Each geological formation was assumed to give rise to characteristic soils with little regard to other factors, so that soils were classed as Chalk soils, Gault soils, Old Red Sandstone soils and so on (Robinson 1943). It is regretted by Muir (1961) that this approach was encouraged by the notes on soils in the early memoirs of the Geological Survey which suggest that the geological maps could serve as the basis of soil maps.

The early workers, who paid little attention to effects on soil formation of factors other than the parent rock, suffered from a limited geographical appreciation of soil variation. This was not the case with Russian scientists led by Dokuchaiev (1846–1903), who in about 1870 introduced the entirely new concept of soil as an independent natural body. It was recognized that a vertical section in the soil had a definite form (morphology) expressed by a succession of distinctive layers or horizons roughly parallel to the surface constituting the soil profile. These horizons resulted from transformation of the original material by the interaction of various soil-forming processes determined by environmental factors. Thus the type of soil profile at any place was determined by climate, vegetation and relief, together with the parent material in which the soil developed and the age of the ground surface.

The Russian school envisaged soils as dynamic bodies developing with time to reach a condition of equilibrium corresponding to a particular vegetational climax (Basinki 1959). 'Mature' soils with well developed horizons contrasted with 'immature' soils with weakly developed horizons differing little from the original material. Dokuchaiev's classification (Table 3.1) has the three classes of normal, transitional and abnormal soils which were shortly replaced by the concepts of zonal, intrazonal and azonal soils (Sibirtsiev 1901). *Zonal soils* were defined as mature soils with profile characteristics reflecting the influence of regional climate and vegetation. *Intrazonal soils* differed from associated zonal soils in having

Table 3.1 Classification of soils by Dokuchaiev, 1900 (after Soil Survey Staff, 1960)

Class A. Normal, otherwise dry land vegetative or zonal soils	
Zones	Soil types
I Boreal	Tundra (dark brown) soils
II Taiga	Light grey podzolised soils
III Forest-steppe	Grey and dark grey soils
IV Steppe	Chernozem
V Desert steppe	Chestnut and brown soils
VI Aerial or desert zone	Aerial soils, yellow soils, white soils
VII Subtropical and zone of tropical forest	Laterite or red soils
Class B. Transitional soils	**Class C. Abnormal soils**
VII Dry land moor-soils or moor-meadow soils	X Moor-soils
IX Carbonate containing soils (rendzina)	XI Alluvial soils
X Secondary alkali-soils	XIV Aeolian soils

characters reflecting the effects of local physiographic factors such as poor drainage or parent materials resistant to zonal processes. *Azonal soils* were considered immature and formed in freshly deposited sediments or ground surfaces newly exposed by erosion.

On the Russian plains these divisions are expressed geographically by the distribution of belts of zonal soils corresponding to major climatic and vegetational zones, each containing local areas of intrazonal and azonal soils. The tri-partite division was subsequently adopted in other parts of the world and has remained a dominant theme of soil classification to this day.

The innovations of Dokuchaiev and his associates stimulated the study of soils in the field in Russia and later in the USA and western Europe, much of the influence of the work in other countries being attributed to the prolific writings of Dokuchaiev's pupil, Glinka, which were widely read in translations.

The early work in Russia provided a framework of soil classes closely related to climatic and vegetational zones on a continental scale. In the USA, by contrast, considerable progress had been made with detailed soil surveys before the development of a classification scheme comprising broad groups. A national soil survey was started in 1899 under Whitney using soil series and subdivisions known as soil types. Physiographic regions were identified as soil provinces and soils from similar parent materials within a province were distinguished as soil series. A soil type was a subdivision of a series based on texture which at that time embraced consistence, aggregation and related properties as well as particle size. It was much later that Marbut, building on earlier proposals by Coffey (1912) and stimulated by the Russian work described by Glinka, produced a national classification in terms of great soil groups.

Marbut dominated developments in soil classification and soil survey in the USA during the first part of this century and, by the 1920s, he had adopted the Russian concept of soils being independent natural bodies. He anticipated the basis of most modern systems of classification when he wrote '. . . the basis of grouping should be the characteristics of the objects grouped. They should be tangible, determinable by a study of the objects themselves and by direct observation and experiment' (Marbut 1922). However, these views were not consistently reflected in his system of classification, presented first in 1927 and developed in more detail in the masterly *Atlas of American Agriculture* (1935). Emphasis was placed on 'normal' or 'mature' soils while soils with weakly developed horizons were excluded from the higher categories. The division at the highest level into Pedocals (soils in which calcium carbonate accumulates) and Pedalfers (in which aluminium and iron accumulate) was widely accepted despite being misleading and has been perpetuated for decades in geographical textbooks. The drawbacks of Marbut's system are discussed by Buol *et al.* (1980) and the SSS (1960).

Marbut's scheme was revised by Baldwin *et al.* (1938) in the USDA Yearbook 'Soils and Men', and further modifications were introduced by Thorp and Smith (1949) (Table 3.2), where it will be noted that the three highest categories (orders) follow the zonal, intrazonal and azonal concepts of the Russian model. Each sub order includes soils with common

Table 3.2 Classification of the soils of the USA (Thorp and Smith 1949)

Order	Sub-order	Great Soil Groups
Zonal soils	1. Soils of the cold zone	Tundra soils
	2. Light-colored soils of arid regions	Desert soils Red desert soils Sierozem Brown soils Reddish-Brown soils
	3. Dark-colored soils of semi-arid, sub humid and humid grasslands	Chestnut soils Reddish Chestnut soils Chernozem soils Prairie soils Reddish Prairie soils
	4. Soils of the forest–grassland transition	Noncalcic Brown or Shantung Brown soils
	5. Light-colored podzolized soils of the timbered regions	Podzol soils Gray Wooded, or Gray Podzolic soils Brown Podzolic soils Gray-Brown Podzolic soils Red-Yellow Podzolic soils
	6. Lateritic soils of forested warm-temperate and tropical regions	Reddish-Brown Lateritic soils Yellowish-Brown Lateritic soils Laterite soils
Intrazonal soils	1. Halomorphic (saline and alkali) soils of imperfectly drained arid regions and littoral deposits	Solonchak or Saline soils Solonetz soils Soloth soils
	2. Hydromorphic soils of marshes, swamps, seep areas, and flats	Humic-Glei soils (includes Wiesenboden) Alpine Meadow soils Bog soils Half-Bog soil Low-Humic Glei soils Planosols Ground-Water Podzol soils Ground-Water Laterite soils Brown Forest soils (Braunerde)
	3. Calcimorphic soils	Rendzina soils
Azonal soils		Lithosols Regosols (includes Dry Sands) Alluvial soils

characteristics related to the conditions of their formation and are divided into great soil groups corresponding approximately to the Russian genetic soil types. Successive lower categories are called soil families, soil series and soil types. These are differentiated chiefly on profile features reflecting local differences in parent material and water regime, and form the basis of units shown on detailed soil maps. Each soil series category was identified by a geographic name and theoretically allowed only a limited range in the characteristics of subsurface horizons; minor variations in the texture of surface horizons were included and distinguished as soil types.

At the conclusion of a review of the development of soil classification systems in Russia and the USA it was noted (Avery 1962):

By 1939 comparable systems of classification were in use wherever soil surveys were being carried out, but whereas workers in Britain and most Commonwealth countries adopted the American soil series and type as basic units of classification [*lower categories of a system*], *in Russia, and in Europe generally, lower categories, in as far as they were recognised, were given connotative names (often very unwieldy) indicating their status as sub-divisions of major soil groups or 'genetic soil types'.*

Before going on to consider more modern systems it is useful to pause to consider the shortcomings of what went before, with particular reference to the USA (Kellogg 1963). The main weaknesses identified are listed and amplified below.

1. Use of environmental factors rather than soil properties in the definition of certain categories, though, as will be seen, this view is not accepted universally.
2. Emphasis on virgin soils.
3. Vague definition of classes.
4. Confused nomenclature.

Though both Dokuchaiev and Marbut emphasized the need to classify soil on the basis of its particular properties, classes were nevertheless defined partly by inferred relationships with soil-forming processes and contemporary environmental conditions. This can be illustrated by suborder designations like 'light-colored podzolized soils of the timbered regions' (Table 3.2). The approach is expressed aptly by the SSS (1960):

Essentially all our previous natural classifications have assumed that the soil scientist knew where he was, geographically, in relation to broad climatic and vegetation zones and to land forms. Such knowledge enters into his view of the soil. It prompts him to look for certain features because he expects them. Inescapably, the genetic features are considered, not simply the morphology of the soil.

Work in the relatively small countries of western Europe in landscapes with a complex of different rocks and landforms found the concept of zonality wanting. In such areas, contrasting with continental plains with a relatively uniform cover of Pleistocene deposits, particularly loess, it was apparent that the importance of parent material and geomorphic factors had been neglected.

A further dimension was added by the recognition of soils which had apparently acquired distinctive characteristics under environmental conditions different from those at the site today. Such soils, called *paleosols* (Yaalon 1971; Catt 1979), throw further doubt on the application of simple geographical correlation. It is now clear that only the soils of the youngest ground surfaces have originated entirely under existing environmental conditions.

The emphasis on virgin soils in natural landscapes remains a feature of certain genetic classification systems but more attention is now paid to cultivated soils in systems designed to aid land use. In the USA – 'cultivated soils have either been ignored or classified on the basis of properties that they are presumed to have had when virgin' (SSS 1960). As will be seen, the correction of this shortcoming is very much at the heart of the new US system.

Recognition of these and other shortcomings has led to a gradual revolution in approaches to soil classification to an extent that references to zonal, intrazonal and azonal soils are mainly to be found in outdated texts, except in Russia which retains its traditional 'geographic-evolutionary' approach. In most western countries the old 'genetic-geographical' systems are being replaced by what have been called 'morphogenetic' systems guided by two chief principles (Avery 1962):

1. *Classes in all categories are defined in terms of measurable soil properties(including features such as average temperature and seasonal moisture regime) which are either a reflection of genesis, or affect the evolution of the soil.*
2. *Since it is recognised that no one character or group of characters is equally significant in all soils, the characters chosen as criteria are variable within each category, the aim being to attain the most harmonious and useful grouping.*

3.2 MODERN SOIL CLASSIFICATIONS

1. USA – Soil Taxonomy

By far the most detailed and comprehensive national system is the *Soil Taxonomy* prepared by the Soil Conservation Service of the US Department of Agriculture and published, after successive approximations, in

1975. Its purpose is made clear in its subtitle – 'A basic system for making and interpreting soil surveys'. Above the level of soil series it makes a complete break with previous systems in design and nomenclature, and introduces carefully defined diagnostic criteria for class differentiation. It is regarded by many as an international reference system (Ragg and Clayden 1973) and has strongly influenced other national schemes, and that developed by FAO/UNESCO for the soil map of the world.

The hierarchical system is based on class distinction by precisely defined diagnostic horizons, soil moisture regimes and soil temperature regimes. Diagnostic surface horizons known as epipedons (Table 3.3) are defined in such a way that their main properties are unchanged by cultivation over short periods. Diagnostic subsurface horizons (Table 3.4) are more numerous, and like the epipedons, vary a good deal in the taxonomic significance assigned to them. Classes of soil moisture regime (Table 3.5) generally determine placement at the second categoric level of suborder. Each table given here provides only the briefest outline definition and users must turn to the original text for guidance. Classes of soil temperature regime, used for defining classes at various categoric levels, also have lengthy definitions and are merely listed below as follows – pergelic, cryic, frigid, mesic, thermic, hyperthermic, isofrigid, isomesic, isothermic and isohyperthermic.

Nomenclature is a central part of any classification because of the need for unambiguous communication. In an hierarchical system nomenclature should also demonstrate the relationship between classes at each categoric level. *Soil Taxonomy* uses a complete set of new names for categories above series level derived mainly from classic Greek and Latin sources. At first sight the nomenclature seems unpalatable but, with application, its merits become evident. Thus the names are connotative and relatively

Table 3.3 Diagnostic surface horizons – epipedons (Soil Survey Staff 1975)

1. **Mollic epipedon**. A thick, dark-coloured surface horizon with a base saturation of 50 per cent or more, and more than 0.6 per cent organic carbon (1 per cent organic matter) after mixing to a depth of 18 cm.
2. **Anthropic epipedon**. A surface horizon with properties of a mollic epipedon except for larger amounts of phosphorus accumulated by long-continued use by man.
3. **Umbric epipedon**. A surface horizon with properties of a mollic epipedon but with a base saturation of less than 50 per cent.
4. **Histic epipedon.** A peaty surface horizon containing 12 to 18 per cent organic carbon depending on clay content that is saturated with water for 30 consecutive days or more in most years, unless artificially drained.
5. **Plaggen epipedon**. A man-made surface horizon 50 cm or more thick produced by long-continued manuring.
6. **Ochric epipedon**. A surface horizon too pale in colour, too low in organic carbon or too thin to be a mollic, anthropic, umbric, histic or plaggen epipedon, or is both hard and massive when dry.

Table 3.4 Diagnostic subsurface horizons (Soil Survey Staff, 1975)

1. **Argillic horizon.** An illuvial horizon in which silicate clay has accumulated to a significant extent.
2. **Agric horizon.** An illuvial horizon below the plough layer in which silt, clay and humus have accumulated due to the effects of cultivation.
3. **Natric horizon.** An argillic horizon with prismatic or columnar peds and with 15 per cent or more saturation with exchangeable sodium.
4. **Sombric horizon.** An illuvial horizon in which humus has accumulated. The illuvial humus is neither associated with aluminium, as in spodic horizons, nor dispersed by sodium, as is common in natric horizons.
5. **Spodic horizon.** A horizon in which active amorphous materials, composed of organic matter and aluminium with or without iron, have precipitated.
6. **Placic horizon.** A thin, black to dark reddish pan cemented by iron, iron and manganese, or an iron-organic matter complex, with a thickness ranging generally from 2 to 10 mm.
7. **Cambic horizon.** An altered horizon, excluding most sandy materials, as indicated by structure, colour, gley features, clay formation or evidence of removal of carbonates.
8. **Oxic horizon.** A strongly altered horizon at least 30 cm thick consisting of a mixture of hydrated oxides of iron or aluminium, or both, with variable amounts of kaolinitic clay and insoluble minerals such as quartz sand.
9. **Duripan.** A horizon cemented by silica such that dry fragments do not slake in water.
10. **Fragipan.** A horizon of high bulk density, brittle when moist and hard or very hard when dry. Dry fragments slake in water.
11. **Albic horizon.** A horizon from which clay and free iron oxides have been removed to the extent that the colour of the horizon is determined by the colour of the primary sand and silt particles rather than by coatings on these particles.
12. **Calcic horizon.** A horizon of accumulation of calcium carbonate (sometimes with magnesium carbonate) 15 cm or more thick with more than 15 per cent calcium carbonate equivalent.
13. **Gypsic horizon.** A horizon of calcium sulphate enrichment 15 cm or more thick with at least 5 per cent more gypsum than the underlying material.
14. **Petrocalcic horizon.** A calcic horizon that is cemented or indurated such that dry fragments do not slake in water.
15. **Petrogypsic horizon.** A gypsic horizon that is cemented with gypsum such that dry fragments do not slake in water.
16. **Salic horizon.** A horizon 15 cm or more thick containing a secondary enrichment of salts more soluble than gypsum.
17. **Sulfuric horizon.** A mineral or organic horizon that has both a pH value of less than 3.5 (1 : 1 in water) and yellowish mottles of jarosite.

short. A formative element from each of the higher categories is successively carried down to and including the family category, soil series retaining their geographic names (Table 3.6).

The categories of the system are order, sub order, great group, subgroup, family and series.

Orders. There are ten orders which, with the exceptions of Aridisols, Vertisols and Histosols, are distinguished principally by the presence or absence of diagnostic horizons (Table 3.7). Thus the unique feature of

Table 3.5 Classes of soil moisture regime (Soil Survey Staff 1975)

These are defined in terms of the ground-water level and in terms of the presence or absence of water held at a tension less than 15 bars in the moisture control section by periods of the year (the moisture control section varies according to particle-size class).

Aquic moisture regime. (L. *aqua*, water)
A reducing regime due to saturation by water for unspecified periods when the soil temperature is above 5°C. For differentiation in higher categories the whole soil must be saturated.

Aridic and torric moisture regimes. (L. *aridus*, dry, and L. *torridus*, hot and dry)
These are used for the same moisture regime but in different categories of the classification. The moisture control section in most years is:
(i) Dry in all parts more than half the time (cumulative) that the soil temperature at a depth of 50 cm is above 5°C; and
(ii) Never moist in some or all parts for as long as 90 consecutive days when the soil temperature at a depth of 50 cm is above 8°C.

Udic moisture regime (L. *udus*, humid)
In most years the soil moisture control section is not dry in any part for as long as 90 days (cumulative).

Ustic moisture regime (L. *ustus*, burnt)
Intermediate between the aridic and udic regime.

Xeric moisture regime (Gr. *xeros*, dry)
This is typical of Mediterranean climates where winters are moist and cool and summers are warm and dry.

Table 3.6 Nomenclature related to a pedon (Pedon 26, Appendix IV, p. 537, *Soil Taxonomy*) (Soil Survey Staff, 1975)

Order	Sp*od*osols
Suborder	Orth*ods*
Great Group	Fragiorth*ods*
Subgroup	Typic Fragiorth*ods*
Family	Typic Fragiorth*od*, coarse-loamy (particle-size), mixed (mineralogy), frigid (temperature regime).

Spodosols is the presence of a spodic horizon. The argillic horizon, a layer of accumulation of silicate clay, is diagnostic of both Alfisols and Ultisols but the latter differ in having a low supply of bases indicative of intense leaching. It should be noted that Mollisols and Aridisols can also have argillic horizons, in the same way that mollic epipedons are not restricted to Mollisols.

Suborders. The orders are divided into 47 suborders (Table 3.7) based mainly on soil moisture regimes believed to influence genesis and to be important to plant growth. Other differentiae were selected to distinguish what seemed to be the most important variants within the order.

Table 3.7 Orders and suborders of soil taxonomy (Soil Survey Staff 1975)

1. **Entisols**. Weakly developed mineral soils
   - 1.1 Aquents – with gleyic (hydromorphic) features
   - 1.2 Arents – others with fragments of soil horizons (disturbed soils)
   - 1.3 Psamments – others in sandy materials
   - 1.4 Fluvents – others in alluvium or colluvium
   - 1.5 Orthents – other Entisols
2. **Vertisols**. Cracking clay soils
   - 2.1 Xererts – with long dry periods
   - 2.2 Torrerts – usually dry
   - 2.3 Uderts – usually moist
   - 2.4 Usterts – others, with short dry periods
3. **Inceptisols**. Moderately developed soils of humid regions normally with a cambic horizon and no spodic, argillic or oxic horizon.
   - 3.1 Aquepts – with gleyic (hydromorphic) features
   - 3.2 Andepts – others mainly in volcanic ash with a low bulk density and an exchange complex dominated by amorphous material.
   - 3.3 Plaggepts – others with a plaggen epipedon
   - 3.4 Tropepts – others of tropical climates, mainly with ochric and cambic horizons
   - 3.5 Ochrepts – others of mid- to high latitudes, mainly with ochric and cambic horizons
   - 3.6 Umbrepts – others of humid mid- to high latitudes, mainly with an umbric epipedon
4. **Aridisols**. Soils of deserts and semi-deserts
   - 4.1 Argids – with an argillic or natric horizon.
   - 4.2 Orthids – others, without an argillic or natric horizon.
5. **Mollisols**. Base-rich soils with a mollic epipedon (mainly of the steppes)
   - 5.1 Albolls – with albic and argillic horizons and affected by fluctuating ground-water
   - 5.2 Aquolls – others with gleyic (hydromorphic) features
   - 5.3 Rendolls – others with a mollic epipedon over extremely calcareous materials
   - 5.4 Xerolls – others with long dry periods
   - 5.5 Borolls – others of cold climates (with a frigid, cryic or pergelic temperature regime)
   - 5.6 Ustolls – others of sub-humid or semi-arid climates
   - 5.7 Udolls – others of humid climates
6. **Spodosols**. Soils with a spodic horizon (most podzols)
   - 6.1 Aquods – with gleyic (hydromorphic) features
   - 6.2 Ferrods – others with little organic carbon in the spodic horizon
   - 6.3 Humods – others with little iron in some part of the spodic horizon
   - 6.4 Orthods – others with a spodic horizon containing aluminium, iron and organic carbon in which no one element dominates
7. **Alfisols**. Soils with an argillic horizon and moderate to high base status.
   - 7.1 Aqualfs – with gleyic (hydromorphic) features
   - 7.2 Boralfs – others of cold climates
   - 7.3 Ustalfs – others of warm sub-humid or semi-arid climates
   - 7.4 Xeralfs – others with long dry periods
   - 7.5 Udalfs – others of humid climates
8. **Ultisols**. Soils with an argillic horizon and low base status
   - 8.1 Aquults – with gleyic (hydromorphic) features
   - 8.2 Humults – other humus-rich soils of mid- or low latitudes
   - 8.3 Udults – others of humid climates
   - 8.4 Ustults – others of warm regions with high rainfall and pronounced dry season
   - 8.5 Xerults – others with long dry periods

Table 3.7 (continued)

9. Oxisols. Soils with an oxic horizon or with plinthite within 30 cm depth
- 9.1 Aquox – with gleyic (hydromorphic) features
- 9.2 Torrox – others of arid climates
- 9.3 Humox – other humus-rich soils of low base status
- 9.4 Ustox – others with long dry periods
- 9.5 Orthox – others of humid climates

10. Histosols. Soils in organic materials (peat soils)

A. Never saturated with water for more than a few days
- 10.1 Folists

B. Saturated with water for six months or more:
- 10.2 Fibrists – mainly composed of little decomposed plant remains
- 10.3 Hemists – mainly composed of partly decomposed plant remains
- 10.4 Saprists – mainly composed of almost completely decomposed plant remains.

Great groups. At this level soils are grouped with the following properties in common.

(i) Kind, arrangement and degree of expression of horizons, with emphasis on the upper sequum in bi-sequal profiles.
(ii) Soil moisture and temperature regimes.
(iii) Base status.

The use of base status can be illustrated in the suborder of Ochrepts, where the great group of Eutrochrepts includes soils containing carbonates or with a base saturation of 60 per cent or more while Dystrochrepts have a lower base saturation. This separation corresponds approximately to that formerly made on base status within brown earths in Britain. The presence or absence of certain diagnostic horizons is also differentiating as in Fragiochrepts, Ochrepts with a fragipan. In framing the great groups, bias in favour of soils known to occupy large areas of the USA is acknowledged.

Subgroups. This category, not present in earlier US schemes, was introduced to define central concepts of great groups, e.g. Typic Dystrochrepts, intergrades to other great groups, e.g. Dystric Eutrochrepts, and intergrades to other suborders or orders. There is also provision for extragrades, transitional to 'non-soil'.

Families. At this level soils are grouped with similar physical and chemical properties that affect their responses to management. As summarized in *Soil Taxonomy*, the main differentiae used to distinguish the 4500 families are particle-size distribution in horizons of major biological activity below plough depth (normally 25–100 cm), mineralogy, temperature regimes and thickness of soil penetrable by roots.

Series. This is the lowest category of the system and about 10 500 had been recognized in 1975. Most were first defined in accordance with earlier classifications and changes in concepts and differentiae have been made progressively over the years. Its purpose, like that of the family, is mainly pragmatic and the classes are closely linked to interpretative uses of the system. Series are differentiated within families on the basis of the kind and arrangement of horizons, the colour, texture, structure and reaction of horizons, and the chemical and mineralogical properties of horizons. They are mainly characters that can be observed or inferred with reasonable assurance.

It should be noted that the categories within the system have been arrived at in an empirical fashion, by selecting the differentiae required to make the desired separation. Likewise the differentiating characteristics are not uniformly applied, or applicable, to all soils at a given categoric level. Thus an aquic soil moisture regime, together with specific morphological features, distinguishes Aquic suborders in all orders except Vertisols, Aridisols and Histosols. The argillic horizon, used to differentiate the orders of Alfisols and Ultisols, is also used to distinguish the suborders of Aridisols.

Soil Taxonomy is a monumental publication extending to over 700 pages and it is only possible to classify a soil with confidence by careful reference to the text. Familiarity with the diagnostic criteria is a first requirement. Thereafter it is often necessary, as will be seen in other systems, to work through the keys which provide the most precise way of placing a soil at a particular categoric level. Keys are provided for orders, suborders and great groups, while subgroups are identified by the ways in which they depart from the definition of the Typic subgroup. The use of keys can be illustrated by reference to the suborder of Ochrepts in which the great groups of Fragiochrepts (with a fragipan) and Durochrepts (with a duripan) are keyed out before Eutrochrepts and Dystrochrepts and may therefore be of any base status.

An introduction to the development of the US system and its application to British soils is contained in a monograph prepared by the Soil Survey of Great Britain (Ragg and Clayden 1973). This also summarizes the reaction to the system (as presented in the approximation preceding the 1975 publication) from a wide range of pedologists. The main criticisms were listed as follows:

1. The rejection of all previous soil names above series and the unpalatability of the new nomenclature.
2. The inflexibility which requires a name to be changed when a revised definition transfers a great group to another order.
3. The insufficient weight attached to genetic properties, which results

in soils shaped by similar processes being placed in different orders, and the genetic heterogeneity of classes.
4. The selection, criticized by some writers in contrast to 3 above, of genetically significant properties for class separation.
5. The use of differentiae based on data unavailable in many parts of the world.
6. The subjective selection of criteria.
7. The attempt to make the classification applicable to all soils of the world (comprehensive).
8. The failure to make the taxa unequivocal.
9. The absence of a group of hydromorphic soils at the highest categoric level.

Some of these are elaborated in subsequent works like Duchaufour (1977) where it is suggested that the apparent precision of the system can be illusory, particularly when the user has inadequate data as for soil moisture regimes. Attention has also been drawn to criteria relying on laboratory analyses, as in the chemical requirements for spodic horizons, when analytical techniques vary considerably between laboratories.

Nevertheless, given sufficient data, *Soil Taxonomy* enables a pedologist to give names to most soils of the world that are both unambiguous and meaningful. For this reason it is widely used as an international reference system. The system has been in use in the USA since 1965 and is being tested and used in several other countries, notably New Zealand. Its introduction must have presented many problems and its usefulness and practicality for the purpose intended can only be assessed after a reasonable trial period.

2. Canada

The concepts developed in the USA have been closely followed in the simpler Canadian system of soil classification (Canada Soil Survey Committee – CSSC – 1978), the latest version of which is presented in a handsome book containing valuable introductory sections for the general reader. After stating the purpose of the system it indicates that the overall philosophy is pragmatic '. . . the aim is to organize the knowledge of soils in a reasonable and usable way. The system is a natural or taxonomic one in which the classes (taxa) are based upon properties of the soils themselves and not upon interpretations of the soils for various uses.' Attributes of the system are listed. Like *Soil Taxonomy*, the selection of differentiae for higher categories indicates a genetic bias, but 'Classification is not based directly on presumed genesis because soil genesis is incompletely understood, is subject to a wide variety of opinion, and cannot be measured simply.'

The Canadian system is hierarchical and the classes are defined on the basis of measurable soil properties. It differs from the US scheme as follows:

1. It is designed for Canadian soils only.
2. There is no suborder category.
3. Solonetzic, Gleysolic and Crysolic soils are differentiated at the highest categoric level (as in some European systems).
4. All horizons to the surface can be diagnostic since 90 per cent of Canada is unlikely to be cultivated.
5. It uses a different nomenclature.
6. There are fewer named diagnostic horizons and they are more tersely defined.

The categories of the system are order, great group, subgroup, family and series, the differentiae at family and series level following closely those of the USA. Table 3.8, taken from the Canadian text, usefully relates the Canadian orders and great soil groups to classes of the US and FAO systems. In using the table it must be remembered that the definitions of horizons and classes differ from one system to another and that the categorical levels of related classes are generally not equivalent. The closest parallel is between the Canadian Organic soils and the US Histosols because pedologists from the two countries worked together on the classification of peat soils.

3. FAO/UNESCO

The US *Soil Taxonomy* is regarded by many as the most valuable international reference system of soil classification. On the other hand, the best picture of the global distribution of soils can be obtained from the *Soil Map of the World* at 1 : 5 000 000 published by FAO/UNESCO (1974–8). Unfortunately the classification used for this project, though largely derived from US system, has a different terminology. The existence of the two systems imposes an unfortunate strain on everybody involved with the subject and particularly on the student concerned with the distribution of world soils. Thus in texts at various levels it has been found advisable to provide names from both systems when labelling soil profiles (Bridges 1978a; Duchaufour 1976; De Bakker, 1979).

The development of the FAO/UNESCO scheme must be seen as a delicate problem in scientific politics to reconcile the differing approaches of powerful countries with modern classification systems applicable to soils beyond their own national frontiers, particularly the USA, Russia and France (Young 1979). An international advisory panel of soil scientists was convened and the system published as a 'legend' (FAO 1974 – Volume 1 of a set of ten forming the complete publication). Among the

Table 3.8 Taxonomic correlation at the Canadian order and great group levels (only nearest equivalents are indicated. CSSC 1978)

Canadian	US	FAO
Chernozemic	Boroll, some Vertisols	Kastanozem, Chernozem, Rendzina, Phaeozem
Brown	Aridic Boroll subgroups	Kastanozem (aridic)
Dark Brown	Typic Boroll subgroups	Kastanozem (typic)
Black	Udic Boroll subgroups, Rendoll	Chernozem, Rendzina
Dark Gray	Boralfic Boroll subgroups, Alboll	Greyzem
Solonetzic	Natric great groups, Mollisol and Alfisol	Solonetz
Solonetz	Natric great groups, Mollisol and Alfisol	Mollic, Orthic or Gleyic Solonetz
Solodized Solonetz	Natric great groups, Mollisol and Alfisol	Mollic, Orthic or Gleyic Solonetz
Solod	Glossic Natriboroll, Natralboll	Solodic Planosol
Luvisolic	Boralf and Udalf	Luvisol
Gray Brown Luvisol	Hapludalf or Glossudalf	Albic Luvisol
Gray Luvisol	Boralf	Albic Luvisol, Podzoluvisol
Podzolic	Spodosol, some Inceptisols	Podzol
Humic Podzol	Cryaquod, Humod	Humic Podzol
Ferro-Humic Podzol	Humic Cryorthod, Humic Haplorthod	Orthic Podzol
Humo-Ferric Podzol	Cryortod, Haplorthod	Orthic Podzol
Brunisolic	Inceptisol, some Psamments	Cambisol
Melanic Brunisol	Cryochrept, Eutrochrept, Hapludoll	Cambisol Eutric Cambisol
Eutric Brunisol	Cyrochrept, Eutrochrept	Eutric Cambisol
Sombric Brunisol	Umbric Dystrochrept	Dystric Cambisol
Dystric Brunisol	Dystrochrept, Cryochrept	Dystric Cambisol
Regosolic	Entisol	Fluvisol, Regosol
Regosol	Entisol	Regosol
Humic Regosol	Entisol	Fluvisol, Regosol
Gleysolic	Aqu-suborders	Gleysol, Planosol
Humic Gleysol	Aquoll, Humaquept	Mollic, Humic Calcaric Cleysol
Gleysol	Aquent, Fluvent, Aquept	Eutric, Dystric Gleysol
Luvic Gleysol	Argialboll, Agriaquoll, Aqualf	Planosol
Organic	Histosol	Histosol
Fibrisol	Fibrist	Histosol
Mesisol	Humist	Histosol
Humisol	Saprist	Histosol
Folisol	Folist	Histosol
Cryosolic	Pergelic subgroups	Gelic
Turbic Cryosol	Pergelic Ruptic subgroups	Cambisol, Regosol, Fluvisol, etc.
Static Cryosol	Pergelic subgroups	
Organic Cryosol	Pergelic Histosol or Pergelic Histic subgroups of other orders	Gelic Histosol

project's stated objectives is to 'promote the establishment of a generally accepted soil classification and nomenclature'. The introduction adds

The legend of the Soil Map of the World is not meant to replace any of the national classification schemes but to serve as a common denominator. Improving understanding between different schools of thought could profitably lead to the adoption of an internationally accepted system of soil classification and nomenclature which would considerably strengthen the status and impact of soil science in the world.

The system has two categories of which the higher (26) is roughly equivalent to the USA great group. Classes are defined by the use of diagnostic horizons many of which mirror those of *Soil Taxonomy*, though generally less rigorously defined. Nomenclature is of mixed origin and the main units have 'traditional' names like Chernozems and Podzols, names adopted from North America like Gleysols, Histosols and Vertisols, and newly coined names like Luvisols and Acrisols. Unhappy slight inconsistencies with *Soil Taxonomy* usage are introduced as in defining Podzols as having a spodic B horizon and using the diagnostic cambic horizon to define Cambisols.

The main categories are listed in Table 3.9 with brief explanations of their meanings. The full definitions are short by the standards of *Soil Taxonomy* but, because of the large number of classes, are repetitive in having to spell out what features are excluded. The example given below indicates that the class of Gleysols does not include all kinds of hydromorphic soils.

Gleysols. 'Soils formed from unconsolidated materials exclusive of recent alluvial deposits, showing hydromorphic properties within 50 cm of the surface; having no diagnostic horizons other than (unless buried by 50 cm or more new material) an A horizon, a histic H horizon, a cambic B horizon, a calcic or a gypsic horizon; lacking the characteristics which are diagnostic for Vertisols; lacking high salinity; lacking bleached coatings on structural ped surfaces when a mollic A horizon is present which has a chroma of 2 or less to a depth of at least 15 cm' [*as in Gleyic Greyzems*].

As noted earlier (p. 62), the most reliable and economical way of identifying a soil is by using the keys in which classes are set apart in a particular order. Thus Fluvisols which include hydromorphic soils in alluvium and colluvium are keyed out before Gleysols, and sandy Arenosols before other weakly developed soils in unconsolidated materials classed as Regosols. Unfortunately space does not allow the keys to be included here.

Those interested in understanding more about the historical develop-

Table 3.9 FAO/UNESCO Soil Map of the World (1974). Abbreviated, non-exclusive definitions of higher classes

1. Fluvisols	Soils in alluvium (or colluvium) (Fluvents)*
2. Gleysols	Hydromorphic soils in non-alluvial materials
3. Regosols	Weakly developed soils in loamy or clayey unconsolidated deposits other than alluvium
4. Lithosols	Shallow soils with hard rock at less than 10 cm depth
5. Arenosols	Soils in unconsolidated sandy deposits other than alluvium
6. Rendzinas	Soils with a mollic A horizon over extremely calcareous material (Rendolls)
7. Rankers	Soils with an umbric A horizon and no subsurface diagnostic horizon
8. Andosols	Soils, mainly in volcanic ash, with a low bulk density and an exchange complex dominated by amorphous material (Andepts)
9. Vertisols	Cracking clay soils (Vertisols)
10. Solonchaks	Saline soils in non-alluvial materials
11. Solonetz	Soils with a natric B horizon
12. Yermosols	Desert soils with a *very* weak ochric A horizon (containing little organic matter) (Typic Aridosols)
13. Xerosols	Desert soils with a weak ochric A horizon (containing more organic matter than Yermosols) (Mollic Aridisols)
14. Kastanozems	Soils with a mollic A horizon and a calcic or gypsic horizon or concentrations of lime (Ustolls, or Chestnut soils of Thorp and Smith 1949)
15. Chernozems	Soils with a mollic A horizon and a calcic or gypsic horizon or concentrations of lime
16. Phaeozems	Soils with a mollic A horizon and no calcic or gypsic horizon or concentrations of lime
17. Greyzems	Soils with a mollic A horizon and bleached coatings on ped faces of subsurface horizons; usually with an argillic B horizon
18. Cambisols	Soils with a cambic B horizon and without horizons diagnostic of other classes
19. Luvisols	Soils with an argillic B horizon of medium to high base status
20. Podzoluvisols	Soils having an argillic B horizon with an irregular upper boundary due to a tonguing E horizon or to ferruginous nodules
21. Podzols	Soils with a spodic B horizon (Spodosols)
22. Planosols	Soils with a gleyed albic E horizon over a slowly permeable horizon
23. Acrisols	Soils with an argillic B horizon of low base satus
24. Nitosols	Soils with an argillic B horizon extending to 150 cm or more
25. Ferralsols	Soils with an oxic B horizon (Oxisols)
26. Histosols	Peat soils (Histosols)

* The classes are correlated with orders or suborders of *Soil Taxonomy* where possible.

ment of the system are advised to turn to earlier reports of the FAO/UNESCO project (e.g. Dudal 1968) which give insights into the conception of classes like Podzoluvisols and Nitisols.

As it stands there is a need to be very familiar with the soils of the world before advancing serious criticism of the classes distinguished but it can usefully be compared with the US system.

1. The classes of Vertisols, Podzols (Spodosols), Ferralsols (Oxisols) and Histosols are near equivalent of US orders.

2. The US Aridisols are divided into Yermosols and Xerosols.
3. The division between Kastanozems, Chernozems and Phaeozems is less prominent in the current US system. (Note that Kastanozems and Chernozems are difficult to distinguish by the definitions provided in the FAO text.)
4. As a consequence of the structure of the FAO system, classes of weakly developed or shallow soils appear to have a higher status, as do saline soils and those containing large amounts of exchangeable sodium (Solonchaks and Solonetz). Partly for the same reason no less than seven classes at the higher categoric level are defined as normally having an argillic B horizon, notably Solonetz, Greyzems, Luvisols, Podzoluvisols, Planosols, Acrisols and Nitosols.

It has been suggested that the most valuable part of the Soil Map of the World Project is the classification system on which it is based (Young 1979). Certainly there is a gross disparity between the precision of the classes and the information on which the map is based. Like *Soil Taxonomy*, the system is widely used for international communication and has the advantages of brevity and the authority of the major international soil science organizations. However, while accepting that the US system is complex it can be questioned whether the FAO criteria are sufficiently well defined for unambiguous placement of soil profiles.

Any comparison of the two schemes needs to take account of the purposes for which they were compiled. It is not surprising that the US system designed to accommodate soil series identified in more than half a century of detailed soil survey differs markedly from the FAO scheme framed by an international committee for a world map at 1 : 5 000 000. It is clear that the subclasses of the FAO system are inadequate for use in soil survey at scales ranging from 1 : 10 000 to 1 : 50 000, and the lack of climatic separations limits their practical significance.

4. USSR

Modern soil classification in the Soviet Union follows the 'ecological–genetic' approach pioneered by Dokuchaiev and Sibirtsiev in relating soil properties to pedogenic processes and soil-forming factors. Its development is traced by Tiurin (1965) and later work is conveniently summarized by Buol *et al.* (1980).

The most recent comprehensive scheme available is that described by Rozov and Ivanova (1967). It is based on soil-forming processes and environmental factors as well as soil properties, so that classes are not identifiable primarily by intrinsic soil characters that can be readily observed or measured. The main reference classes are 110 genetic soil types, conceived at roughly the same level of generalization as the great soil groups of *Soil Taxonomy*. Each soil type is defined by specified bioclimatic and hydrologic features, and by certain morphological characters of the soil profile. The names are based largely on topsoil colour combined with 'zem' meaning land, as in Chernozem and Krasnozem.

The classification is developed upwards into orders and classes on a co-ordinate system and downwards into subgroups, genera, species, varieties and subvarieties. The authors note that, while the classification of the lower categories is well developed, the proposals above soil type level are more tentative; the framework adopted is just one of several approaches proposed by Soviet scientists. A complex scheme is necessary that '. . . should be constructed in multi-dimensional space but we are still at the stage of approaching its development' (Rozov and Ivanova 1967).

The genetic soil types are arranged in accordance with three co-ordinate axes as shown in Table 3.10 which shows the arrangement for one ecological–genetic group. The three axes are as follows.

1. Ecological–genetic or bioclimatic classes based on climatic indices, weathering cycle and type of biological cycle.
2. Four genetic soil orders based on moisture regime. 'Automorphic' or non-hydromorphic includes the soils regarded as 'zonal'.
3. Five biophysicochemical orders based on characteristics of organic matter decomposition, soluble salts, degree of base saturation and cation composition, and 'the general make up of the soil profile'. As the definitions of these orders are somewhat long and complicated they are replaced by the letters A to E in Table 3.10. The tabular scheme is given in full in *Soil Taxonomy* with approximate equivalents in the US system.

Genetic soil types are divided into subtypes on the basis of 'genetic processes and phenomena' which include form of organic matter accumulation, degree of podzolization, depth to carbonates or soluble salts and criteria related to depth of freezing and thawing and soil temperature.

Genera are identified within subgroups by a 'set of genetic soil properties that depend on characteristics of the soil-forming parent material, the chemical content of ground-water or characteristics preserved from a previous phase of soil formation'.

Soil species are distinguished within genera by 'quantitative indices describing the degree of expression of the fundamental processes responsible for the development of the soil,' as, for example, the thickness of a particular horizon.

Varieties are identified by particle-size and *subvarieties* on the petrographic characteristics of the parent material, as in the example given – medium podzolic, illuvial-humic loamy sands or *ancient alluvial sands*.

Table 3.10 Classification of the soils of the USSR (Rozov and Ivanova 1967) (a part of the co-ordinate system)

1 Ecological–genetic (bioclimatic groups and classes)	2 Genetic orders	3 Biophysicochemical orders *Divisions based mainly on type of organic material, absorption complex and soluble salts*				
		A	B	C	D	E
Soils of Brown Earth – Forest subboreal regions $\Sigma T^{\circ} > 10^{\circ}C = 1800–3800$ Precipitation (annual) 450–1000 mm Moisture coef. > 1 Weathering siallitic with clay formation Biological cycle – moderately retarded calcium-nitrogen	Automorphic	Brown Forest (Brown Earths) Podzolic Brown Earths	Brown Rendzinas	Soil types of these biophysiochemical orders are not represented in this ecological-genetic class		
	Weakly hydromorphic-alluvial	Alluvial Sod of Brown Earth zone	Meadow Chernozem-like of prairies			
	Semi hydromorphic	Gley Brown Forest (Gley Brown Earths) Podzolic Brown Earth Gley Alluvial Sod Gley of Brown Earth zone				
	Hydromorphic		Meadow Dark of prairie zone Alluvial moist-Meadow of prairie zone			
		Alluvial Bog of Brown Earth zone and prairie zone				

5. Western Europe

This outline of some of the modern schemes of soil classification in Western Europe is suitably introduced by a summary of the 'natural system' of Kubiena (1953) published in *The Soils of Europe*. The introduction includes a general grouping of soils by their profiles (Table 3.11) which gives an overall view of how soils can be classified by the degree of profile evolution. However, such a grouping is not regarded by Kubiena as an adequate foundation for a natural system since it is based on a 'single' characteristic, albeit one that correlates with many others.

The bulk of the book is devoted to a 'general key, with descriptions of the soils, arranged after the natural system, proceeding from the simplest to the most complex'. Kubiena's concept of a natural system (p. 60) is one in which soils are ordered after all their characteristics and also after their interrelationships, though the importance of certain characteristics varies for different soil formations.

The classification is hierarchical with seven categories and a primary

Table 3.11 General grouping of soils by their profiles (Kubiena 1953)

1. (A)C – Soils	With soil life, but without macroscopically distinguishable humus layers, and only with an upper layer colonized by organisms (with or without a plant root layer). *Sub-aqueous*: Underwater-raw soils (e.g. red deep sea clays, coral reefs, marine marl and marine chalk). *Semi-terrestrial*: Raw gley soils. *Terrestrial*: Raw soils (e.g. nival raw soils of the Alps, arctic raw soils, desert soils, white rendzinas).
2. AC – Soils	With distinct humus horizon, but without B horizons. *Sub-aqueous*: Underwater humus soils (e.g. dy, gyttja, sapropel, reed peat). *Semi-terrestrial*: Humus-gley soils (e.g. anmoors, gleyed grey warp soils, mull gley soils). *Terrestial*: Rendzina-and ranker-like soils (e.g. rendzinas, chernozems, rankers, para-chernozems).
3. A (B)C – Soils	With pronounced B horizons which however are not real illuvial horizons built up by peptizable substances, but whose origin, in the first place, is due to deep reaching weathering with sufficient aeration and oxidation. *Terrestrial*: Brown earth and red earth like soils (e.g. brown earths, brown and red loams, red earths, terra rossa).
4. ABC – Soils	With B horizons which are at the same time developed illuvial horizons, i.e. having a strong enrichment of peptizable substances. *Terrestrial*: Bleached soils (e.g. podzols, bleached brown and red loams, soloti).
5. B/ABC – Soils	With strong enrichment of illuvial substances transported to the surface layer in peptized state by intensive capillary rise and irreversible precipitation. *Terrestrial*: Rind and surface crust soils.

separation is made into the divisions of Subaqueous soils, Semiterrestrial or Flooding and Groundwater soils, and Terrestrial soils. The 17 classes at the next categoric level, including brown earths, pseudogleys and podzols, have parallels at a similar level in several later systems. Detailed description of 269 subtypes, varieties and subvarieties are given and include information on physical, chemical and biological characteristics, water relations and micromorphology. The emphasis placed on humus forms, for which the book gives a separate key, has been followed particularly in France. The system provides classes for fossil and relic soils (paleosols) developed under former climatic conditions, and comparable classes are perpetuated in current German schemes. The work pioneers the application of soil micromorphology, the study of undisturbed thin sections of soil, now regarded as an indispensable tool for the identification of certain diagnostic features in soil classification (Bullock 1974, SSS 1975).

Kubiena's system was the major classification of European scope of its period and as such had a considerable influence. For example, in the first post-war classification of British soils (Avery 1956) the concepts are in general accord with those of Kubiena and major significance was attached to humus forms. Unfortunately the classes were defined by 'central concepts' in a somewhat qualitative manner so that unambiguous application to particular soils can be difficult. The system has been criticised for having no place for *sols lessivés* (Udalfs) and the limited attention paid to cultivated soils is also a serious shortcoming.

France

The French have been most influential in developing ideas on classification in western Europe on the 'genetic model' and these have been ably disseminated by the prolific writings of Duchaufour (1960, 1965, 1970, 1976). The latest official French system (Table 3.12) was published by the *Commission de Pédologie et de Cartographie des Sols* (CPSC 1967) and was developed from that of Aubert and Duchaufour (1956). The system is not restricted to the soils of France and has been applied by the *Office de la Recherche Scientific et Technique Outre-Mer* (*ORSTOM*) in overseas countries where France has close historical ties, but is less used for international communication than the US and FAO systems.

The classes of the French classification are not defined precisely. At the highest categoric level the twelve *Classes* are distinguished by: (i) degree of profile development; (ii) nature of alteration products; (iii) type of humus form and distribution of organic matter; (iv) features due to gleying or salinity (Aubert 1965). Particular significance in some classes is attached to humus form since it is considered to condition the evolution of the mineral soil below. This was more explicit in an earlier scheme (Aubert and Duchaufour 1956) where *sols brunifiés* were termed *sols évolués à mull* and *sols podzolisés* as *sols évolués à humus brut* (raw humus). Unlike *Soil Taxonomy*, but in keeping with most European schemes, a class of hydromorphic soils is set apart at the highest categoric level and includes peat soils (Histosols). More prominence is also given to Andosols while the class of Vertisols corresponds to the order in the US system.

Subclasses are separated mainly on a climatic basis (Table 3.12). At the third categoric level, groups are defined by morphological characters resulting from pedogenetic processes. Thus within the subclasses of *sols brunifiés* soils with or without an argillic horizon are distinguished as *sols lessivés* (Udalfs) and *sols bruns* (Ochrepts). This is in marked contrast with the US system where the presence of an argillic horizon distinguishes soils at the highest categoric level.

A new comprehensive system being developed by ORSTOM was outlined by Segalen *et al.* (1978). Like the US system, it comprises a hierachy of classes defined by intrinsic soil properties, but the approach entails logical (rather than pragmatic) ranking of criteria and more importance is attached to mineralogical composition than to diagnostic horizons, particularly the mollic epipedon and the argillic horizon.

No summary of the French work on soil classification would be complete without further reference to the enormous contribution made by Duchaufour. He focussed attention on the process of clay translocation (*lessivage*) and the developmental sequences leading to soils with marked Bt horizons of clay enrichment (argillic horizons). At the same time his work deals in detail with the contrasting environmental conditions required for the development of podzolised soils.

In a recent book, which replaces and expands the editions of *Précis de Pédologie* published in 1960, 1965 and 1970, Duchaufour (1977) elaborates on a revised version of a former French system which he calls a '*classification ecologique*'. This is admirably supplemented by his *Atlas Écologique des Sols du Monde* (1976, English translation 1978) which illustrates the classification by descriptions, analytical data and photographs of typical profiles and correlates them with the US and FAO systems.

The objective of the Atlas is to show that a soil cannot be adequately defined without reference to the environment of its formation. It advocates an ecological approach to the study of soils and the need for genetic classification to take account of environment and process as well as soil features *per se*. Duchaufour admits that an ideal system based on these principles is impossible for the following reasons.

1. Processes of ecological evolution are not well understood.
2. The rigid framework of any classification system creates artificial components with little allowance for intergrades.

Table 3.12 Classes and subclasses of the French classification (CPCS 1967)

1. **Raw mineral soils** (*sols mineraux bruts*)
   1.1 of erosion or deposition (*non climatiques*)
   1.2 of cold deserts
   1.3 of hot deserts
2. **Weakly developed soils** (*sols peu évolués*)
   2.1 with permafrost
   2.2 rankers (*sols humifières*)
   2.3 of dry climates
   2.4 of erosion or deposition (*non climatiques*)
3. **Vertisols** (*vertisols*)
   3.1 with reducing conditions and no external drainage
   3.2 with external drainage
4. **Andosols** (*andosols*)
   4.1 of cold climates
   4.2 of tropical climates
5. **Calcimorphic soils** (*sols calcomagnésiques*)
   5.1 calcareous soils (*sols carbonatés*)
   5.2 calcimorphic soils with little or no free carbonate (*sols saturés*)
   5.3 gypsiferous soils
6. **Isohumic soils** (*sols isohumiques*)
   6.1 of humid climates
   6.2 of very cold climates
   6.3 with rain in the cool season
   6.4 with rain in the hot season
7. **Browned soils** (*sols brunifiés*)
   7.1 of humid temperate climates
   7.2 of continental temperate climates
   7.3 of boreal climates
   7.4 of tropical climates
8. **Podzolized soils** (*sols podzolisés*)
   8.1 of temperate climates
   8.2 of cold climates
   8.3 with gley (hydromorphic) features
9. **Ferruginous soils** (*sols à sesquioxydes de fer*)
   9.1 of tropical climates
   9.2 fersiallitic soils
10. **Ferrallitic soils** (*sols ferrallitiques*)
    10.1 with high base saturation (40–80 per cent)
    10.2 with moderate base saturation (20–40 per cent)
    10.3 with low base saturation (< 20 per cent)
11. **Hydromorphic soils** (*sols hydromorphes*)
    11.1 organic (peat) soils
    11.2 humic hydromorphic soils
    11.3 non-humic hydromorphic soils
12. **Halomorphic soils** (*sols sodiques*)
    12.1 with non-degraded structure
    12.2 with degraded structure

3. In some environments soil formation can be related to more than one evolutionary process.

Nevertheless it is suggested that the rigid compartments of hierarchical systems should be replaced by evolutionary sequences established in relation to various environmental factors as used in the Atlas.

Germany

Soil classification in West Germany is chiefly associated with the name of Mückenhausen (1962, 1975, 1977), though the framework of the national classification was prepared by a working party of the German Society of Soil Science. Unfortunately none of Mückenhausen's books has been translated into English but useful summaries of the West German approach can be found in Mückenhausen (1965) and in a review by Greene (1963). The scheme has been altered little since its introduction in 1962 and, among current national systems, it most reflects the influence of Kubiena. It has seven categories, with a four-fold primary division into Terrestrial soils, Hydromorphic soils, Subhydric soils and Peat soils. While Mückenhausen, like Duchaufour, places emphasis on genetic relationships, the classification clearly reflects the origin of German soil science as a branch of geology in the importance placed on parent rock, and the significance attached to layers formed or deposited during landscape evolution in Pleistocene and Holocene times.

The flexibility of the German language facilitates the naming of intergrade subtypes in a manner which reflects their interrelationships. Thus *Podsol-Braunerde* forms a subtype of brown earths and the subtype of *Braunerde-Podsol* is distinguished within podzols. However, it is questionable whether the diagnostic criteria provided enable consistent identification of some subtypes, particularly in land with upper horizons mixed by cultivation. As noted by Greene (1963), the 'orderly procedure keeps in mind the gradual development of soil profiles and the gradual transitions from one kind of soil to another', but '. . . the system seems to meet intellectual rather than practical needs'.

The lowest category of the system, the soil form, is defined by profile type, particle-size class and parent material, and is roughly equivalent to the soil series of the USA. It has the advantage of being a more flexible concept that enables the nature of soil maps to be specified in connotative terms while avoiding the correlation problems that arise from geographically named soil series.

The classification developed in East Germany (Ehwald 1968; Ehwald *et al.* 1966) merits separate treatment as it illustrates the 'co-ordinate' approach whereby the soil form is specifically defined by combining the name of a soil type with a lithological class called a *Sippe*. The types and subtypes are defined on the basis of pedological characteristics and are derived from those of West Germany but, as explained by Lieberoth (1969), there are fewer classes. Those defined by lithology, like the clayey soils distinguished as *Pelosole*, were excluded, and classes containing soils that resemble each other after cultivation were amalgamated. In addition, the number of intergrade subtypes was reduced.

The classification of lithological types is essentially separate. The classes are defined on the nature and origin of the material so that stony soils from hard rocks are distinguished from those in loess, alluvium and colluvium, and other separations based on broad particle-size groupings and carbonate content. Emphasis is also placed on whether the material is homogeneous or stratified.

Soil forms are thus identified by a connotative lithological prefix to the type name. Examples of some main soil forms of the brown earths are given below (Lieberoth 1969).

Fels-Braunerde – shallow brown earth on solid rock at less than 40 cm depth.
Berglehm-Braunerde – brown earth in the weathering and redistribution products of solid rock 40 to 120 cm thick, over rock.
Sand-Braunerde – brown earth in a layer of sand more than 80 cm thick.

The East Germans seem to have developed a more practical system than in West Germany with less emphasis on the characteristics of uncultivated soils. In contrast with other schemes outlined, it attempts to make a clear-cut separation between pedological and lithological characters for class differentiation.

Netherlands

The Dutch soil classification is of particular interest since it illustrates a system developed for a limited but special range of soils and landscapes. In the foreword to the English version of De Bakker (1979), Dudal writes:

> *Soil science in the Netherlands puts strong emphasis on the relationship between parent material and soil formation, and between physiographic conditions and land use. This approach . . . is quite understandable in a country where soils have to a great extent developed from alluvial and aeolian materials of recent geological origin. Dutch soil scientists have paid much attention to pedogenesis in fresh sediments, known as 'initial soil formation' or 'ripening', and to groundwater as a soil forming factor. Furthermore human influence on soil genesis, in this land of man-made soils, has been thoroughly investigated.*

The multicategoric, morphometric system (De Bakker and Schelling 1966) in current use has replaced the physiographic scheme of soil landscapes (Edelman 1950) and was developed by the Soil Survey Institute as a basis for soil mapping at 1 : 50 000. The scheme shares features of other classifications, particularly that of the USA, which was evolved over a similar period, but has a distinct national character.

The classification gives precise definitions of differentiating criteria used to define 5 orders, 13 suborders, 25 groups and 60 subgroups. The orders are as follows.

1. Peat soils – with peaty material more than 40 cm thick within 80 cm of the surface.
2. Podzol soils – with a 'prominent podzol B horizon' and an A1 horizon less than 50 cm thick (mainly in cover sands).
3. Brick soils – with a 'brick layer' (argillic horizon) starting within 80 cm depth (mainly in loess and river-terrace deposits).
4. Earth soils – with a surface 'mineral earthy layer' significantly darkened by humus. 'Thick earth soils' with an A1 horizon thicker than 50 cm occur where generations of farmers have applied bulky organic-rich top dressings (*Plaggen* soils).
5. Vague soils – with weakly developed A horizons which are pale and contain little organic matter. They embrace most of the soils of the polders and dunes. A suborder of initial (unripened) Vague soils is included.

Some features of the system are listed below.

1. A weathered B (cambic) horizon is not diagnostic so there are no equivalents of Ochrepts or brown earths *sensu stricto*.
2. Relatively permanent features of surface horizons are used widely for class identification.
3. It is designed for well-sorted parent materials and no provision is made for stony materials or bedrock.
4. No distinction is made between gley and pseudogley soils as in the French system. ('Pseudogley features' are restricted to some Brick soils; otherwise gleying is entirely related to a fluctuating groundwater table.)
5. The significance attached to the depth of 'ripening'.
6. Unlike the US scheme, base status is not used for differentiation and the presence or absence of carbonates is diagnostic only at low categoric levels.
7. The classification of peat soils does not follow the North American systems in the use of fibre content.

A much fuller understanding of the soils can be gained from De Bakker's *Major Soils and Soil Regions in the Netherlands* in which the soils of the major soil geographic districts are illustrated by 32 coloured plates of soil profiles supported by descriptions and analytical data. These profiles are classified by three Dutch systems and seven others including most of those discussed in this chapter.

Belgium

The soil classification used in the soil survey of Belgium illustrates the use of a co-ordinate scheme to define soil series, the basic units of the system (Tavernier and Marechal 1962). They are defined by: (i) particle-size class (textural) groupings with a separate provision for organic materials; (ii) drainage classes based on morphological properties with definitions related to texture; and (iii) profile development. Each soil series is designated by letter symbols derived from the three separate axes as in the examples below.

Series Aba: loam with textural B horizon without gleying.
Series Aca: moderately gleyed loam with textural B horizon.
Series Sbc: dry loamy sand, with discontinuous textural B horizon.
Series Zcg: moderately dry sand, with a well developed humus and/or iron B horizon.

The designations indicate clearly how a particular series is defined and which characters are shared with others.

Great Britain

Soil surveys in Britain based on soil profile observations began in the 1920s (Muir 1961) and, as earlier in the USA, soil series were identified by mapping at widely scattered localities during the pre-war period before the development of a classification framework. A scheme of major groups and subgroups was subsequently compiled by a committee of soil surveyors that included five major groups of brown earths, podzols, calcareous soils, gley soils and organic soils, together with a group designated 'undifferentiated alluvium' which was strictly a cartographic unit (Clarke 1940). The shortcomings of the system became apparent with the subsequent growth of survey and supporting investigations and proposals for modification (Avery 1956, 1965; Mackney and Burnham 1964) closely reflect the influence of other West European schemes. However, the original system with minor modifications continued in use until the *Soil Survey of England and Wales* adopted the outline scheme published by Avery (1973) in the *Journal of Soil Science*. For the first time in Britain this uses specific soil properties to define class limits as in the systems developed in the USA and the Netherlands for a similar purpose, that is '. . . to provide a nationally uniform, systematic basis for legends of soil maps made to aid land use . . .'

The system has the four categories of major group, group, subgroup and series. Classes in the two highest categories are listed in Table 3.13.

Table 3.13 Classification of soils in England and Wales – Major groups and groups (Avery 1980)

Major Group	Group
1 Terrestrial raw soils	1.1 Raw sands
	1.2 Raw alluvial soils
	1.3 Raw skeletal soils
	1.4 Raw earths
	1.5 Man-made raw soils
2 Raw gley soils	2.1 Raw sandy gley soils
	2.2 Unripened gley soils
3 Lithomorphic soils	3.1 Rankers
	3.2 Sand-rankers
	3.3 Ranker-like alluvial soils
	3.4 Rendzinas
	3.5 Pararendzinas
	3.6 Sand-pararendzinas
	3.7 Rendzina-like alluvial soils
4 Pelosols	4.1 Calcareous pelosols
	4.2 Non-calcareous pelosols
	4.3 Argillic pelosols
5 Brown soils	5.1 Brown calcareous earths
	5.2 Brown calcareous sands
	5.3 Brown calcareous alluvial soils
	5.4 Brown earths (sensu stricto)
	5.5 Brown sands
	5.6 Brown alluvial soils
	5.7 Argillic brown earths
	5.8 Paleo-argillic brown earths
6 Podzolic soils	6.1 Brown podzolic soils
	6.2 Humic cryptopodzols
	6.3 Podzols
	6.4 Gley-podzols
	6.5 Stagnopodzols
7 Surface-water gley soils	7.1 Stagnogley soils
	7.2 Stagnohumic gley soils
8 Ground-water gley soils	8.1 Alluvial gley soils
	8.2 Sandy gley soils
	8.3 Cambic gley soils
	8.4 Argillic gley soils
	8.5 Humic-alluvial gley soils
	8.6 Humic-sandy gley soils
	8.7 Humic gley soils
9 Man-made soils	9.1 Man-made humus soils
	9.2 Disturbed soils
10 Peat soils	10.1 Raw peat soils
	10.2 Earthy peat soils

Following modern definitional systems, classes in higher categories are differentiated mainly by the use of diagnostic horizons and by the composition of the soil material, and several of the differentiating criteria are adopted from the US system.

The desire to maintain continuity with previous work can be recognized by the correspondence of the brown, podzolic, gley and peat major groups with those of the 1940 classification. The system introduced major groups for terrestrial and *gley raw soils* with little or no pedogenetic alteration, lithomorphic soils to include rankers and rendzinas, *man-made soils* for those strongly modified by man and including the Dutch 'Plaggen' soils, and *pelosols* for non-hydromorphic clay soils as in the West German system. At group or subgroup level clayey (pelo-) and sandy classes are distinguished from loamy soils. The reasons for these introductions were 'firstly to create a manageable number of classes more homogeneous for practical objectives than those based on pedogenic horizons alone, and secondly because extreme lithologies tend to modify or obscure the expression of such horizons' (Avery 1973). In addition soils in recent alluvium are distinguished from otherwise similar soils.

The scheme differs from most other major systems by separating three classes of hydromorphic (gley) soils (2, 7 and 8) at the highest categoric level. The gley soils of Western European systems are distinguished as ground-water gley soils, while those with impeded drainage or excess surface wetness identified in France and Germany as *Pseudogley* and *Stagnogley* are designated surface-water gley soils.

The classification has been thoroughly tested by the Soil Survey of England and Wales over the last five years. Classes in the sub-group category were used to characterize map units on the 1 : 1 000 000 soil map of England and Wales (Avery *et al.* 1975) and the system as a whole has been applied with minor variations in legends of 1 : 25 000 and 1 : 250 000 maps from 1975 onwards. A short monograph has now been prepared for users of the system containing specifications of horizons and other differentiating features used to define classes, and definitions of categories above soil series level (Avery 1980).

The classification systems used by the other soil survey organizations in the British Isles are not detailed in separate publications but are described in the various memoirs. The *Soil Survey of Scotland* has used a slightly modified form of the simple British scheme (Clarke 1940) consistently through its mapping programme (Bown and Heslop 1979), while in Eire use is made of an amended form of the US system before the introduction of the new taxonomy (Finch 1971).

6. Australasia

Australia

Soil classification and soil survey in Australia is reviewed in the opening chapters of *A Handbook of Australian Soils* (Stace *et al.* 1968) where it suggests that classification developed along similar lines as in Europe and North America. The soil map and accompanying bulletin of Prescott included a classification into major soil groups which has only recently been superseded. He recognized the zonal relationships of soils to existing climate and vegetation but noted that these relationships were obscured over large areas of old landscapes with lateritic soils developed under former humid tropical climates. The system distinguished major soil groups based mainly on soil colour and kind of profile, related to present or past climate.

Prescott's system was modified and extended by Stephens (1962) following the modern trend of basing groupings on actual soil characteristics such as colour, texture, structure and consistence and the nature of boundaries between horizons. Account is also taken of the presence, form and amount of organic matter, calcium carbonate, gypsum, soluble salts, clay skins, concretionary material and the nature of the parent material. The principal and most used category was the great soil group though these are arranged under Undifferentiated and Differentiated Solum Classes, soil orders and suborders and subdivided into families, series, types and phases. It is interesting to note that in the main Solum Class of differentiated soils, there is a primary division into pedalfers and pedocals, following the terminology developed in the USA by Marbut (p. 64), and reflecting the importance of arid and semi-arid soils in which calcium carbonate tends to accumulate. The divisions are defined as follows.

Pedalfers. 'Soils in which lime carbonate does not accumulate in any part of the profile, but such carbonate as may have been present in the parent material is continually in process of disappearance from the soil profile'.

Pedocals. 'Soils in which, regardless of the presence or absence of lime carbonate in the parent rock, lime carbonate (and/or sulphate) has accumulated in the soil during the progress of soil-making and as a result of the soil-forming processes.'

Stace *et al.* (1968) present an amended and simplified system (Table 3.14) which forms the basis of the description of soils in the Handbook. The great groups are now ordered to indicate progressive increase in degree of profile development and leaching.

This brief summary of Australian work would be incomplete without reference to the factual key for the recognition of Australian soils (Northcote 1971), based on specified soil properties with little emphasis on genetic connotations to define 'principal profile forms'. It does not describe or name central class concepts and thus breaks away from the traditional approach. The relationships of the classes to the Australian great soil

Table 3.14 Australian Great Groups in order of profile development and degree of leaching (Stace *et al.* 1968)

1. No Profile Differentiation	1. Solonchaks
	2. Alluvial soils
	3. Lithosols
	4. Calcareous sands
	5. Siliceous sands
	6. Earthy sands
2. Minimal Profile Development	7. Grey-brown and red calcareous soils
	8. Desert loams
	9. Red and brown hardpan soils
	10. Grey, brown and red clays
3. Dark Soils	11. Black earths
	12. Rendzinas
	13. Chernozems
	14. Prairie soils
	15. Wiesenboden
4. Mildly leached soils	16. Solonetz
	17. Solodized solonetz and solodic soils
	18. Soloths
	19. Solonized brown soils
	20. Red-brown earths
	21. Non-calcic brown soils
	22. Chocolate soils
	23. Brown earths
5. Soils with predominantly sesquioxidic clay minerals	24. Calcareous red earths
	25. Red earths
	26. Yellow earths
	27. Terra rossa soils
	28. Euchrozems
	29. Xanthozems
	30. Krasnozems
6. Mildly to strongly acid and highly differentiated	31. Grey-brown podzolic soils
	32. Red podzolic soils
	33. Yellow podzolic soils
	34. Brown podzolic soils
	35. Lateritic podzolic soils
	36. Gleyed podzolic soils
	37. Podzols
	38. Humus podzols
	39. Peaty podzols
7. Dominated by organic matter	40. Alpine humus soils
	41. Humic gleys
	42. Neutral to alkaline peats
	43. Acid peats

groups and FAO classes are given by Northcote *et al.* (1975) in describing the soils occurring in the *Atlas of Australian Soils* (Northcote *et al.* 1960–68). Stace *et al.* (1968) claim that the key 'represents a considerable advance in operational efficiency due to greater precision in definition of classes in terms of observable features' and the representative profiles of great soil groups given in the Handbook are coded by the system.

New Zealand

Soil classification was developed independently in New Zealand to serve the purposes of the Soil Bureau, a division of the Department of Scientific and Industrial Research (DSIR). As in the USA, the work has been approached from two directions (Taylor and Pohlen 1968). First, soil types were identified by soil survey and grouped into soil series and families. Second, broad soil groups were recognized and subdivided according to soil-forming processes.

Soil groups were conceived in acordance with the zonal system of Taylor and Cox (1956) in which intrazonal soils were associated with particular parent materials like volcanic ash and pumice, and azonal soils included gley soils, organic soils and alluvial soils. The genetic classification introduced by Taylor and Pohlen (1968) did not receive general acceptance and, following a visit from Guy Smith the principal author of *Soil Taxonomy*, the US system was adopted in 1977 for use over a trial period. Experience of its application has already suggested the need for modifications to accommodate some New Zealand soils and the practical problems in everyday survey work have been noted. In particular there is the difficulty experienced in many countries of the overlap of established categories at series level with the subgroups of the implanted system.

7. Numerical classification

Numerical taxonomy, defined as 'the grouping by numerical methods of taxonomic units based on their character states' (Sneath and Sokal 1973), has developed rapidly in recent years in line with computer techniques. It aims to create classes and demonstrate their relationships objectively rather than intuitively as discussed in Chapter 4. Thus the recognition of patterns in soils and soil properties implicit in classification is done numerically with the aid of computers (Rayner 1969). Numerical systems of soil classification are still in their infancy but are likely to mature quickly with the increasing desire and need to quantify soil data and automate its manipulation (de Gruijter 1977; Webster 1977b).

Sneath and Sokal (1973), who deal with biological systematics, describe the sequence of operations in numerical taxonomy as follows – 'organisms and characters are chosen and recorded; the resemblances between organisms are calculated; taxa are based upon these resemblances; and last, generalizations are made about the taxa'. This methodology cannot be applied directly to soils because, as noted earlier, the choice of individuals is arbitrary and the selection of characters is complicated by the division of soil profiles into horizons of varying thickness so that characters are commonly measured at standard depths (Rayner 1969).

Let us assume that a set of data on soil properties obtained from field observations or laboratory measurements is available. Multivariate methods such as principal components analysis or principal co-ordinate analysis as described by Webster (1977b) can be used to examine relationships between individuals and should reveal the inherent structure of the population. If the units sampled are allocated to a 'traditional' classification system multiple discriminant analysis offers possibilities of improvement by identifying the best discriminants between classes and the formation of more homogeneous classes. That is to say it can suggest ways to maximise differences between classes and minimize differences within classes, albeit within the necessarily finite sample of the population included in the study.

Otherwise the data can be used in an attempt to construct a non-hierarchical system of classes within which variation is minimal. 'The methods involve creating a classification, evaluating some function of within-class variation as a criterion of goodness, transferring individuals between groups, and evaluating the criterion afresh. The process is repeated until no further improvement seems possible' (Webster 1975). This technique is regarded by Webster as the most promising for soil classification, since soil populations rarely form the clusters compatible with a hierarchical system (p. 61).

Developments in numerical taxonomy have progressed with the introduction of other numerical methods in soil science (Ch. 4) and the literature suggests that the Netherlands Soil Survey Institute is in the forefront in applying the new techniques. The advantages over traditional methods include the ability to integrate data from a variety of sources, greater efficiency by the use of automation, the flexibility to produce keys and associated descriptions and maps by data processing systems, and greater sensitivity in delimiting classes. However, numerical classification faces similar problems to traditional systems in dealing with the vertical anisotropy of soil properties, the selection of characteristics and the need to include large number of both soil individuals and attributes in the analysis if the bias inherent in the initial selection is to be reduced to an acceptable level. The application of numerical methods has as yet been limited but it seems likely that, if married to conventional techniques, useful improvements over purely traditional methods will result.

CONCLUSIONS

This introduction to soil classification has traced its development from systems with a strong reliance on environmental factors to morphogenetic systems based on the selection of intrinsic soil features to define classes at each categoric level. The extent to which genetic relationships should underlie soil classification has been earnestly debated and there are clearly wide differences of approach, though Duchaufour (1977) suggests these may be less fundamental than they appear. Development has been aided by improved methods of field and laboratory diagnosis, and the growing discipline of micromorphology pioneered by Kubiena has made a significant contribution. Various investigations have assisted in the identification of materials that have undergone pedogenic alteration under previous environmental conditions and soils that have been profoundly influenced by human activity. Numerical taxonomy will undoubtedly develop with the increasing desire to quantify soil data but at the moment is still a research tool and no national system has yet been produced.

Classifications for soil survey purposes have been designed to relate cultivated soils to their virgin or semi-natural counterparts and emphasis has been placed on the properties of relatively permanent subsurface horizons not readily altered by man. Little has been said about soil classification in relation to soil mapping and the use of classes to define map units at different scales. These topics are dealt with elsewhere where the techniques of soil survey are described.

Soil classification reflects knowledge of different soils and the degree of understanding of the processes that formed them. It will continue to stimulate research to improve our knowledge of soil distribution and pedogenetic processes. Hopefully work in the future will extend across national boundaries with the aim of achieving an acceptable international classification system. It is feared that despite the steps made in the USA and by FAO there is a long way to go before this end is achieved.

4

HANDLING SOIL SURVEY DATA

C. C. Rudeforth

Scientists in several countries are attempting to integrate a wide range of soil and related information by setting up *Soil Information Systems* (SIS). These should contribute increasingly towards recognizing the best uses for different kinds of land. As is described in previous chapters, soil surveyors record, classify and display the distribution of soils. In each of these tasks, modern data handling methods, particularly by computer, enable the large amounts of information accumulated to be processed. Easy access to computers has further encouraged surveyors to store more data which hitherto was either unrecorded or only noted in the field on books or maps for personal reference when compiling the work for publication. The surveyor himself would be well aware of the reasons why the data were collected and their reliability. Once collected and stored mechanically, however, others have much more chance of using the information and it has become important that the nature and limitations of different forms of sampling should be well understood and that handling methods and result reporting should be suited to the nature and purpose of the original observations. This chapter outlines ways of collecting different kinds of soil information and describes how it is organized, stored, retrieved and displayed for various purposes. It concludes with brief references to some soil information systems being developed in different countries.

4.1 SAMPLING DESIGNS

Information about soils is usually obtained on a three-dimensional basis. In this context sampling refers not only to the actual collection of soil material for physical or chemical testing, but includes collection of any information about soil and site. Information about soil distribution is collected in at least six ways, each needing a different approach to data handling. Free, fortuitous, probability, nested, traverse and representative sampling designs are described.

Free sampling

Free sampling allows a surveyor to choose where he wishes to observe the soil and gives him the freedom to change his choice as he proceeds. For many years and in many countries experienced surveyors have conventionally used free sampling to locate boundaries between different soil classes (see p. 47). There are no constraints on the surveyor, except perhaps of total time available, which restrict his choice of sites for examination. He uses his experience to predict where boundaries are likely to occur and is free to examine (i.e. sample) the soil where he thinks he can most economically check his predictions. He is likely to sample much more frequently where soil classes do not relate in a simple manner to visible landscape features. Not only does this result in a somewhat uneven spread of observations, but the observations are likely to be significantly biased towards difficult boundary cases so that soils occurring close to boundaries can be over represented in numbers of observations (Webster 1977b). Records of such data need careful handling, as will be shown on page 104. It is wrong, for example, to average values for depth of soils in a unit where a disproportionate number of these values were measured at sites chosen for plotting boundaries around shallow variants.

Fortuitous sampling

Digging soil profiles and trenches can be time-consuming and hard work, and a surveyor may well think himself fortunate when he finds a ready-made soil exposure. He will often take pains to describe features such as lateral variability in more detail and make observations to greater depth than would otherwise be practical. Exposures may be natural as in river banks, or artificial as in quarry faces, road cuttings or excavations for building foundations. Many such exposures are located in particular parts of the landscape. Quarries are located to yield particular rock materials, but will rarely represent all materials in proportion to their spatial occurrence. Glacial till may remain unquarried while neighbouring good quality glacio-fluvial sand may be taken for building purposes. Likewise, road or rail cuttings often occur in rocks which form hills and are relatively rarely in clay vales. It is thus important to appreciate that soil data collected by fortuitous sampling may differ from those obtained by other sampling methods and may therefore require different forms of data handling.

Probability sampling

Probability sampling requires that every spot in the landscape should have an equal chance of being chosen. This is more easily said than done, but four main schemes approximate to this ideal: random; grid; stratified-random; and stratified-grid. Random samples are generally located by specifying the co-ordinates of points within a survey area from random number tables. In practice, sample sites chosen at random are often spread rather unevenly across the landscape, giving a pattern which is not very satisfactory for display on maps because unsampled areas may be confused with areas genuinely lacking the particular property (Fig. 4.1a). This problem is overcome by grid sampling. Sites for grid sampling are located at regular intervals such as at the intersections of the National Grid shown on British Ordnance Survey maps or simply by overlaying a sheet of squared paper on a map or aerial photograph of the district to be sampled. Such sampling has the theoretical disadvantage that it could seriously misrepresent the soils of a district if regular soil variations coincide with the orientation and spacing of the grid. Thus narrow strips of alluvial soil in Fig. 4.2 occupy 3 km^2 and yet are unrepresented on the 1 km grid through which they run. Stratified-random [Fig. 4.1(b)] and stratified-grid [Fig. 4.1(c)] schemes overcome these objections to a considerable extent but are still imperfect. The former scheme has one or

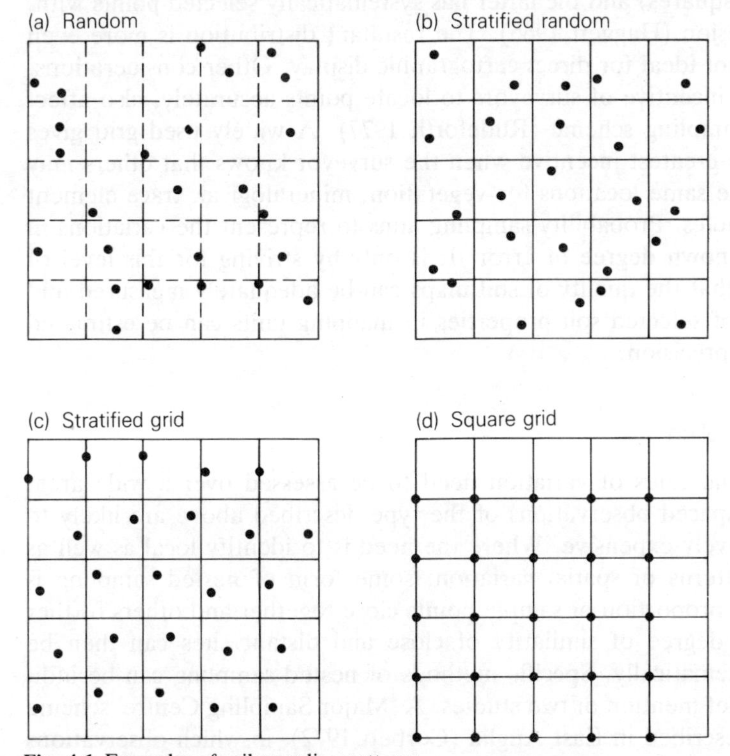

Fig. 4.1 Examples of soil sampling patterns.

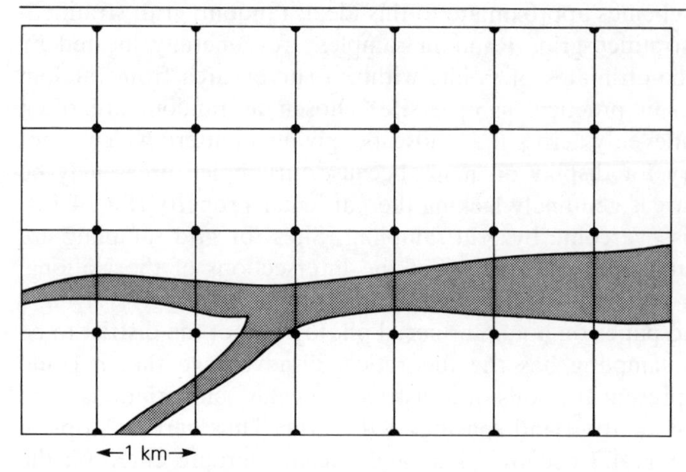

Fig. 4.2 An alluvial soil remaining unsampled between 1 km grid points although it covers over 3 km^2

more points chosen at random within predefined landscape divisions (often grid squares) and the latter has systematically selected points within each division (Haggett 1965). The resultant distribution is more even but is still not ideal for direct cartographic display. Other considerations, such as the incentive of surveyors to locate points accurately, also affect choice of sampling scheme (Rudeforth 1977). A widely used grid gives perhaps the greatest incentive when the surveyor knows that others may well visit the same locations for vegetation, mineralogical, trace element or other studies. Probability sampling aims to represent the variations in soils with known degree of error. It is only by striving for this level of objectivity that the quality of soil maps can be adequately measured and the values of selected soil properties in mapping units can be estimated with stated precision.

Nested sampling

Where spatial rates of variation need to be assessed over a wide area, very close spaced observations of the type described above are likely to be prohibitively expensive. Where the need is to identify local as well as broader patterns of spatial variation, some form of nested sampling is used with a proportion of sample points close together and others further apart. The degree of similarity of close and distant sites can then be studied systematically. Specific methods of nested sampling can be indicated by brief mention of two studies. A 'Major Sampling Centre' scheme has been described in East Anglia (Corbett 1972), in which observations are clustered as in Fig. 4.3. The scheme has a symmetry that makes it suitable for terrain without a strongly oriented 'grain', a situation which

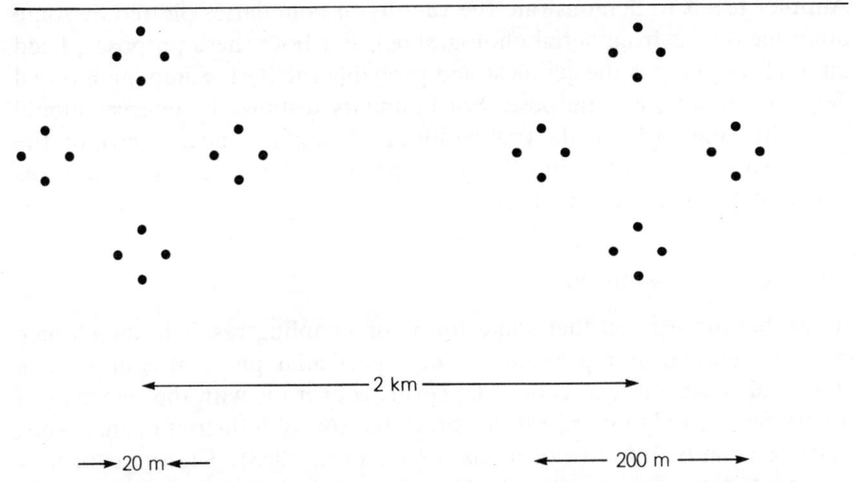

Fig. 4.3 'Major sampling centre' scheme of sampling for finding variation within facets (20 m interval), between facets (200 m interval) and across landscapes (2 km interval) (After Corbett 1972)

contrasts with Pembrokeshire, which has a marked alignment of geological strata and parent materials. In this latter area, a simpler method was chosen by the present author, who selected a 41 km traverse through National Grid intersects across the county roughly at right-angles to the rock strike. Supplementary records were made at 1 m, 10 m and 100 m from each km intersect directed so as to fall in the same field and in the same map unit. The supplementary records were taken due north of the intersects if all the sites were in the same field and unit, otherwise due east; or, failing this, south or west until the conditions were fulfilled. The order of frequency of variations was thus examined within map units down to 1 m spacing using only 123 profiles.

Sampling is also commonly nested within map units of limited extent which prove to have been undersampled during a general survey. Nested samples can be located systematically by superimposing a fine grid over the maps: first the number of intersects (I) falling in the map unit of interest is counted and divided by the number of samples (S) needed to bring the total to an acceptable level (e.g. 20); then, starting at random within the unit, every (I/S)th intersect is chosen for sampling. Unlike the first two forms of nested sampling chosen to reveal rates of spatial variation, the last is really a form of 'Representative Sampling' as explained later.

Traverses

A primary purpose of traverse sampling is to produce sections to illustrate soil variation across a piece of country in a chosen direction.

Another use is to demonstrate the validity of boundaries plotted by some other means (as from aerial photographs). For both these purposes, fixed interval sampling is the simplest and probably the best. Sampling interval depends on scale and purpose. For boundary testing, the interval should be of the same order as the true width of a boundary line as drawn on the map. An interval of about 20 m is appropriate for testing field maps drawn at 1 :25 000 (see p. 101).

Representative sampling

It can be appreciated that some forms of sampling result in data which can be regarded as representative of a particular piece of country in a statistical sense. In this section, however, concern is with the selection of a very few or only one site to illustrate the soil of a district or map unit. Selection can be intuitive or modal (Protz *et al.* 1968). The surveyor may select intuitively in several ways. He may have built up in his mind during the course of mapping a strong image of what the soil is like in a map unit. He may then take considerable trouble to search for a profile which fits this mental picture, rejecting many profiles he finds during his search as in some way deviant from the 'ideal'. This is 'ideal profile' sampling. The very fact that he has felt obliged to reject many profiles before finding one which fits his image closely enough is itself evidence that such an 'ideal' profile might in fact be something of a rarity and therefore not truly representative of much of the unit. Jansen and Arnold (1976) warn that data from such profiles (typifying pedons) are unsuitable for defining ranges in map units because the pedons are selected on the basis of their profile characteristics. Another approach is to select what looks like a typical part of a map unit from landscape appearance and accept whatever profile there is at that point. This is 'pseudo-random' selection. If there is choice of site within the 'pseudo-random' selection and the final site is chosen for convenience, this is called 'convenient spot' sampling.

Modal selection is more objective but still has limitations. The simplest form is to select a soil which has a single property (e.g. thickness to rock) which is within the most frequently occurring band in the map unit. Suppose a grid survey revealed that soil thicknesses between 50 and 60 cm occurred most frequently, so the profile selected should be one of those recorded with a thickness between these limits. If many profiles have thicknesses between the same limits, choice of a representative soil can be reduced by narrowing the limits or by specifying other properties that should be modal. The more closely the surveyor chooses to specify the limits of a 'modal' soil and the more characteristics he requires to be 'modal', the fewer soils he will find in his sampling that meet his specification. It is questionable therefore whether one or only a few profiles can ever really 'represent' the soils of a map unit. Profiles presented in soil survey accounts are perhaps better regarded as examples illustrating particular points about map units rather than being representative of them. Care is required when choosing ways of using data from these profiles.

Sampling depth

Emphasis has so far been placed on choice of sample sites across the landscape, but as soils are three dimensional bodies depth must be considered. Two main choices are available, fixed depth sampling and sampling soil horizons. Systematic sampling at fixed depths is compatible with fixed interval (grid) sampling and is independent of any judgements of the nature of soil horizons. It enables a surveyor to produce maps of topsoil and subsoil quality without making subjective decisions about horizons. Nevertheless, adequate profile descriptions should always accompany fixed depth sampling so that differences and similarities can be attributed where appropriate to horizons. Considerable problems of interpreting fixed depth sampling could arise if contrasting horizons are sampled in different profiles at the same depth because the horizon boundary undulates across the sampling depth. In such cases, where bimodal distributions are apparent in some properties, separation of samples according to horizon designation is appropriate and feasible even where the sampling is at fixed depth. Sampling by horizons alone, on the other hand, often gives different numbers as well as different depths of samples. This may be useful for detailed study of the processes of soil formation and plant root environments, but requires careful handling during a survey and especially in subsequent data manipulation.

4.2 KINDS OF DATA AND ITS STORAGE

After the discussion of sampling strategies in the preceding section, attention can now focus on the data collected. This topic was introduced in Chapter 2. The main categories are recorded as: presence or absence of an attribute; grades of ordered classes (ranked data); and measured values. Examples of attributes often recorded simply as present or absent are of particular soil horizons such as an organic topsoil. Kinds of rock may also be recorded in this way such as the presence of limestone, and of fauna such as earthworms. Grades of ordered classes include ratings commonly given for soil properties such as stoniness, particle size classes and soil water state. Measured values are perhaps most useful, but they can be difficult or expensive to obtain. Chemical analyses are often presented in this form. Simpler and cheaper examples are soil thickness and slope angle measurements.

Soil descriptions and analytical results are generally kept in their original field record form or as typed copies. Other notes are commonly made on field maps or aerial photographs. The descriptions are usually stored by grid reference to their location or by order of acquisition. This enables records from particular areas to be retrieved directly, but other groupings of the descriptions such as those of a particular soil class are not nearly so easily made. Some form of coding and a method of distinguishing records of different kinds are needed to pick out groupings of interest. It is not generally practicable to code all the data collected at every site. The coded information is therefore used partly as a key for drawing attention to records for which more data are available. Alternatively the coded data themselves may suffice to answer users' questions.

Coded information is commonly stored on punched cards and/or computer. Punched feature-cards have proved useful even where computers are available (Rudeforth and Webster 1973). A feature-card is ruled with a numbered grid with cell size 2 to 3 mm. Each card represents an attribute or feature and each cell a soil sample (Fig. 4.16) If the sample has the attribute represented by the card a hole is punched in the cell corresponding to the sample. When several feature-cards for different attributes have been punched they can be superimposed to show samples with a particular combination of properties.

Computer storage of data is achieved by the input of standard 80 column holerith cards but may be keyed in directly, and considerable efforts are now being made in many countries to collect data in forms directly readable by computer. One or several 80-column cards represent a sample and each card, when fed into the computer, produces a row of a data table (Table 4.1). From such a table it is possible to pick out automatically values in any column or row and so to identify any site or sample with an attribute or combination of attributes. It is also possible to retrieve, decode and print the information stored for individual sites.

Table 4.1 Part of a data file in fixed format from a survey of minor elements in soils of Pembrokeshire

Site No.	Grid	Ref.	D pH Zr	Y	Sr	Rb	Pb	Br	As	Ga	Zn	Cu	Ni
342	172	224	1063278	25	71	78	93	50	10	16	53	31	22
353	173	225	1059269	17	80	48	19	44	21	13	65	29	31
353	173	225	4068377	18	91	60	14	25	12	11	63	36	36
344	174	224	1061230	19	81	58	31	43	30		66	30	24
344	174	224	3072221	20	77	52	25	29		12	51	28	35
364	174	226	1057192	22	75	39	14	24	44	10	54	21	19
365	174	226	5065267	24	84	60	9	10	10	12	49	24	28

Columns show: sample site numbers; Ordnance Survey National Grid Reference Numbers (eastings and northings); Mean depth (cm) of sampling a soil layer 10 cm thick; soil pH value; the total concentrations of 11 elements in mg kg^{-1}

4.3 DATA HANDLING METHODS

The main subject of this chapter is a discussion of the ways in which the different kinds of data are handled. The sampling design associated with each method is noted.

Editing and merging data (all samples)

It is essential to check that data are correctly coded and complete. Errors arise for several reasons: wrong assessment in field or laboratory, bad writing, careless copying, card punching or direct keying for computer storage. Most computer departments have verifying facilities which reduce the numbers of card-punching or keying errors. The data are keyed a second time and the results compared with those of the first attempt. Any discrepancy is highlighted automatically, and the operator can make the appropriate correction. For this reason, errors are more likely to arise from the earlier stages, and are often difficult to detect. For the present purpose concern is not with the preventive measures to reduce these errors, but the ways in which they can be corrected when detected.

As indicated above, most survey data are stored as tables with a 'fixed format'. The position of a number or letter describing a particular property is fixed in each row. A table with data for many sites can be stored in a computer file. When the file is printed out, gaps and unusual values are relatively easy to see and can, if necessary, be detected by simple instructions to the computer itself which will then list samples with discrepant values.

Editing then consists of making the appropriate corrections to the data file by instructions on punched cards or directly using a terminal keyboard. One method is by duplicating most values on a punched card, inserting a changed value when the appropriate column is reached, then continuing duplication. The data on the altered card replaces the original data in the file using editing instructions supplied by the computing centre. Editing by terminal keyboard is similar.

A soil surveyor may also need to link data from one file to another: a file of field data to one of laboratory data. For example, he may wish to compare soil colour value from one file with organic matter content from another. The merging procedure is to read site numbers and the required data from each file in turn and produce a new file containing all the data needed for the comparison.

Sorting sites into classes (all samples)

Allocating profiles to classes according to their soil and site characteristics is one of the commonest activities of a soil surveyor in his attempt to simplify and describe the variety of soils. For the relatively few profiles chosen to illustrate soil reports a surveyor can reasonably look at each and match it against the classification criteria to make the allocation. But with increasingly complex classifications using a multiplicity of criteria, and many more profiles, automatic allocation saves much time and repetitive work. Furthermore, changes in classification for different purposes can easily be accommodated and the sites reallocated according to the new class limits with the minimum of further effort.

A general scheme for allocation by computer is outlined in Fig. 4.4. There are six stages, A to F. Stage A reads from a data file and stage B (optional) prints it out. Stage C sets limits of properties acceptable in

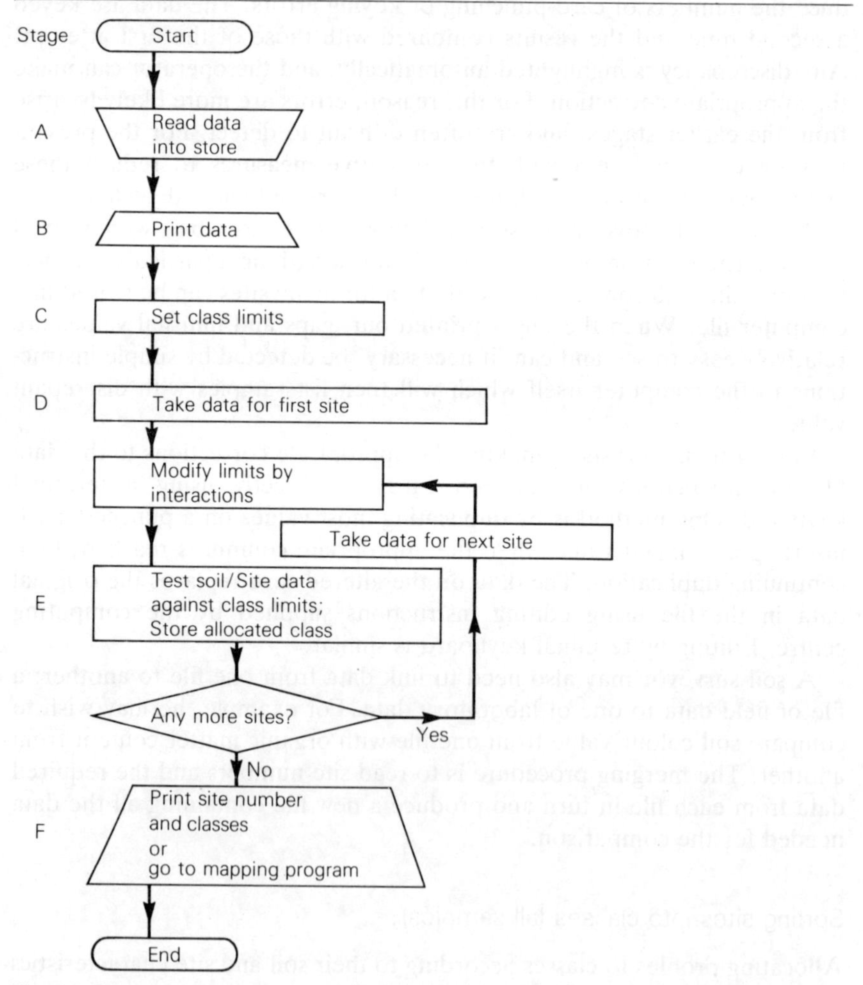

Fig. 4.4 Generalized flow chart for soil or land classing by computer (after Rudeforth 1975)

each class. Stage D scans the data for each site and places values for each property in store, then alters property limits by interactions. This gives extra flexibility by changing the limits previously set for some properties according to values of others found at the site to be classed. Stage E allocates each site to a class, and the final stage (F) prints site numbers (or grid reference) followed by class designation, or stores the results for automatic mapping. Cipra *et al.* (1969) describe an alternative but fundamentally similar approach which emphasizes the use of decision tables for keying out world soil orders.

Classing by cluster analysis and related techniques (Probability and systematic traverse samples)

There is a range of computer techniques for creating classes from objectively sampled data; these include ordination, hierarchical classification and ways of analysing dispersion of sites based on many properties which lead to systems of 'optimal classification'. The objective is to divide soils into groups which are as homogeneous as possible internally, and which are as distinct from one another as possible, without reference necessarily to arbitrary divisions that may be thought useful for a particular purpose. Computing strategies are inevitably complex and cannot be described here in detail. They are described by Arkley (1976) and reviewed comprehensively by Webster (1977b), but a brief description of the objectives of three main approaches is appropriate.

Ordination. It is a simple matter to place a soil in order of the value of a single property. This becomes difficult when many soil properties are being handled. To simplify the problem to manageable proportions the data can be reorganized into a few principal components which account for as much variation in the properties as possible. The projection of points on to the axis of the first principal component then gives the desired simple ordering. The method can be extended for plotting points in the plane of the first two principal components or for three dimensional plots using also the third principal component.

Hierarchical classification. The range of a single soil property can be divided at critical or convenient points, and the sample divided accordingly. The range of several soil properties could be divided simultaneously and still give manageably few groups, but far too many groups result if many properties are used. The need is for a system for placing individual samples into small groups, the small groups into larger groups and so on with, hopefully, relatively few of the soil samples lying at the boundaries between groups compared with those falling comfortably within them. Rayner (1966) describes how similarity between soils can be scored and pays particular attention to sorting soil horizons as well as whole profiles to produce a classification tree or dendogram, while Grigal and Arneman (1969) stress the sequence of horizons (Fig. 4.5). Webster (1977b) enlarges on the alternative strategies producing such hierarchies.

Optimal classification. Webster goes on to describe how soils can be allocated to classes based either on the natural clustering of the data alone, or partly on the intuitive recognition of a core sample of a few profiles unequivocally belonging to each desired class. Such a procedure of 'seeding' to guide classification has advantages in reducing computing time and store space needed to approximate to the best obtainable or 'optimal' classifications. It also allows a surveyor to use his experience and intuition to choose broad classes while the computer handles much data and makes the routine decisions. 'All the hard decisions, the borderline issues that take a disproportionate amount of human's time and attention, can be made by computer' (Webster 1977b).

Correlation and confidence (probability and systematic traverse samples)

One of the most valuable uses of automatic data handling for soil survey is in extending the predictive power of any one surveyor by allowing him use of the recorded experience of others – provided that the sampling was unbiased. A surveyor builds confidence in his predictions, for example of soil depth from site shape when a particular relationship is observed many times. The more the relationship is found to hold (e.g. shallow soils on convex sites) the more confident he can be about it. Confidence will be increased in his work if his colleagues' records show the same. A relationship noted once might or might not be generally true, found 8 times out of 10 indicates at least that it sometimes holds and could generally be true, but found 800 times out of 1000 gives considerable confidence that the relationship generally holds, but exceptions are to be found in about a fifth of the cases. If the sampling were biased (e.g. sites chosen as part of a study of shallow soils occurring on convex slopes, while ignoring most of those that happened to be deeper) we should have no confidence in the proportions of each that happened to have been recorded.

Many relationships have been demonstrated between soil properties, sometimes with a view to predicting soil behaviour and crop yields, but care should always be taken in interpreting such results to ensure that the population we wish to apply them to is the same as or equivalent to that originally sampled. Thus McKeague *et al.* (1971) show how the closeness of relationships among soil properties, and hence the feasibility of predicting unmeasured properties from available data, varies according to the population of samples involved. They conclude that useful relationships do exist for certain soils between visible properties and prop-

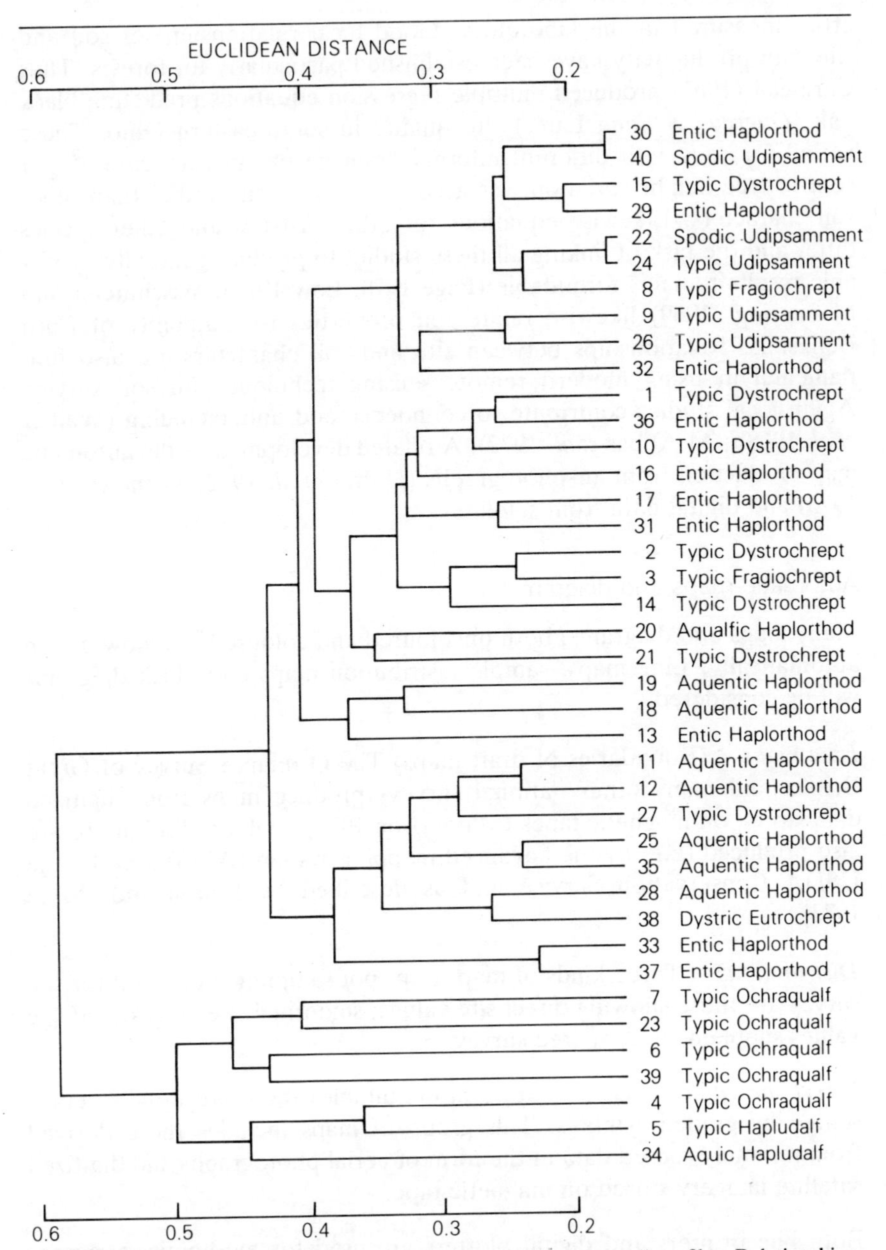

Fig. 4.5 Dendrogram showing relationship among 40 Minnesota soil profiles. Relationships among profiles based on similarity in properties and in sequence of component horizons using Euclidean distance to measure similarity. Dendrogram based on all 22 measured soil properties (after Grigal and Arneman 1969)

erties measured in the laboratory. Good local relationships of soil and site with productivity have been established particularly for forests. Thus Carmean (1967) produced multiple regression equations predicting black oak (*Quercus velutina* Lam.) site quality in south-eastern Ohio. These varied for soils with different internal drainage but in each case 75 per cent of tree height variation was accounted for by site rather than other soil characters. Likewise equations for other districts and other species differ and the task of linking all these studies to produce generally applicable results remains formidable (Page 1970; Low 1976). Wischmeier and Mannering (1969) likewise relate soil properties to erodibility of Corn Belt soils. Relationships between site and soil characters are also fundamental in using modern remote sensing techniques for soil survey. Again local studies contribute to confidence and understanding (Walker *et al.* 1968b; Al-Abbas *et al.* 1972). A related development is the automatic map production from air-photographs (Hoffer *et al.* 1972; Mathews *et al.* 1973) and digital data from satellites.

Automatic maps and diagrams

Many maps and diagrams, both uncoloured and coloured, are now drawn automatically. Line maps, sample distribution maps and block diagrams will be considered.

Line maps. (Boundaries of draft maps) The Ordnance Survey of Great Britain and many other national surveys produce maps from digitized data stored on magnetic tapes (Margerison 1978). Coloured soil maps are also produced (e.g. by the advanced mapping system (AMS) used by the USDA Conservation Service and as described by Linton and White 1973).

Distributions. Three kinds of map from spot sampling most used for soil survey are those showing direct site values; smoothed site values; and site values supplemented by free survey.

Direct site values (Grid samples most suitable) these are usually represented by graded symbols. This group of maps includes those derived from remotely sensed data in the form of aerial photographs and digitized satellite imagery stored on magnetic tape.

Both line printers and digital plotters are used for automatic mapping. Line printers are widely used (as in SYMAP) for 'grey scale' maps with darkness of symbol roughly related to values to be plotted. They have the advantage of speed and the maps are relatively cheap to produce compared with maps drawn by digital plotter. However, they cannot be used so conveniently for line drawing or for symbols of varying size. For some purposes, e.g. in trace element mapping, such systems have their limitations. Ideally, these complex maps are best in colour, but they can be very expensive (Lowenstein and Howarth 1972; Applied Geochemistry Research Group 1978; Thornton and Webb 1979). Areas of unusually high or unusually low values are easily perceived by focussing attention on contrasting colour patterns for each. Monochrome grey scale mapping displays high values (darkest symbols) fairly well, but the pattern of lightest symbols (low values) is not so easy to see and can sometimes be confused with gaps representing missing data. The problem is to gain the advantages of coloured maps in showing patterns of extreme values at a glance while using only black and white. This is solved (Rudeforth 1977) by taking fundamentally contrasting symbols (X and O) to represent the upper and lower ends of the scale and diminishing their size to a small dot about the middle of the scale. The diminution of size achieves two objectives, namely the smooth and logical transition between the two contrasting symbols, and the reasonable impression that missing values would be assumed to have about average values. The most balanced effect is achieved when the X symbols have double line thickness (Fig. 4.6)).

An example of grey scale automatic mapping from multispectral aerial photography in a test area of Indiana is compared with a soil map made by conventional field survey (Fig. 4.7) (Kristoff and Zachary 1974). Bashir *et al.* (1978), while presenting a coloured map of part of the Sudan from satellite sensing, also show how computers can now produce uncoloured maps with complex symbols of any design (Fig. 4.8).

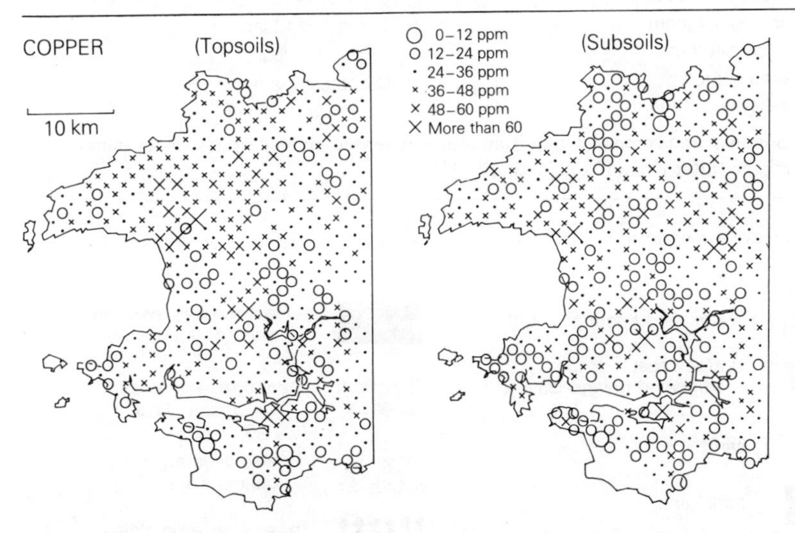

Fig. 4.6 Distribution of copper at sites in west and central Pembrokeshire. Symbols indicate values at individual sites: O = low values, X = high values

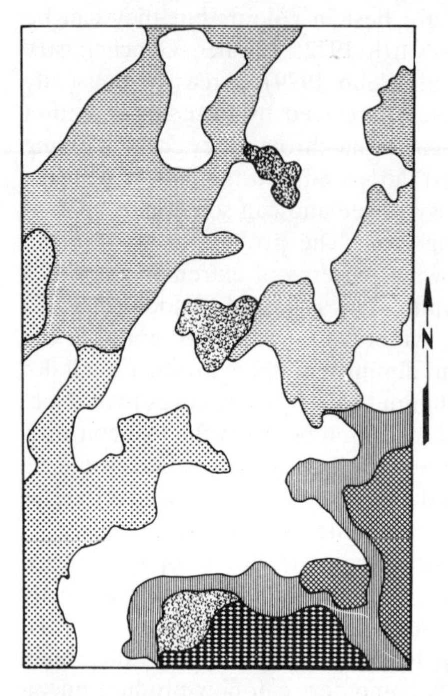

Ragsdale silty clay loam
Brookston silty clay loam
Brookston silt loam
Toronto silt loam
Crosby silt loam
Celina silt loam
Reeseville silt loam

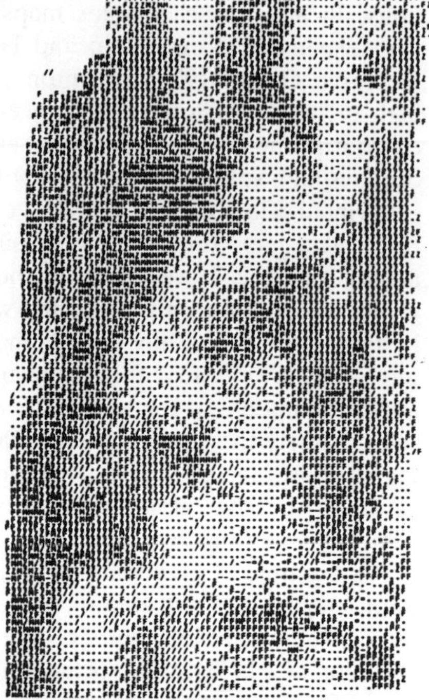

M Ragsdale silty clay loam
Z Brookston silty clay loam
F Brookston silt loam
/ Toronto silt loam
– Crosby silt loam
= Celina silt loam
· Reeseville silt loam

Fig. 4.7 Soil map and computer map from remotely sensed multispectral scanner data in part of Indiana (after Kristoff and Zachary 1974)

Key for Figure 4.8 opposite.

water
water (and floating wood)
sparse vegetation on truncated upland clay
sand
dark sand
sparse tree vegetation on vertisols
sparse tree vegetation on vertisols
vertisols (dark swelling clays in floodplain)
very sparse vegetation on upland clays
sparse vegetation on upland clays
trees (dense on clays)
dense woody vegetation, trees in floodplain

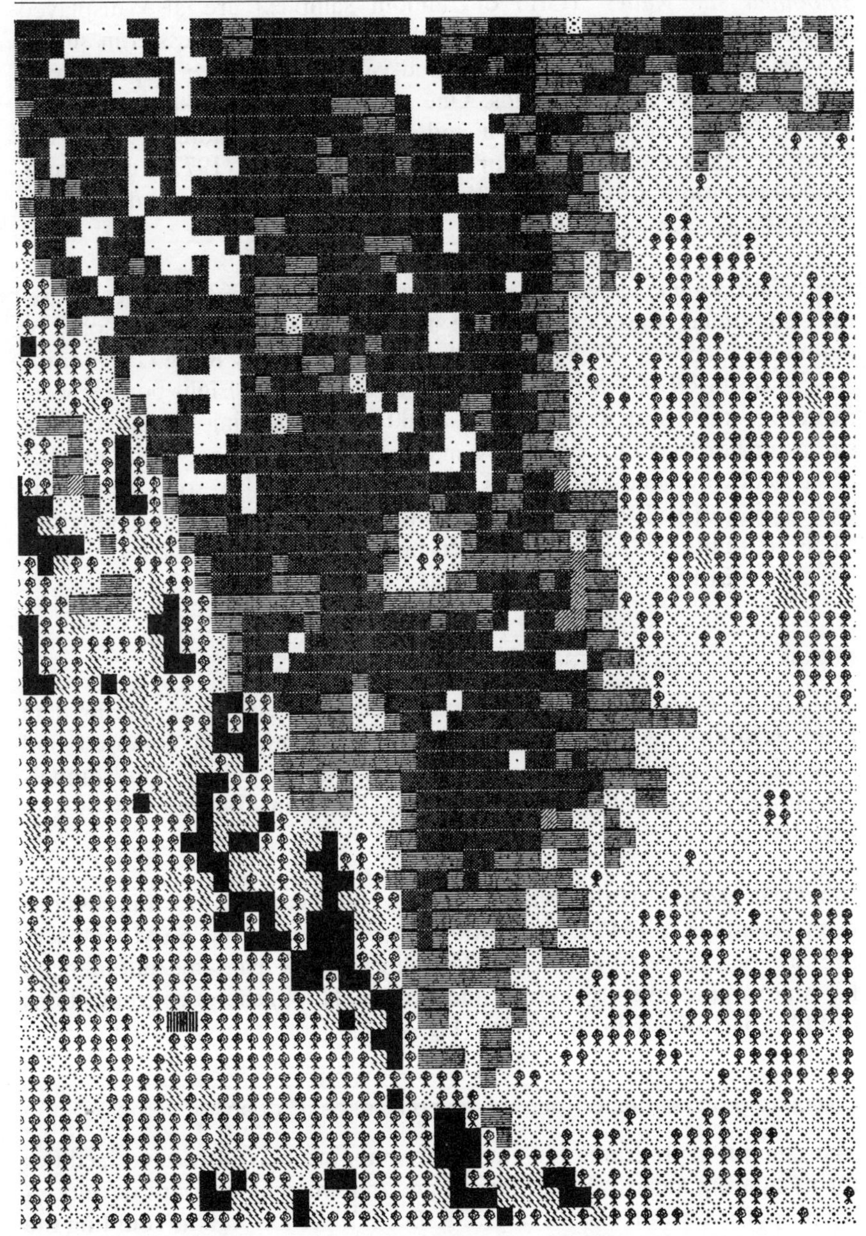

Fig. 4.8 Computer-implemented classification of landscape northeast of Jebel ed Dair and northwest of Jebel Dumbeir, Sudan. (After Bashir *et al* 1978)

Smoothed site values (Grid or random samples) are also displayed on grey scale or coloured maps, but the data are preprocessed in such a way as to average out extreme isolated values to produce maps showing more clearly the general trends. One procedure is to make the value at any one point the average of it and the eight surrounding points. The SYMAP package has options for smoothing. Webster (1977a) used canonical correlations to show smoothed trends. Smoothed and unsmoothed SYMAPs are compared in Fig. 4.9.

Site values supplemented by free survey (Grid plus free samples) Carefully combined, these can produce useful results suitable for display by automatic mapping. Such a system has been devised by Ragg (1977) using a program GRID CAMAP (Finch and Hotson 1974). Eight borings were made per km^2, four were sited on a 500 m grid, leaving the surveyor to site a further four in a subjective manner. In complex areas the number of subjectively sited points was increased, whereas in areas where the distribution of soil types was simple the density was reduced. Numerous single or multi-factor maps can then be printed. Three simple examples are shown in Fig. 4.10.

Interpretive maps (Soil map data)

The USDA Conservation Service has developed methods for storing data from existing soil maps and printing more generalized coloured and monochrome maps automatically for special interpretations. The computing strategy is described by Nichols (1975) and an example is shown in Fig. 4.11.

Block Diagrams. (Probability samples) These can enhance impressions given by maps. SYMVU, an option associated with the SYMAP package, has been used to good effect to produce block diagrams automatically. An example given by Davies and Roberts (1978) shows concentrations of Pb in part of north-east Wales (Fig. 4.12).

Calculating properties of map units and classes (Probability samples)

The computer programs for automatic mapping may also have facilities for calculating mean values, maxima and minima and standard deviations for properties within classes (e.g. GRID CAMAP) as well as counting the proportions of classes in map units. Other programs have been written specially for this purpose and results presented as in Table 4.2

---

* variate: a set of measured values of a soil property in which there is more or less random variation. The 'soil property' in this case is complex and derived from a number of measured values during the process of canonical variation.

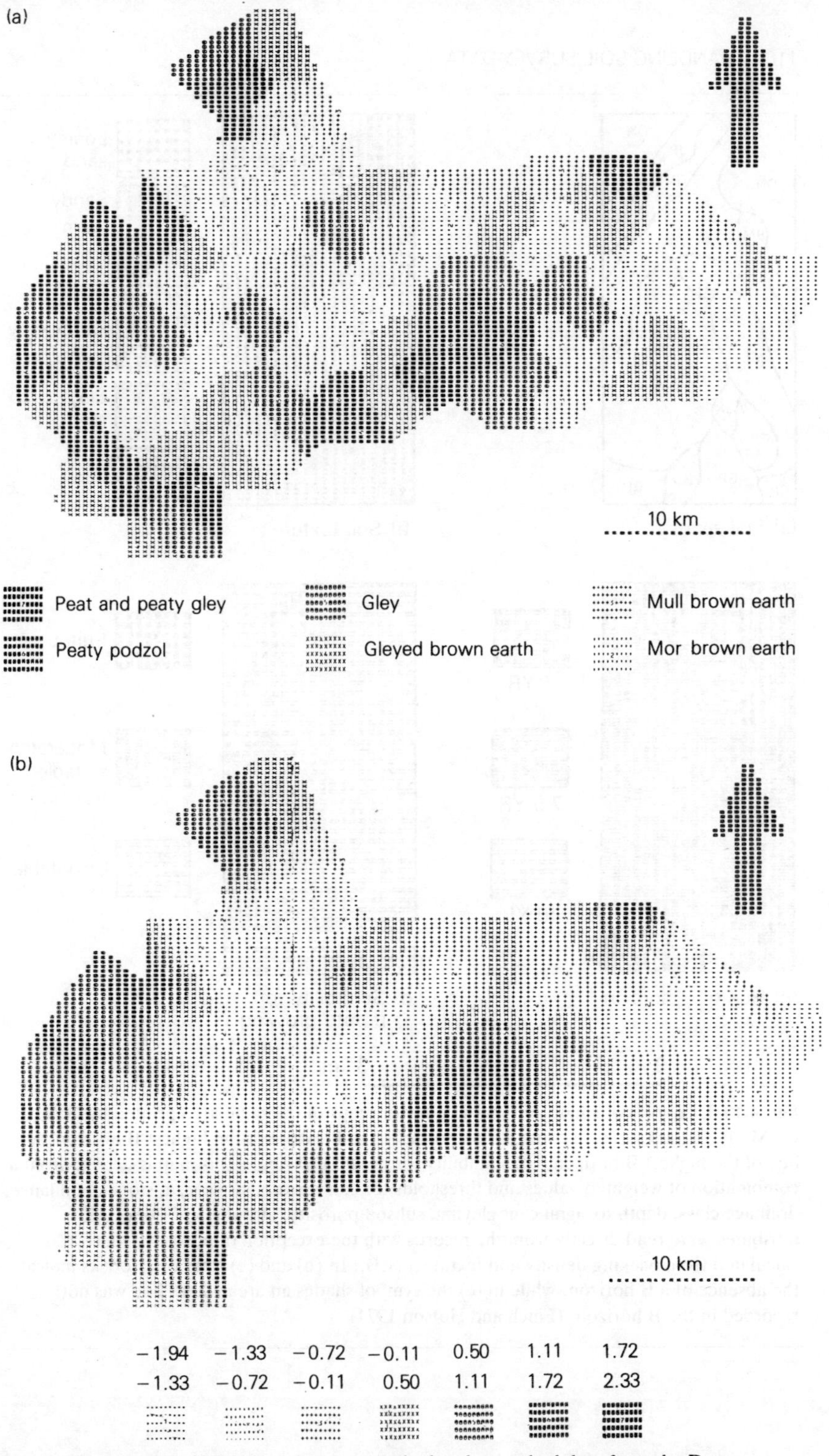

Fig. 4.9 Examples of SYMAPs using unsmoothed and smoothed data from the Dee catchment in north-east Wales. (a) Soil classification (unsmoothed) (b) First canonical variate* (smoothed) (after Webster 1977a; Rudeforth and Thomasson 1970).

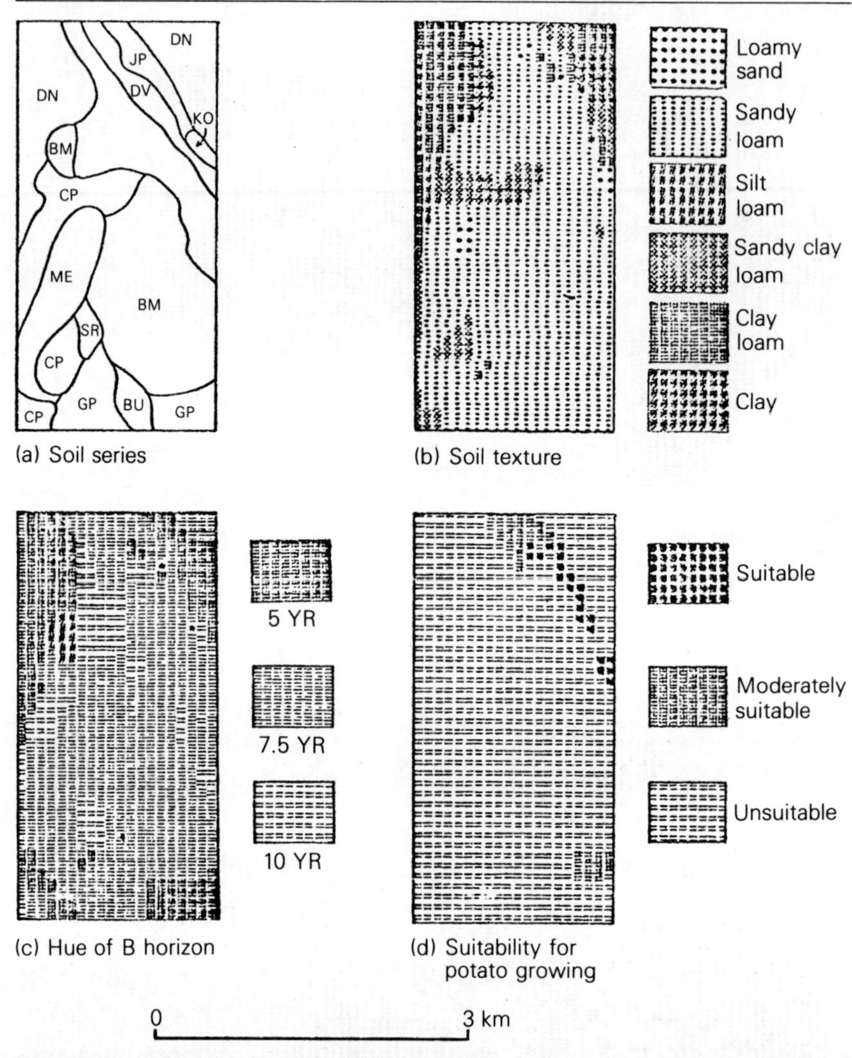

Fig. 4.10 Examples of single and multiple factor maps of the same area printed by GRID CAMAP. (a) soil series in conventional form; (b) soil texture of the highest B horizon; (c) hue of the highest B horizon; (d) suitability for potato growing. (This map is derived from a combination of weighted values and thresholds for soil texture, stone size, stone abundance, drainage class, depth to significant gleying, subsoil porosity, altitude and slope. All attributes were read directly from the records with the exception of porosity, which was calculated from packing density and texture.) N.B.: In (b) and (c) stars or asterisks indicate the absence of a B horizon, while in (c) the symbol shades an area where hue was not recorded in the B horizon. (Finch and Hotson 1974)

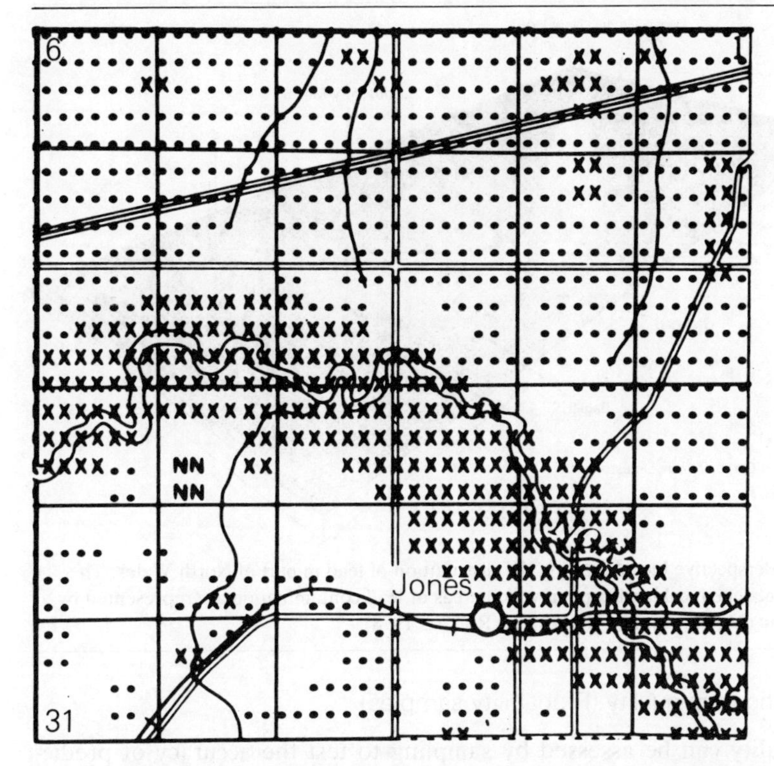

Fig. 4.11 A 93.2 km^2 (36 miles^2) segment of a computer-generated interpretative map. Limitations for dwellings without basements in Oklahoma County, Okla. Cells were 16.20 ha (40 acres). Severe limitations – XX; Moderate limitations – ..; Slight limitations – blank; Not surveyed – NN (after Nichols 1975)

Compiling map keys (Probability samples)

Where map units have been drawn independently of sampling, probability samples from each unit may be used for estimating the proportions of different soil classes. One of the striking features of compiling map keys in this way is that attention is drawn to the true complexity of soils in each unit, while at the same time enabling the surveyor to generalize and simplify as required. For example, the Manod map unit in Pembrokeshire (Table 4.3) contains mainly brown podzolic soils (76 per cent of the 231 sample spots) and yet several other classes are present. Such a unit would nevertheless be regarded as 'simple' rather than complex, because only one soil group occupies most of the land (brown podzolic soils 86 per cent) and contrasting soils (stagnogleys and gleys) together only occupy a few per cent. The Nercwys complex from the same survey clearly contains similar numbers of soils, but no one soil type is dominant.

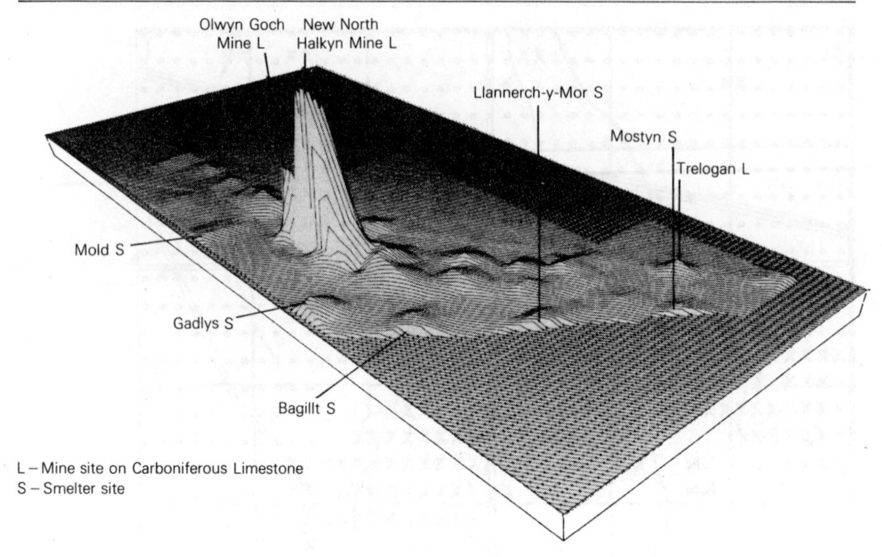

Fig. 4.12 Perspective block showing the distribution of lead in part of North Wales. The figure is a geochemical landscape in which areas of lead contamination are represented by topographic elevations (after Davies and Roberts 1978)

Comparing map quality (Probability samples)

Map quality can be assessed by sampling to test the accuracy of predictions, map unit purity, intricacy and placement of boundaries and variance within and between map units (Rudeforth 1977). The proportion of sample points where soil class is correctly predicted from the map legend is fairly straightforward to assess and needs relatively little data handling. For map unit purity a scoring procedure like that used by Bie and Beckett (1971) can be used. For each map, the units are listed with the fraction of the survey area that each covers (W). The stated fraction of each named profile class within each map unit is then listed as legend purity (L). Actual purity (A) is recorded as the proportion of probability samples fitting the legend description. The score for each profile class is then the product of W and the lesser value, L or A. The score for each map unit is the sum of scores for its constituent profile classes, and the score for the map is the sum of the map unit scores. From tables built up in this way, for each map unit (j) in which a single class exceeds a stated percentage, the proportion of the map covered by the map unit is noted (Wj). The sum of Wj values then gives the proportion of the whole map covered by units of the stated purity. Intricacy can be measured by counting the number of boundaries on each map crossing the lines of a superimposed grid. Boundary placement is tested along traverses by closely spaced observations. The resulting data can be handled by multi-

Table 4.2 Computer printout showing total selenium contents of soils (mg/kg) in part of South Wales. The unit numbers refer to soil series classes. The upper and lower acceptable limits indicate the normal range and attention is drawn to soils with a substantial proportion of samples outside this range. (Unit 2254 topsoil; unit 1806 subsoil).

									Acceptable Limits					
property	sample	unit	mean	sd	sem	max	min	n	lower limit	no.	%	upper limit	no.	%
selenium-tot	topsoil	305	0.69	0.27	0.06	1.3	0.3	23	0.2	0	0	2.0	0	0
selenium-tot	topsoil	1237	0.63	0.84	0.16	5.0	0.2	29	0.2	0	0	2.0	1	3
selenium-tot	topsoil	1806	0.35	0.15	0.07	0.6	0.2	4	0.2	0	0	2.0	0	0
selenium-tot	topsoil	208	0.41	0.19	0.06	0.8	0.2	10	0.2	0	0	2.0	0	0
selenium-tot	topsoil	516	0.57	0.26	0.07	1.1	0.2	13	0.2	0	0	2.0	0	0
selenium-tot	topsoil	2254	2.59	1.01	0.23	5.4	0.8	20	0.2	0	0	2.0	13	65
selenium-tot	topsoil	1735	2.00	0.23	0.12	2.4	1.8	4	0.2	0	0	2.0	1	25
selenium-tot	topsoil	1941	1.20	0.57	0.33	1.7	0.4	3	0.2	0	0	2.0	0	0
selenium-tot	subsoil	305	1.40	2.42	0.52	9.0	0.2	22	0.2	0	0	2.0	2	9
selenium-tot	subsoil	1237	0.65	1.05	0.20	6.0	0.1	28	0.2	1	4	2.0	1	4
selenium-tot	subsoil	1806	0.33	0.21	0.12	0.6	0.1	3	0.2	1	33	2.0	0	0
selenium-tot	subsoil	208	0.41	0.18	0.06	0.8	0.1	10	0.2	1	10	2.0	0	0
selenium-tot	subsoil	516	0.32	0.13	0.04	0.6	0.1	12	0.2	1	8	2.0	0	0
selenium-tot	subsoil	2254	0.84	0.83	0.19	4.0	0.1	20	0.2	1	5	2.0	1	5
selenium-tot	subsoil	1735	0.63	0.33	0.19	1.1	0.4	3	0.2	0	0	2.0	0	0
selenium-tot	subsoil	1941	0.70	0.36	0.21	1.2	0.4	3	0.2	0	0	2.0	0	0

Table 4.3 Part of a map legend for Pembrokeshire compiled automatically from grid samples. (a) Manod unit (from 231 sample profiles); (b) Nercwys complex (from 21 sample profiles)

(a)	Brown (non-humic) ranker	6%
	Stagnogleyic ranker (drift on rock)	0%
	Brown earth	2%
	Stagnogleyic brown earth	0%
	Brown podzolic soil	76%
	Humic brown podzolic soil	1%
	Stagnogleyic brown podzolic soil	8%
	Humic stagnogleyic brown podzolic soil*	1%
	Cambic stagnogley soil	3%
	Cambic stagnohumic gley soil	0%
	Typical alluvial gley soil	1%
(b)	Brown earth	11%
	Stagnogleyic brown earth	29%
	Humic brown podzolic soil	10%
	Stagnogleyic brown podzolic coil	5%
	Humic stagnogleyic brown podzolic soil*	5%
	Ferric stagnopodzol	10%
	Cambic stagnogley soil	5%
	Podzolic stagnogley soil*	5%
	Gambic stagnohumic gley soil	10%
	Podzolic stagnohumic gley soil (with E)*	10%

variate methods (Webster and Wong 1969; Webster and Cuanalo 1975) or more simply by studying how soil classes and some individual soil properties vary across boundaries.

Variance of measured soil properties accounted for by mapping is estimated by comparing 'total' and 'within unit' variance (Snedecor and Cochran 1967). Webster and Beckett (1968) applied this concept to soil maps and reported large variances within units for some properties and much smaller variances for others in relation to their mean values. Similarly differing amounts of variation for other properties were recorded by Rudeforth (1977, 1978), showing that soil maps could be particularly useful in Wales for predicting properties such as degree of gleying. By contrast, mapping accounted for less than 20 per cent of soil thickness and stoniness variations. The actual properties best accounted for will differ markedly according to terrain.

Assessing spatial rates of variation (Nested traverse sampling)

In the Pembrokeshire study described on page 101 soil classes were the same at 80 per cent of sites 1 m distant, 59 per cent at 10 m distant and 37 per cent at 100 m distant within map units. Allowing for close intergrades, the classes were definitely different at 5 per cent of sites 1 m apart, 10 per cent of sites 10 m apart and 39 per cent of sites 100 m apart. Such figures help to guide decisions about sampling intensity, and in this case imply that observations should be spaced at substantially less than 100 m intervals if an accuracy of 85 per cent in mapping soil classes is to be achieved. An alternative way of assessing spatial rates of variation is to devise a strategy for estimating the percentage of total variation of properties taken separately or together which can be accounted for within the first 1 m, 10 m, 100 m, etc.

Printing profile descriptions from coded data (All samples, but mainly representative)

Webster *et al.* (1976) have described a computer program DECODE for translating coded soil profile data descriptions into text. Coding follows the scheme of the Soil Survey of England and Wales (Hodgson 1974) and is completed by the surveyor on forms such as shown in Fig. 4.13 and punched on to a set of cards. Figure 4.14 illustrates a form of computer printout, but better output devices are now available which print lower case characters and produce descriptions which appear identical with normal typescript.

Assessing and mapping potential cropland (Grid samples)

Land suitable for specific crops is identified and mapped by extending the computer methods described for allocating and mapping land classes from grid data (Rudeforth 1975). Sites growing a particular crop at the time of survey were selected to find the range (maximum and minimum) of each soil property compatible with the crop. All the sites, irrespective of cropping, are then rescanned and those with properties outside the range found for the crop excluded. The remainder are plotted as land potentially capable of growing the crop. However, if by chance during a survey a site was described with one or more properties well outside the normal range acceptable for a crop (e.g. a very shallow soil in the middle of a field of deeper soil), then it would be wrong to include all such sites as potential cropland. To cope with this situation, the data for the sites actually growing the crop are rescanned after the whole range is established and sites with extreme values excluded. In this way 'subminima' and 'submaxima' are found and the process repeated several times to build up a table (Table 4.4). The line of subminima and submaxima can then be chosen according to the percentage of sites to be excluded from the whole range. Maps are thus made showing land most central to the range, which may well be the better land for the crop within the survey area in many cases. Figure 4.15 shows clearly how the distribution of the better cropland (limits less 4 per cent) differs with the nature of the crop.

Profile no. SK 12/1670	SERIES: MARCHINGTON	SUBGROUP: TYPICAL STAGNOGLEY
Grid ref. 41156 32695	Described by: R. JONES & D. HALL	LOCALITY: DAISY BANK FARM NEWBOROUGH

General & site description

Grid reference (to 10 metres) Eastings	Northings	Continuation	Date Day	Month	Year	Elevation (metres)	Slope (°)	Aspect (as a bearing)	Form	Soil erosion and deposition	Rock outcrops	Land Use	Soil Surface Form	Condition	Thickness of L layer (cm)	Rock Lithology Type	Subtype	Structure	Hardness	Colour Hue	Value	Chroma	No. of horizons described	Spare 1	Spare 2	Spare 3	Spare 4	Spare 5	Spare 6	Spare 7	Spare 8	Spare 9	Spare 10	Supp. Info.
41156	32695	01	4	10	74	143	6	102	0	0	0	2	N	N	0	208	**	3	1	6	6	2	05											

Profile description

Grid reference (to 10 metres) Eastings/Northings	Continuation	Lower average depth (cm)	Colour (Moist soil; Ped face; Air-dry; Rubbed; Main mottle — Hue, Value, Chroma), Mottle, Organic matter status, Particle-size class, Stone, Soil-water state, Ped, Voids, Consistence, Root, Nature of O (peaty) hor., Carbonates, Nodules, Coats, Bdy., Spare 11, Spare 12	Supp. Info.	Horizon notation	Sample number
4115632695	10	25	432 *** 442 433 443 2322 0 18 3111203NN 4 246 22N1 33*022 132 N 0 441* 11		Ap	
4115632695	11	50	544 *** *** 554 653 3332 0 18 3211203NN 4 246 31N1 33*033 222 N 0 442* 21 23		Eg	
4115632695	12	85	663 652 672 456 357 4*31 0 22 N0NNNNNN 2 326 31N1 33*033 212 N 1 442* 21 31		Btg	
4115632695	13	105	662 *** 672 556 358 3533 0 18 N0NNNNNNN 2 222 323N2 32*022 N0N N 1 441* N0 4*	1	BCg	
4115632695	20	130				
4115632695	83	1	ALSO LIGHT OLIVE BROWN (2·5Y 5/4) MOTTLES;			
4115632695	83	2	POCKETS OF SILT LOAM.			
4115632695	90	1	SHALE IS MOTTLED STRONG BROWN (7·5YR5/6 AND 5/8)			
4115632695	90	2	AND LIGHT OLIVE BROWN (2·5Y 5/4)			
4115632695	99	1	THE PERMANENT GRASSLAND IS MAINLY PERENNIAL			

Diagram

Notes

Soil Survey of England and Wales

Soil description card

Fig. 4.13 A standard A4-sized 80-column recording pro-forma completed in the course of routine survey (after Webster *et al.* 1976).

GRID REFERENCE: 4115632695
DESCRIBED BY:
PROFILE NO: SK 12/1670
DATE: 4-10-74
WEATHER
LOCALITY:
SERIES:
ELEVATION: 143 M O.D.
SLOPE AND ASPECT: 6 DEG SW STRAIGHT
SOIL EROSION OR DEPOSITION: NONE
ROCK OUTCROPS: NONE
LAND USE AND VEGETATION: PERMANENT GRASSLAND
SOIL SURFACE:
THICKNESS OF LITTER LAYER: NONE
THE PERMANENT GRASSLAND IS MAINLY PERENNIAL
RYEGRASS WITH SOME WHITE CLOVER, COCKSFOOT AND WEEDS.

HORIZON DEPTH (CM)	DESCRIPTIONS OF HORIZONS
0-25	VERY DARK GREYISH BROWN (10YR 3/2) (RUBBED 10YR 3/3, DRY 10YR 4/2) SILTY CLAY LOAM; COMMON DISTINCT FINE BROWN TO DARK BROWN (10YR 4/3) MOTTLES WITH CLEAR EDGES; A FEW MEDIUM ROUNDED QUARTZITE STONES: VERY MOIST; MODERATELY DEVELOPED MEDIUM SUBANGULAR BLOCKY; MEDIUM PACKING DENSITY; SLIGHTLY POROUS, VERY FINE MACROPORES; MODERATELY FIRM SOIL STRENGTH; MODERATELY FIRM PED STRENGTH; MODERATELY STICKY; MODERATELY PLASTIC; MANY VERY FINE FIBROUS ROOTS; NON-CALCAREOUS; A FEW FERRI-MANGANIFEROUS SOFT CONCENTRATIONS; SHARP SMOOTH BOUNDARY.
25-50	OLIVE BROWN (2.5Y 4/4) (RUBBED 2.5Y 5/4) SILTY CLAY LOAM; MANY PROMINENT FINE OLIVE (5Y 5/3) MOTTLES WITH CLEAR EDGES; COMMON MEDIUM ROUNDED QUARTZITE STONES; VERY MOIST; MODERATELY DEVELOPED MEDIUM SUBANGULAR BLOCKY; HIGH PACKING DENSITY; VERY SLIGHTLY POROUS, VERY FINE MACROPORES; MODERATELY FIRM SOIL STRENGTH; MODERATELY FIRM PED STRENGTH; VERY STICKY; VERY PLASTIC; COMMON FINE FIBROUS ROOTS; NON-CALCAREOUS; COMMON FERRI-MANGANIFEROUS SOFT CONCENTRATIONS; A FEW CLAY COATS; ABRUPT IRREGULAR BOUNDARY.
50-65	PALE OLIVE (5Y 6/3) (RUBBED 10YR 5/6, DRY 5Y 7/2) SILTY CLAY; VERY MANY PROMINENT STRONG BROWN (7.5YR 5/7) MOTTLES WITH SHARP EDGES; STONELESS; MOIST; MODERATELY DEVELOPED COARSE PRISMATIC WITH OLIVE GREY (5Y 5/2) FACES; HIGH PACKING DENSITY; VERY SLIGHTLY POROUS, VERY FINE MACROPORES; MODERATELY FIRM SOIL STRENGTH; MODERATELY FIRM PED STRENGTH; VERY STICKY; VERY PLASTIC; A FEW FINE FIBROUS ROOTS; VERY SLIGHTLY CALCAREOUS; COMMON FERRI-MANGANIFEROUS SOFT CONCENTRATIONS; A FEW CLAY COATS; CLEAR SMOOTH BOUNDARY.
65-105	LIGHT OLIVE GREY (5Y 6/2) (RUBBED 2.5Y 5/6, DRY 5Y 7/2) SILTY CLAY LOAM; MANY PROMINENT COARSE STRONG BROWN (7.5YR 5/6) MOTTLES WITH DIFFUSE EDGES; STONELESS; MOIST; WEAKLY DEVELOPED MEDIUM PRISMATIC; MEDIUM PACKING DENSITY; MODERATELY POROUS, FINE MACROPORES; MODERATELY FIRM SOIL STRENGTH; MODERATELY WEAK PED STRENGTH; MODERATELY STICKY; MODERATELY PLASTIC; NO ROOTS; VERY SLIGHTLY CALCAREOUS; A FEW FERRI-MANGANIFEROUS SOFT CONCENTRATIONS; GRADUAL BOUNDARY. ALSO LIGHT OLIVE BROWN (2.5Y 5/4) MOTTLES; POCKETS OF SILT LOAM.
105-130	LIGHT OLIVE GREY (5Y 6/2) SOFT FRACTURED SILTY SHALE. SHALE IS MOTTLED STRONG BROWN (7.5YR 5/6 AND 5/8) AND LIGHT OLIVE BROWN (2.5Y 5/4).

Fig. 4.14 Decoded output from the data in Fig. 4.13. Note that the comment 'Ryegrass with . . . weeds' describing the grassland derives from an additional card not shown in Fig. 4.13 (after Webster *et al.* 1976)

Successful application of this method depends on a sufficiency of sites actually under the crop at the time of survey. With fewer than about 20 sites there is a danger that the range is only poorly represented, but this will depend on the degree of similarity of the sites under the crop. If they are very similar and form a distinctive part of the area, contrasting with other parts, than fewer sites might be judged acceptable.

Results will vary according to the kind of number of properties chosen

Table 4.4 Ranges of scaled site and soil parameters for early potatoes and barley in Pembrokeshire (after Rudeforth 1975)

Column		A	B	C	D	E	F	G	H	I	J	K	L	M	N	O	P	Q	R	S	T	U	V	W	X	Y
Early Potatoes	*Minima*	1	0	2	32	1	5	3	0	4	0	1	38	20	0	0	0	0	0	0	0	0	0	1	0	1
		1	0	2	35	1	6	5	0	5	0	1	38	45	0	0	0	0	0	0	0	0	0	1	1	1
		1	0	2	36	1	6	5	0	5	1	1	38	45	0	0	0	0	0	0	0	0	0	1	1	1
		1	0	3	40	1	6	6	0	5	1	1	38	60	0	0	0	0	0	0	0	0	0	1	1	1
		1	0	3	40	1	6	6	0	5	1	1	38	65	0	0	0	0	0	0	0	0	0	1	2	1
	Maxima	3	0	9	90	3	12	12	12	8	14	1	50	380	1	9	9	6	18	5	6	25	20	9	16	3
		2	0	8	90	3	10	10	10	8	10	1	46	335	1	8	7	4	5	4	6	5	11	8	15	2
		2	0	8	90	3	10	10	10	8	9	1	45	290	0	7	5	3	3	3	5	4	6	8	12	2
		2	0	8	86	2	10	10	10	8	8	1	45	280	0	6	5	3	3	3	5	4	4	4	11	2
		2	0	7	76	2	10	10	10	8	7	1	45	240	0	6	4	3	2	2	4	3	4	3	11	1
Barley	*Minima*	1	0	1	16	0	1	0	0	0	0	1	38	20	0	0	0	0	0	0	0	0	0	1	0	1
		1	0	2	20	1	3	0	0	0	0	1	38	30	0	0	0	0	0	0	0	0	0	1	0	1
		1	0	2	20	1	5	0	0	0	0	1	38	35	0	0	0	0	0	0	0	0	0	1	0	1
		1	0	3	22	1	5	0	0	1	0	1	38	35	0	0	0	0	0	0	0	0	0	1	0	1
		1	0	3	25	1	6	0	0	1	0	1	38	35	0	0	0	0	0	0	0	0	0	1	0	1
	Maxima	4	2	10	90	3	10	13	14	8	15	1	60	855	3	12	6	7	8	9	7	12	13	9	16	6
		4	0	9	90	3	10	10	13	8	9	1	60	745	2	5	5	6	7	6	7	10	6	8	15	3
		3	0	8	84	3	10	10	10	7	8	1	60	730	1	5	5	5	5	6	7	7	6	8	15	2
		2	0	7	80	2	10	10	10	7	7	1	58	720	0	4	5	4	4	6	7	5	4	8	15	2
		2	0	7	80	2	10	10	10	7	7	1	58	710	0	4	4	4	4	5	6	4	3	8	15	1

Key to columns:

A *Drainage class* (from profile morphology) (1 = good; 4 = poor).

B *Flooding* (0 = none; 2 = estimated one year in three)

C *Structure rating* (1 = good; 6 to 10 = poor to very poor).

System outlined in Rudeforth and Bradley, 1972; details from author on request.

D *Soil Thickness* in cm (ibid.).
E *Stoniness* (0 = stoneless; 1 = slightly stony; 2 = stony; 3 = very stony).
F *Texture* 0–25 cm (1 = sand; 3 = loamy sand; 5 = sandy loam; 6 = loam; 10 = silty clay loam; 12 = sandy clay; 13 = silty clay; 14 = clay).
G *Texture* 25–50 cm.
H *Texture* 50–75 cm.
I *Moisture Deficit* (number of years in 10 needing irrigation based on Apr.–Sept. soil moisture deficit of 3 in. calculated from climatic data).
J *Gradient* in degrees.
K *Climatic region* (1 = West Britain).
L *Rainfall* (annual average in inches).
M *Altitude* (feet above sea level).
N *Erosion risk* (0 = none; 1 = very slight; 2 = slight; 3 = moderate).
O–V *Topex* N., NE., etc., to NW. (Topographical exposure, measured by angle to the horizon at 8 points of the compass).
W *Slope shape* (provisional) (1 = convex; 2 = plane; 3 = concave; 4–9 = compound shapes and facets).
X *Aspect* (1 = N; 5 = E; 9 = S; 13 = W; etc.)
Y *Air-photo unit* (1 = smooth, apparently dry slopes; 2 = apparently drier intergrade; 3 = apparently wetter intergrade; 4 = mottled, apparently wet slopes; 6 = valley complex).

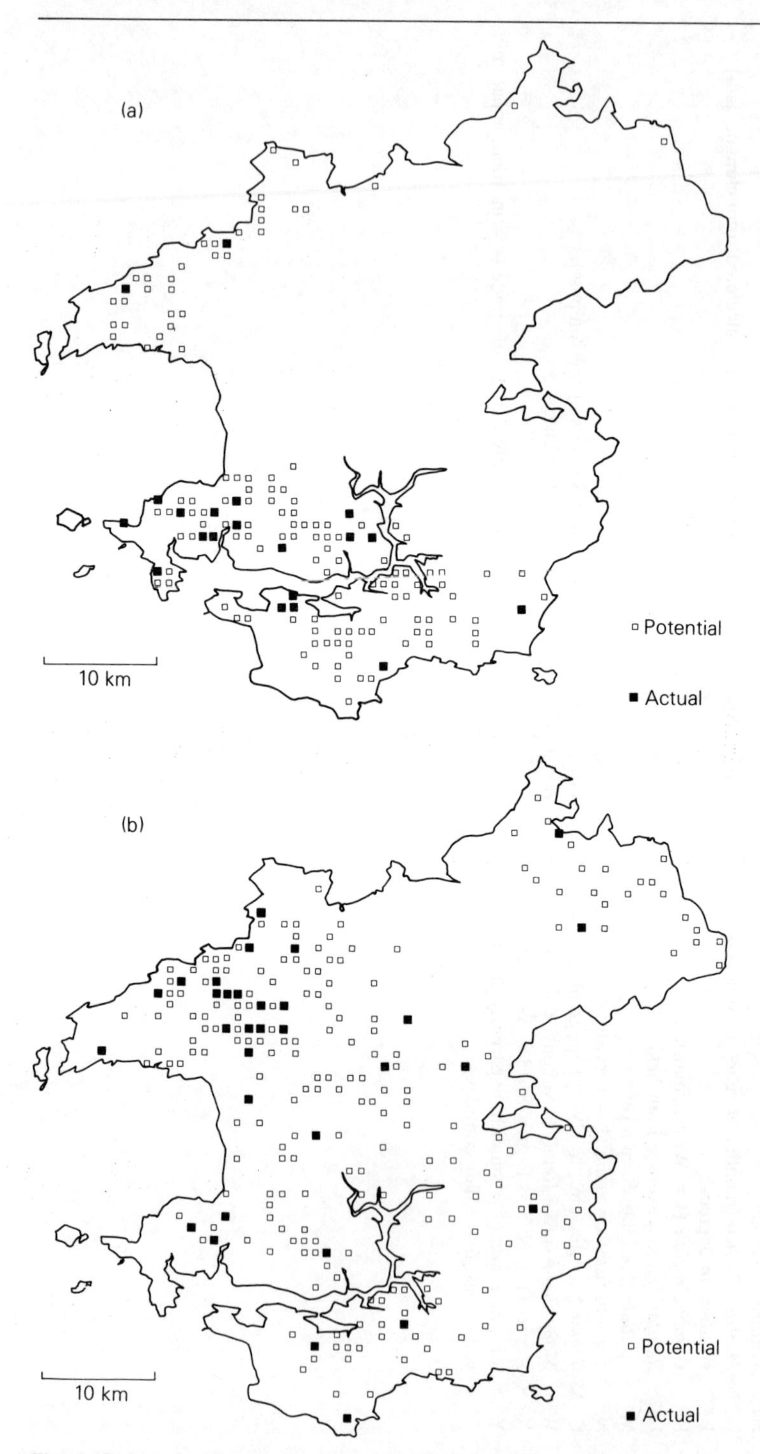

Fig. 4.15 Actual and potential cropland in Pembrokeshire. (a) early potatoes; (b) barley (after Rudeforth 1975)

to describe each site, but fairly consistent results are obtainable from about 25 of the main soil and site factors even if some relevant factors are omitted because data are unavailable. Some shortcomings of the method are discussed by Rudeforth (1975), but the procedures may well prove useful in indicating the limits of a specified acreage of the best land. If, for example, demand were such that 100 km^2 of a crop were needed then a simple set of graphs could show the soil-site limits of the best land of that amount. Encouragement might then be given to growers with land within the range so that maximum environmental efficiency is achieved. Though by no means perfect, a system of this kind could prove better than using arbitrary limits as is done for example when allocating hill land subsidies in Britain.

The potential occurrence of other soil/site related properties (Grid samples)

The technique developed for potential cropland can also be used for predicting the occurrence of individual soil properties. For example, while it may not be possible from air-photo pattern alone to predict whether a wet hollow in hilly country contains peaty soils, a computer scan of all similar sites could quickly reveal the altitude and rainfall, in relation to shape and size of catchment where peaty soils have been found and where, therefore, others are likely to occur. This way of combining information could be valuable for producing soil probability maps from air-photos (Rudeforth 1978) and considerable success has already been achieved in the automatic interpretation of remote sensing (Weismiller *et al.* 1977). Webster (1977a) describes a general statistical method for assessing the level of relationships between soil and environmental variables by canonical correlation.

Feature punched cards (All samples, but particularly useful for square grid samples)

Where computers are available, it may be thought surprising that any other form of data storage and manipulation is useful. However, punched feature-cards have several very useful properties in relation to soil survey; they are also appropriate for offices without immediate access to a computer (see p. 104). The merits of feature-cards in the context of the questions most commonly asked of soil surveys are described by Rudeforth and Webster (1973).

Records from the sample spots are numbered systematically by geographic location and stored in numerical sequence. The information on each record is analysed into a number of simple attributes or 'features', which are then used for indexing. For each feature there is a feature card bearing a matrix of numbered positions corresponding to record numbers in the store. Each card is punched in the positions of all records that possess the feature. To identify and retrieve records, feature-cards are used either singly or superimposed in combination. When applied to grid surveys each feature-card constitutes a map as indicated in Fig. 4.16, and the pattern of punched holes coinciding through several cards gives a map of sites having the combination of properties represented by those cards (Fig. 4.17).

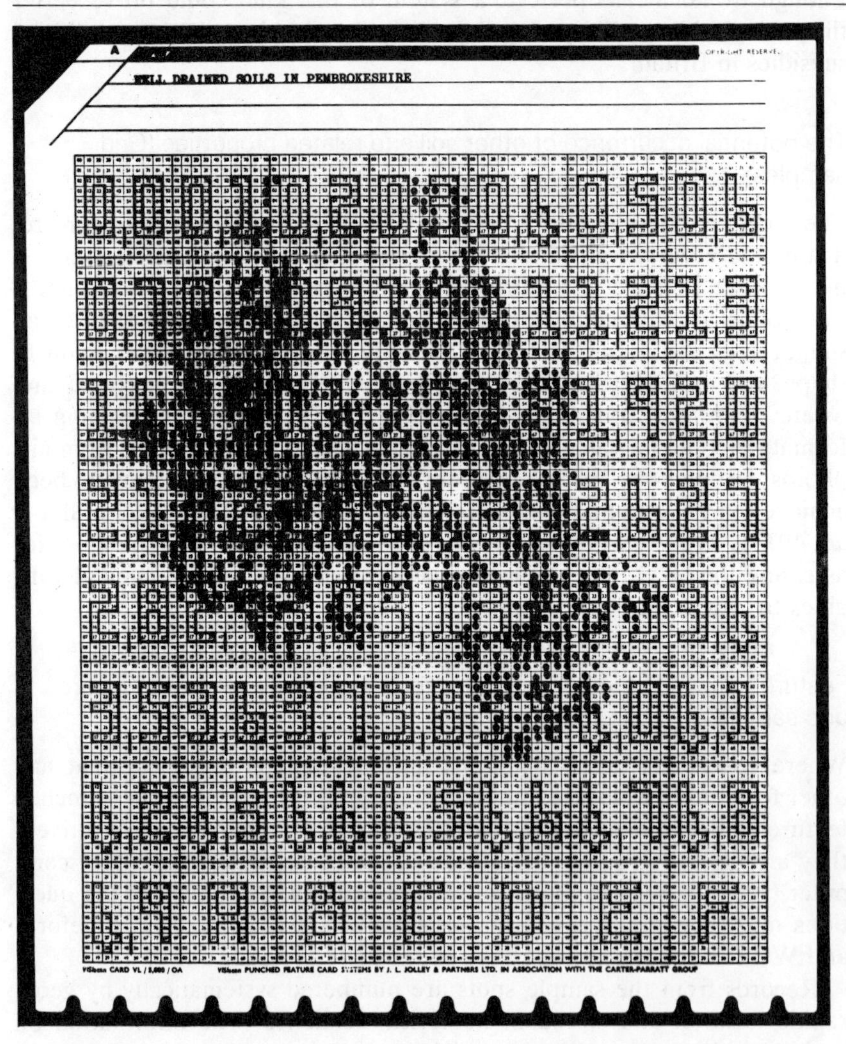

Fig. 4.16 Part of 5600-position feature card as used for surveys in mid- and west Wales.

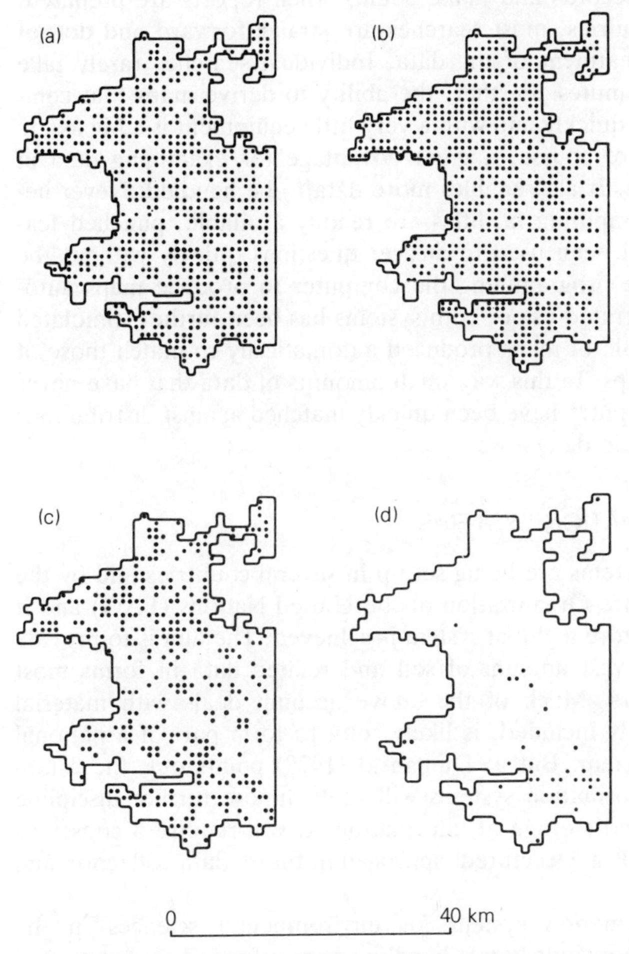

Fig. 4.17 Maps derived from superimposed punched feature cards, showing sites with different combinations of properties in west and central Pembrokeshire. (a) Soil at least 50 cm thick; (b) Soil freely drained; (c) Soil at least 50 cm and freely drained; (d) Soil at least 50 cm thick, freely drained and at less than 30 m above sea level (after Rudeforth and Webster 1973)

Compared with automatic methods, a feature-card system has both advantages and disadvantages. The advantages in any particular application depend on the quantity of data, the frequency with which records are consulted, the complexity of questions asked and the speed with which answers are needed. Feature-cards have been used successfully to answer local questions and accommodate the thousand or so records that accumulate each year in a soil survey district office. Outside enquiries are usually simple and answered swiftly. The index is used frequently by the survey to retrieve records and make counts when reports are prepared. As for outside enquiries, most searches are straightforward and do not require complex treatment of the data. Individual searches rarely take more than a few minutes. Further, the ability to derive maps that combine attributes very quickly and with a very little equipment is a big asset.

A computer is, of course, a great advantage for examining data in more complex ways. It also enables more data to be handled. Nevertheless, even when computing facilities are readily available, punched feature-cards have still been used to answer questions quickly without the need to spend time programming the computer to produce maps automatically. The integrated use of both systems has been further stimulated by adjusting the scale of maps produced automatically to match those of the feature card maps. In this way small amounts of data that have never been stored by computer have been quickly matched against distributions retrieved from a main data store.

4.4 SOIL INFORMATION SYSTEMS

Soil information systems are being set up in several countries and by the Food and Agriculture Organization of the United Nations (FAO) and it may not be long before a global system is achieved. The aim is to provide easy access to the vast amount of soil and related data in forms most appropriate to users. Much of the above account deals with material which, if not already included, is likely soon to form part of a national soil information system. But as Dumanski (1978) points out, the establishment of soil information systems will itself impose further discipline on the collection and storage of information. It will require a conscious commitment towards a 'structured' approach in future data collection and analyses.

A general information system for environmental sciences in the Netherlands aims to include a data handling system for soil survey providing interactive data retrieval, original and derived soil maps for different purposes and plots of original field observations (Schelling and Bie 1978). Similar steps are being taken in England and Wales, where it is planned also to provide a catalogue of the most common named soils and their chief characteristics in a data bank with dial-up access from remote telephone terminals.

The Canadians have in CanSIS an integrated computerized system (started in 1972) intended to enhance the ability of soil scientists to make decisions from their data and increase the scope of decisions that can be made by both scientists and other users (Dumanski *et al.* 1975; Dumanski 1978; McCormack *et al.* 1978). Standardizing methods of data collection is an important internal objective. In the United States, a similar system, RAMIS, was started in 1976, in which the soil data files form the central core. Information is retrieved by classification, soil properties, location and performance, and the system is itself used for management and scheduling of soil surveys.

No integrated, computerized soil information system yet exists at national level in Australia, although in common with many other countries local and regional systems have been developed. The FAO soil information system is conceived within a general system dealing with natural resources, which includes climatic data, grassland, forest and genetic resources as well as productivity information and assessments of soil degradation and land potential. Farm management, pesticide monitoring and agricultural data are also considered.

A first step has already been made in creating an international soil data processing network sponsored by the Agency of Cultural and Technical Cooperation in Paris. A central data bank is planned to access French, Belgian, Canadian and other data banks.

Looking to the future, Schelling and Bie (1978) list requirements for second generation systems and point out the need to use soil data with other environmental data on a world-wide basis. They foresee the establishment of an international Earth Science Data Centre which would provide a central, world-wide system, globally accessible through remote terminals, that will allow soil scientists everywhere to benefit from the progress made, and to contribute to the further development and application of soil science. A user of such remote data would be wise to ensure that the information he asks for was collected in a way appropriate to his needs.

5

MODELS AND SPATIAL PATTERNS OF SOILS

R. J. Huggett

Most studies of soil geography have theoretical underpinnings. The digging of a soil pit, the describing of soil horizons, the measuring of soil properties are of little importance in themselves unless they can be used to prove or refute some hypothesis or theory about the soil. Even those facets of soil geography which seem purely descriptive usually make assumptions about how the soil forms somewhere along the line. In the modern idiom, theoretical frameworks are referred to as models. This chapter will present principles and examples of the conceptual, statistical and mathematical modelling and analysis of the soil system. First of all, the notion of the soil as a system will be presented and, in this setting, two different, conceptual ways of analysing the soil system – Jenny's (1941, 1961) functional analysis approach and Simonson's (1959) system process approach – will then be discussed. Secondly, a special examination of the soil as a spatial system will be made, attention being drawn to soil system units and to soil landscapes. Thirdly, statistical analysis of the soil system, which is based on field observation and involves finding a statistical relation defined by 'best fit' parameters, will be critically explored. Fourthly, the mathematical analysis of the soil system, which is rooted in theory and takes the form of a set of general equations which may be applied to specific cases by fitting parameters from field or laboratory data, will be explained for both soil profiles and soil landscapes. And lastly, the prospect for soil system modelling will be evaluated.

5.1. THE SOIL AS A SYSTEM

Inputs, outputs and storage

As early as 1927, long before the word system in its more modern connotation had entered the geographer's vocabulary, Vernadskii had noted that the soil system is a state in the geochemical cycle. Jenny (1941), in his classic book, the thesis in which will be discussed later, also referred to the soil as a system. The soil system (Fig. 5.1) is a set of interrelated components and is dynamic (see Huggett 1980). The weathering of rock, rainfall and leaf fall all add materials to the soil. Plant uptake, throughflow and erosion remove material from the soil. For the soil system to persist, incoming material must at least replace outgoing material; this fact was recognized by Nikiforoff (1959), who likened the situation to a section of an aggraded stream between two bends: water enters from upstream, water leaves downstream, but between the two bends nothing is lost and work is done. The corresponding 'stream' in the soil system is the collection of surficial materials which constitute the soil. In the words of Buol *et al.* (1973), 'a soil is an evolving entity maintained in the midst of a stream of geologic, biologic, hydrologic, and meteorologic material'. Coexisting with the material soil system is an energy exchange and storage system (Yaalon 1960). Thermal energy (heat) is stored in the soil. The soil system gains heat from incoming solar radiation, terrestrial radiation emitted by the atmosphere, possibly from incoming soil materials and from exothermic reactions; it loses thermal energy in emitting radiation, by conduction out of the system, in outgoing soil materials and in endothermic reactions. Potential energy of a chemical or elevational nature is also stored, imported and exported. Energy transfers in the soil are brought about by heat conduction, by convection associated with water and air movements and by translocation of materials. Energy transformations in the system occur in chemical alterations, biological activity, wetting and drying, freezing and thawing and evaporation and condensation in the soil atmosphere.

Steady state in soils

Associated with the concept of soil as a system is the idea of dynamic equilibrium or steady state. A steady state soil, or soil component, is one which remains roughly the same, even though matter and energy may move in or out, be produced or be consumed. A soil system in a steady state may be stable, its stability being maintained within a more pervasive geochemical flux (Fig. 5.1) by regulatory or homeostatic processes. A change in an external driving force, such as radiation and precipitation, which might tend to perturb the system, should be absorbed and the soil system returned to its steady state. For example, in certain situations, the rate of lowering of the soil–rock interface may be roughly equal to the rate of lowering of the land surface so that soil depth remains virtually constant while the surface and bedrock are lowered through hundreds of metres (Kirkby 1971). The resistance of a system to change is called homeostatis, and it relates to what Bryan and Teakle (1949) styled pedogenic inertia. A system possessing inertia only responds on an exter-

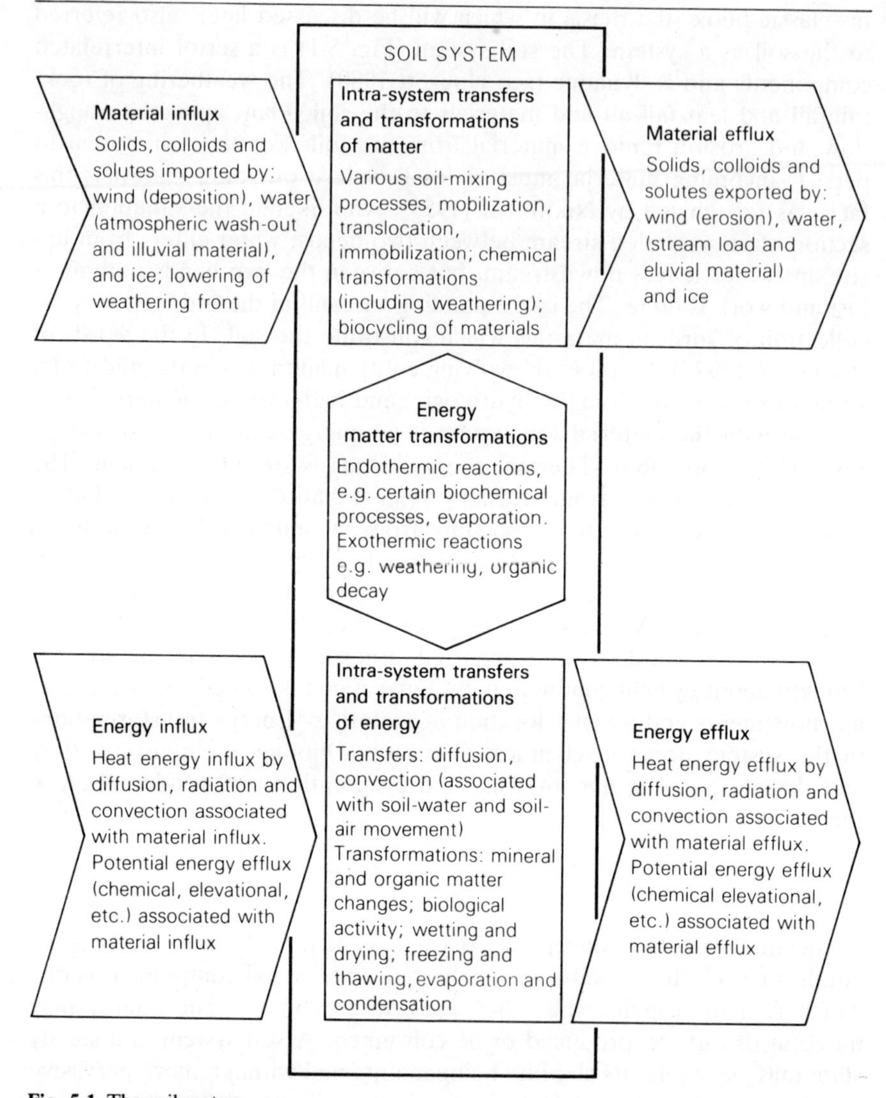

Fig. 5.1 The soil system.

nal change when a threshold value in the degree of that change has been passed. The time taken for a system to respond to a change is known as the reaction time. A threshold value having been passed, the soil may tend towards a new steady state, the time to do so being known as the relaxation time of the system. Many of the driving forces of pedogenesis either show recurrent or intermittent variations or are impulsive, and a small lag (a relaxation period), between change in a driving force and response in the soil system, often exists. A good example is the soil temperature wave, both diurnal and annual, which invariably lags behind the atmospheric temperature wave.

Soil-forming processes operate at different rates and Yaalon (1971), applying the concept of equilibrium and dynamics, distinguished three groups of processes. Firstly, reversible, self-regulating processes which rapidly attain a steady state (between a 100 to 1000 years or less) and are therefore subject to rapid alteration (they have short relaxation times) when environmental conditions, and thus the driving forces of pedogenesis, change. For example, Jenny (1941) has shown that the nitrogen content of a soil may reach an equilibrium or steady-state level in about 80 years, the actual steady-state nitrogen content being determined by the rate of nitrogen fixation and the rate of nitrogen loss. Other examples include the development of mollic horizons, salic horizons, slickensides, mottles, gilgai features and some spodic horizons. Secondly, the processes and properties in a state of near-equilibrium or metastability, in which the soil system, as a result of the slow rate of change (between about 1000 and 10 000 years), is assumed to be in a steady state. This was first realized by Jenny (1941), who said that, unless there is an accurate record which extends over centuries or millennia, it will be uncertain whether true equilibria or merely slow reaction rates are prevalent. Examples include the development of some spodic horizons, cambic horizons, fragipans, argillic horizons, histic horizons and gypsic horizons. Using the terminology introduced into geomorphology by Chorley (1966), both groups of processes so far discussed might be termed timeless, in that they show the time-independent steady state of the open soil system. However, the third group of processes, which Yaalon termed irreversible, self-terminating reactions and derived properties, might, to use again Chorley's words, be termed timebound. These processes, which proceed in a definite direction, include the transformation of primary minerals and other weathering processes in which losses are greater than gains. In the opposite situation, where there is an excess of influx over efflux, as in an enclosed basin in an arid or semi-arid environment, features such as desert incrustations occur. Examples of persistent features include oxic horizons, plinthite, petrocalic horizons, gypsic crusts and natric horizons. Yaalon (1971) also drew attention to the fact that, when determining the existence of a steady state, the size or dimensions of the system are important. For example, he noted that, although a pedon may not attain a steady state, it may constitute part of a catena or some other larger system which does, when viewed over a certain time scale, approximate to dynamic equilibrium.

The functional approach to the soil system

Towards the end of the last century, Dokuchaev suggested the following equation to represent soil formation

$$s = f(pm,c,b,a)$$

where *s* is soil, where *pm* is parent material, *c* is climate, *b* is biosphere, *a* is age of the land (time factor) and *f* is some function. This form of symbolism was extended by Jenny in 1941. Jenny's original concept of the soil states that the soil is a function of five soil-forming factors, namely, climate, organisms, relief, parent material and time. Using the designation *cl* for climate, *o* for organisms, *r* for relief, *p* for parent material and *t* for time, Jenny wrote

$$s = f(cl,o,r,p,t, \ldots)$$

where the dots stand for additional soil-forming factors not yet identified. It was not until 1961, in a paper which updated his concept, that Jenny formulated his ideas in what might now be regarded as a systems framework. Looking at the function of the soil-forming factors in an open system, Jenny reduced the five factors to three 'state factors' of the soil system. In essence, Jenny's revised proposition may be written $s = f(l_o,P_x,t)$. The first state factor, l_o, is 'the initial state of the system, or its assemblage of properties at time zero when pedogenesis starts', and this may include parent material, relief and some organic material. Climatic and organic factors are grouped as the second state factor, P_x; these control the supply and loss of energy to and from the soil system and Jenny termed them the external flux potentials of the system. The third state factor is time, *t*.

The influence of a single factor on soil can be examined by seeing how soil, or a soil property, varies as the state factor varies and all other state factors are held constant. Soil nitrogen content varies with climate in a range of situations where parent material, topography and biotic factors are similar. The basic equation in this case becomes

$$s = f(cl)_{o,r,p,t}$$

which Jenny called a climofunction. Similarly, where the effect of relief on soil is singled out we have a topofunction, $s = f(r)_{cl,o,p,t}$; where the effect of parent material is isolated we have a lithofunction, $s = f(p)_{cl,o,r,t}$; and biofunctions and chronofunctions would be defined using the same logic. The purpose of this kind of modelling is to see how a soil is influenced by a single soil factor, to establish univariant functions; it is therefore known as the functional analysis approach to the soil system and it will be discussed more fully in a later section.

The impact of Jenny's concept has been felt in virtually all branches of pedological research and thinking; it has also reached plant ecology (see Major 1951; and Perring 1958). The main results and achievements of Jenny's functional approach over the last 30 or so years have been brought together in the book by Birkeland (1974).

Jenny's concept has been criticized on many fronts. Some of the criticisms were tidying-up operations; they picked out the weak points in the argument and put them in order. Thus Stephens (1947) was concerned because in reality state factors vary at the same time; he proposed partial differential equations to allow for this. Crocker (1952) tried to justify the use of the biotic factor as an independent variable. Other critics were more derisory. Crowther (1953) said that Jenny's equations were insoluble, a point averred by Kline (1973). Runge (1973), though accepting that Jenny's expression can be differentiated in theory to determine the rate of change of soil development due to various factors, explained that in practice this cannot be done for parent material because it consists of discrete units to which differential rates of change are not applicable. A similar argument can be made for climate and organisms because they are of such a composite nature that data suitable for the solving of the differential equations are difficult to collect. Not that there has been a lack of effort to collect data, as a perusal of Låg's work (Låg 1968, 1974) will show. However, Jenny was aware of the difficulties of obtaining numerical solutions to his equations, especially solutions which genuinely held all variables constant bar one. His aim was not to explain soil process but to provide a vehicle for studying the relationship between soil and the factors responsible for its formation and there is no doubt that Jenny's functional system has proved a milestone for this sort of study.

Runge (1973) modified Jenny's equation to produce what he described as a more workable model of soil development. In essence, Runge's model states that soil development, s, is a function of organic matter production, o, the amount of water available for leaching (defined as precipitation less evaporation), w, and time, t

$$s = f(o, w, t)$$

As it is more specific than Jenny's formulation, it should be easier to implement; but because it excludes, or at least unrealistically diminishes, the parent material factor, Yaalon (1975) suggested its utility would probably be limited to soils formed in loess, till and other unconsolidated deposits where leaching is far and away the dominant process of soil horizon differentiation, and dismissed it as merely a new verbal dressing of the Jenny original.

The process approach to the soil system

Neither Jenny's nor Runge's equations focus directly on soil processes. An approach which stresses soil processes was put forward by Simonson (1959). His generalized theory of soil genesis described four groups of physical, chemical and biological processes common to all soils: additions of organic and mineral material as solids, liquids and gases; their removal; their transfer or translocation; and their transformation. The changing balances between these processes differentiate one soil from another. For instance, mineralization and humification of plant litter engage more or less the same processes of transformation in all environments but differing process rates may lead to different end-products. Buol *et al.* (1973) categorized processes of soil formation using Simonson's general scheme. Enrichment, deposition on the soil surface and littering are additions; leaching, and surface erosion are removals; eluviation, lessivage and pedoturbation are examples of transfers; and humification, mineralization and weathering involve transformations.

Yaalon's (1971) model of soil dynamics, which is essentially an extension of Simonson's approach to include time, considers the importance of the size or dimensions of the soil system.

5.2 THE SOIL AS A SPATIAL SYSTEM

Soil system units

Implicit in the models described thus far is the fact that as soil-forming factors vary from place to place, so will soil types; but allusion to the spatial arrangement of soil and soil processes is not made. The soil system unit to which all the models relate is the soil profile or its three-dimensional equivalents the pedon and soil tessera (Jenny 1958), which refer only to the soil body, and the eco-tessera (Jenny 1965), which includes associated plants and animals as well as the soil body. These four units are conceptually the same, more or less, and are defined as elongated cores through the living skin of the earth, which extends from the soil surface to the unaltered parent materials, and must have a horizontal area small enough to minimize spatial variation within them (SSS 1975). These basic soil units may be grouped into higher-order units which occupy portions of the landscape: there are taxonomic units, such as the polypedon and soil series, which are established on the basis of within-unit uniformity; and there are functional units, such as the soil catena (or toposequence) and soil-landscape (Fig. 5.2). It is to the functional units of soil systems that discussions will now shift.

Milne (1935a, b) conceived the catena as a unified theoretical framework for the functional aspects of soil formation on hilly terrain. Subsequently, Morison *et al.* (1948) proposed that a soil catena, in analogy with an individual soil profile, contains eluvial, colluvial and illuvial sections which are linked to one another by subsurface down-slope movement of water or throughflow; this notion was later reaffirmed by Blume (1968) and Glazovskaya (1968). The down-slope concatenation of soils is produced by selective transport of soil materials by throughflow, in the same way that the soil profile is produced, in part, by selective transport

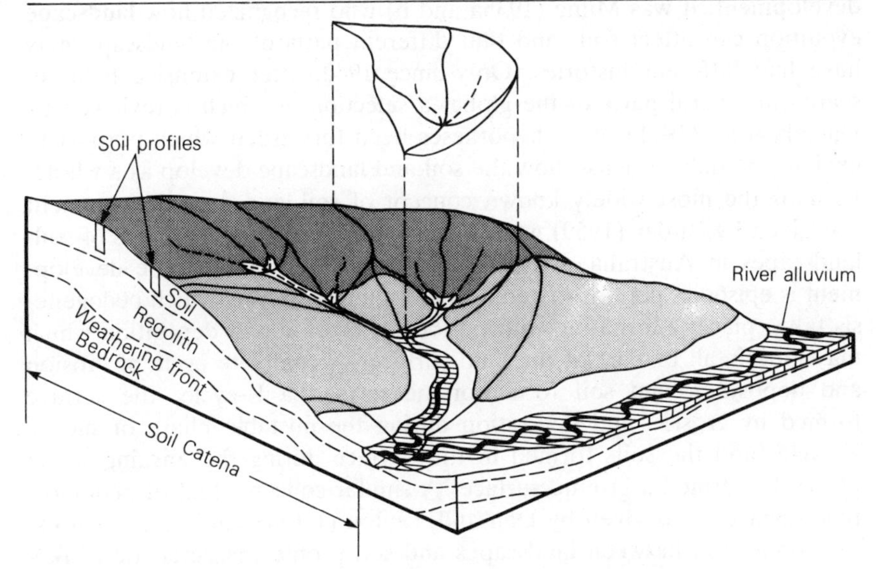

Fig. 5.2 Units of soil and soil-landscape. The different shadings represent soil series.

of soil materials in vertically percolating water. This has prompted Blume and Schlichting (1965) to observe that the valley soils of a landscape can be thought of as the B horizons of hill soils.

Although the catena concept is useful, it fails to give full prominence to the three-dimensional character soil processes. To overcome this weakness, the valley basin or erosional drainage basin, which accommodates catenary relations as well as the three-dimensional character of soil and soil processes, may be adopted as a basic functional unit of the soil system as Huggett (1973) and Vreeken (1973) suggested.

Soil-landscape systems

Although three-dimensional, functional soil units have only recently been proposed, it has long been recognized that soils are part of landscapes (Kellogg 1949; SSS 1951) and so have a spatial expression. The development of a soil landscape or soilscape, a contracted form coined by Walker *et al.* (1968a and c), is seldom solely the formation of a skin of soil at the land surface: the landscape in which soil forms also changes. Soil processes and geomorphological processes go hand in hand within the unitary soil-landscape system (Huggett 1975).

Although early researchers had realized that geomorphology and soils were in some way related, and indeed Neustruev (cited by Rode 1961) had written of the direct link between the Davisian cycle of erosion and soil development, it was Milne (1935a and b) who recognized how landscape evolution can affect soils and that different parts of the landscape may have had different histories. Only since 1940, after extensive field research in several parts of the globe, a selection of which is reviewed by Daniels *et al.* (1971), have hypotheses been forwarded which purport to explain, partially at least, how the soil and landscape develop as a whole. Perhaps the most widely known concept of soil-landscape development was given by Butler (1959) who, having carried out many studies of soil-landscapes in Australia, hypothesized that soil and landscape development is episodic, periods of geomorphic stability, during which pedogenesis takes place, alternating with periods of erosion and deposition which may modify all or part of the land surface. A complete cycle of erosion and deposition and soil formation he termed a K-cycle; the surface formed by erosion and deposition during the unstable phase of such a K-cycle, and the soils formed in this surface during the ensuing stable phase, he termed a ground surface. A similar concept, that of pedomorphic surfaces, was given by Dan and Yaalon (1968) (Table 5.1) to stress the connection between landscapes and soil profile characteristics. Likewise, Conacher and Dalrymple (1977) proposed a nine-unit, landsurface model as an appropriate framework for pedogeomorphic research (Table 5.2). Ruhe and co-workers in Iowa, whose field work focussed mainly on the relationship between geomorphic surfaces and soils, especially on time relations based on a radio-carbon chronology, have made many notable contributions to concepts of soil-landscape development. Ruhe and Walker (1968) and Walker and Ruhe (1968a) developed hillslope models of landscape systems, both open systems in which drainage basins are part of a more extensive drainage network, and closed systems in which drainage is in a closed basin. Within low-order open drainage basins they distinguished a number of geomorphic landscape components (divide, inter-fluve, side-slope, head-slope, nose-slope), the head- and side-slopes being divided into Ruhe's (1960) slope profile components (summit, shoulder, backslope, footslope, toeslope) (Fig. 5.3). In closed basins they defined peripheral slopes which link divides with a central depository. Generally speaking, the landscape components develop in the following manner: summits and shoulders remain relatively unmodified by erosion; hillslopes (head- and side-slopes) become progressively truncated upslope by erosion; part of the debris eroded from the hillslopes may be temporarily stored as valley fill (alluvium) before being removed by a stream or, if it be a closed basin, the eroded debris remains in the toeslope positions. Thus, different landscape elements have different histories and are of different ages. Accordingly, soil age in the drainage basin will also vary. Vreeken (1973) termed this age-transgressive character of landscape elements and soils developed in them the historical nature of soil-landscape relationships. At any given time, however, topography influ-

Table 5.1 Topofunctions of soil properties or features in various landscapes, numerically solved by curve-fitting and regression analysis. (from Yaalon, 1975).

Pedomorphic surface (soil landscape)	Soil property tested Y	Relief determinants X	Function calculated	Reference
Closed bog watershed on drift, Iowa	grain size indices, surficial sediment thickness	distances from summit	polynomial	Walker (1966)
Loess watersheds, Iowa	grain size indices	slope gradient (convex segment)	linear	Ruhe and Walker (1968)
	grain size	distance from summit and slope gradient (convex segment)	multiple linear	
	solum thickness; depth to maximum clay	slope gradient (convex segment)	polynomial, composed of linear subsets	
Closed bog watershed on drift, Iowa	grain size indices, organic matter content; depth to carbonates	distance from summit	polynomial	Walker and Ruhe (1968)
Drift landscape, Iowa	A horizon thickness	distance from summit, grid co-ordinates	polynomial	Walker *et al.*, (1968b)
Loess landscape, Iowa	A horizon thickness, mottling, pH		trend analyses	
Drift and loess landscapes, Iowa	depth of A horizon, mottling, carbonates	slope gradient, elevation and aspect	multiple linear	Walker *et al.*, (1968c)
Chalk and calcareous grit, England	organic carbon, nitrogen and pH of surface soil	slope gradient	linear, separately for upper and lower slopes	Furley (1968)
Weathered gneiss, North Carolina	solum thickness	distance from summit and slope gradient	multiple linear	Ahnert (1970)
Surficial sediment on till, Iowa	grain size indices, organic carbon content, base saturation	distance from summit	polynomial, separately for convex and concave slopes	Kleiss (1970)
Loess watershed, Iowa	clay and organic carbon content	slope gradient	linear, separately for upper and backslopes	Vreeken (1973)
Calcrete on chalk, Israel	Nari (calcrete) thickness	slope gradient, distance from summit and aspect	multiple curvilinear	Yair *et al.* (in preparation)

Table 5.2 A summary of definitions and diagnostic criteria used in the nine-unit landsurface model. (Adapted from Conacher and Dalrymple, 1977, Table I and Fig. 1.)

Land-surface unit	Abbreviated definition	Predominant and/or distinguishing pedological criteria	Predominant and/or distinguishing contemporary pedogeomorphic processes
1	Interfluve with predominant pedogeomorphic processes being those resulting from up and down soilwater movements	Soil development *in situ*; seasonal poorly drained with relatively shallow profiles and marked lateral uniformity in physical and morphological properties.	Vertical pedogenetic processes
2	Responses to mechanical and chemical eluviation by lateral subsurface soil water movements either predominate or serve to distinguish this unit from other units on the catena	Gleying above iron pans; reduced porosity and increased compaction in Eb as compared with underlying Bt horizons, and in the uppeı as compared with lower parts of (B) horizons; mottling, especially small ochreous mottles in A horizons; Mn concretions and concentrations	Mechanical and chemical eluviation by lateral subsurface soil water movement
3	Convex slope element where soil creep is the predominant process producing lateral movement of soil materials	Substitution of soil material from upslope by creep and probably surface and subsurface water movement; better drained than unit 2	Soil creep; terracette formation; processes resulting from subsurface soil water movement
4	Slope greater than 45° characterized by processes of fall and rock-slide	Soil formation restricted to lithosols	Fall, slide, chemical, and physical weathering
5	Response to transportation of a large amount of materials downslope, relative to other units, by flow, slump, slide, raindrop impact, surface wash and man's cultivation practices	Where wash and creep predominate: A horizon does not differ in thickness by more than 10 per cent and does not thicken downslope by more than 5 per cent. Where other processes of mass movement predominate: contrasting areas of deep and shallow soils	Transportation of material by rapid mass movement and/or surface water action; processes resulting from subsurface water movement; terracette formation
6	Response to colluvial redeposition from upslope	Heterogeneous soil mantle containing additions from upslope; possible occurrence of discontinuous and disturbed paleosol horizons and of fragipans and duripans in more humid areas	Colluvial redeposition by mass movement and surface water action; transportation; processes resulting from subsurface soil water movement

Land-surface unit	Abbreviated definition	Predominant and/or distinguishing pedological criteria	Predominant and/or distinguishing contemporary pedogeomorphic processes
7	Response to redeposition from up-valley of alluvial materials	Addition of alluvium, generally poorly drained with deep Al horizons and frequent occurrence of relatively continuous paleosol horizons; occurrence of fragipans and duripans	Alluvial redeposition; processes resulting from subsurface soil and groundwater movement
8	Channel wall, distinguished by lateral corrasion by stream action	Intermittent regosol formation	Corrasion, slumping, fall
9	Stream channel bed, with transportation of material down-valley by stream action	Absence of soil formation	Transportation of material down-valley by stream action; periodic aggradation and corrasion

ences pedogenesis in such a way as to induce areal differentiation of soil properties within a drainage basin. Vreeken (1973) termed this over-all topographical influence on soil properties within a drainage basin the temporal nature of soil-landscape relationships.

5.3 STATISTICAL ANALYSIS

A large group of empirical studies have arisen from Jenny's functional approach to studying the soil system and it is convenient to discuss them from this viewpoint. Analysing the soil system using Jenny's approach involves finding quantitative solutions to the functional relationships which were given in his formulation: soil properties are related to single soil-forming factors to give univariant solutions – lithofunctions, topofunctions, climofunctions, biofunctions and chronofunctions. Statistical curve fitting is usually the basis of the analysis, the mere establishment of correlation coefficients being generally an inadequate procedure in this case. The curves are the solutions and may be derived graphically by fitting a best-fit curve by eye or numerically by running a regression to obtain a mathematical description of a curve thought to describe the data.

Topofunctions

Many investigators have established relationships between topographical measures, usually slope angle, distance from watershed, or slope curva-

(a)

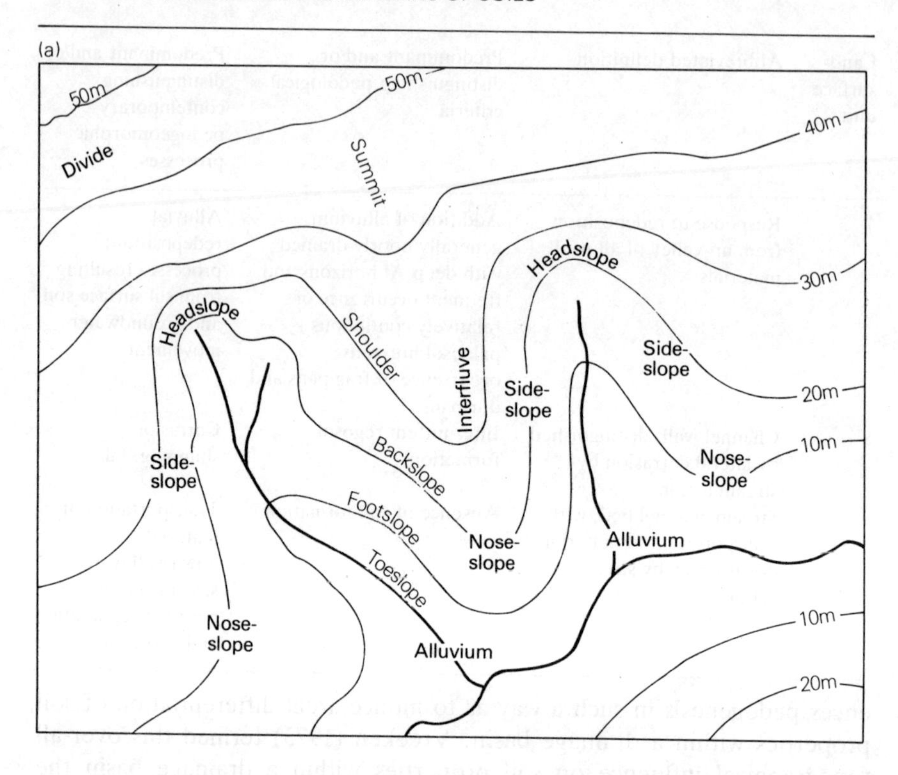

(b)

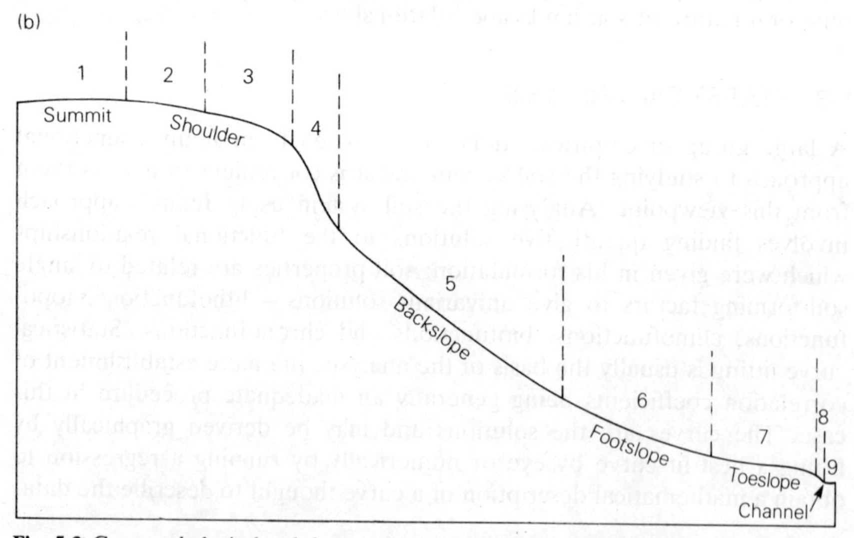

Fig. 5.3 Geomorphological and slope profile components of a drainage basin. The terminology follows that given by Ruhe and Walker (1968). The numbers on Fig. 5.3(b) are the units in the nine-unit landsurface model of Conacher and Dalrymple (1977)

ture, and soil properties. Early work concentrated on relations along a catena; Aandahl (1948) and Troeh (1964) considered relations in a three-dimensional landscape. A variety of soil properties, ranging from grain size characteristic to nitrogen content, have been tested against a smaller variety of relief measures; some examples are shown in Table 5.1. An interesting piece of research by Anderson and Furley (1975), who worked on catenae in the Berkshire and Wiltshire chalk downs, sought to establish the interrelationships amongst selected surface soil properties and topographical measures (slope gradient and slope length) through the use of principal components analysis. Analysis revealed a consistent pattern in the distribution of soil properties over five slope transects. The first component of the pattern, which accounted for between 50 and 60 per cent of the total variance in soil properties, was related to organic matter and soluble constituents: properties associated with organic matter (carbon content, nitrogen content, exchangeable potassium and moisture loss) diminish fairly evenly down-slope, whereas properties associated with soluble constituents (pH, carbonates, exchangeable calcium, sodium and magnesium) increase down-slope. The second component, accounting for 13 to 18 per cent of the variance, was interpreted as a particle-size factor, the pattern of which showed an abrupt increase in finer soil material immediately down-slope of the maximum gradient in the transect, giving a marked discontinuity in the pattern over the slope. In an earlier paper (Furley 1968) it had been suggested that some slopes could be divided into two sections; an upper, generally convex section where net erosion is greater than net deposition; and a lower, generally concave section where net deposition predominates; the zone of interaction between the two sections being known as the junction. The results from the five chalk land transects show that, with the exception of fine materials, soil properties alter gradually over the slope and the transition from net erosion to net deposition in the surface soil layer is diffuse.

Yaalon (1975) argued that, with computer methods facilitating the process of curve fitting, and with better data being collected on the nature and development of slopes, generally valid rules can be expected to emerge from topofunction studies. Anderson and Furley (1975) have proposed some general trends of soil properties across slopes (Fig. 5.4). It is doubtful if the mere acquisition of more and more topofunctions will eventually enable general statements to be made. Rather, it would seem more fitting to use topofunctions as a yardstick against which to test the output from soil simulation models of the kind mentioned later in this chapter.

Care must be exercised in interpreting topofunctions. Daniels *et al.* (1971) gave an example of a toposequence developed in loess in Harrison County, Iowa. They found that certain soil properties appear to be related to slope gradient – solum thickness and depth to carbonates de-

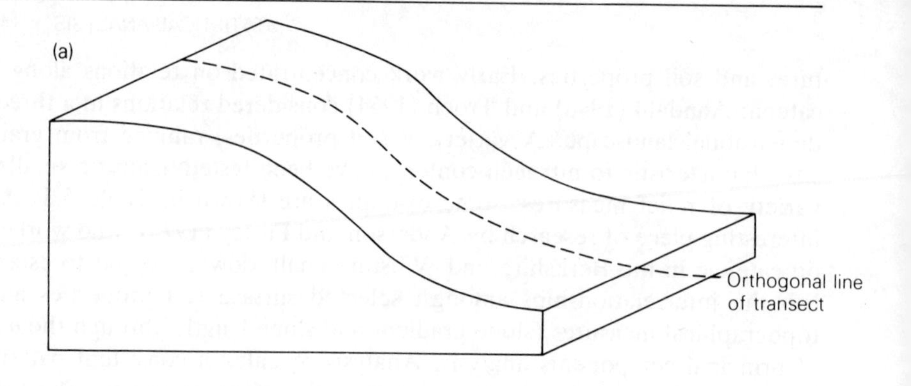

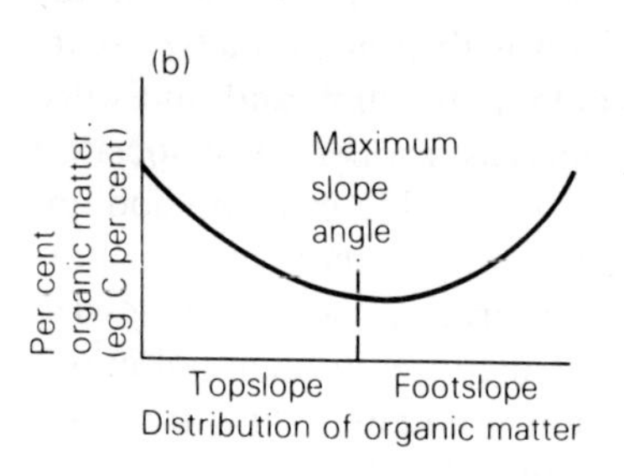

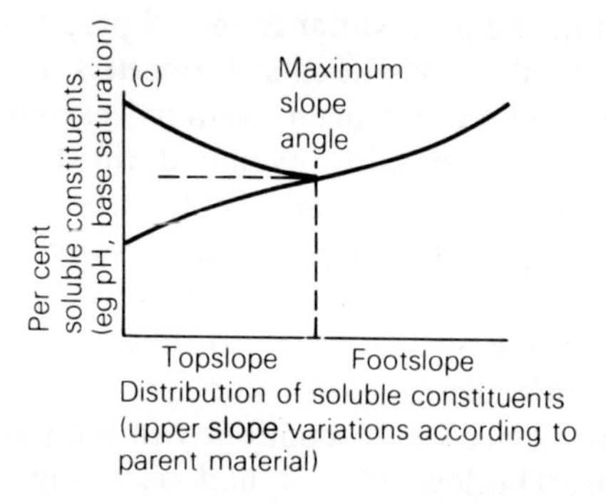

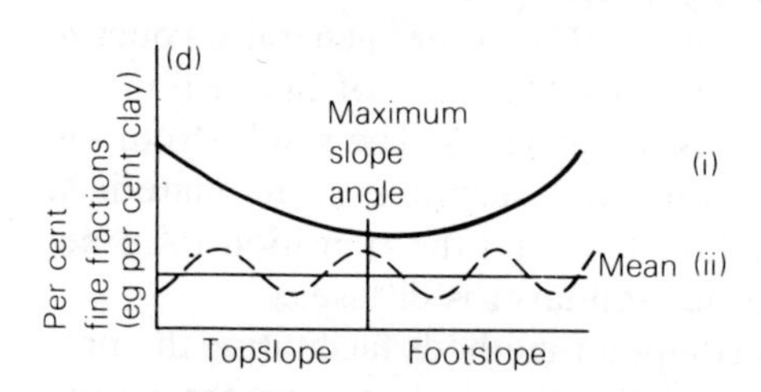

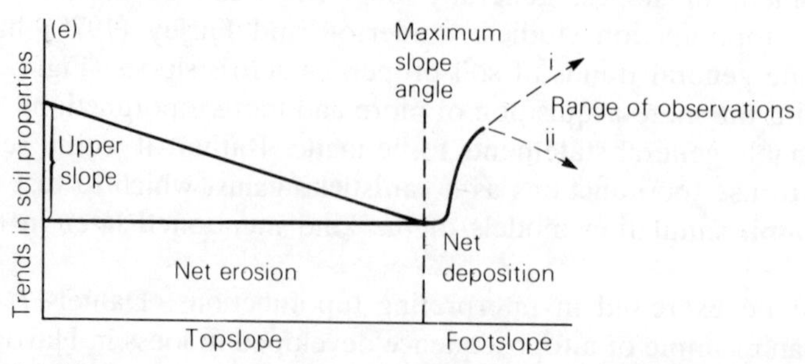

Fig. 5.4 (a) Convex-concave slope form typical of chalk slopes in southern England. (b to d) Models illustrating the distribution of individual soil properties over a slope of form (a). (e) Discontinuous distribution patterns of soil properties (particularly the fine fractions) (from Anderson and Furley 1975, Fig. 1)

crease as slope gradient increases. However, a careful study of the geomorphological history of the loess deposits revealed that some of the soils have developed in a calcareous loess during the last 1800 years while others formed during the last 15 000 years in a non-calcareous loess and in places were eroded 1800 years ago. So the apparent topofunction, in fact, results from a combination of slope gradient, character of parent material, and length of weathering. Vreeken (1973, 1975b), too, has shown how varying ages of soils in drainage basins complicates the interpretation of soil–landscape relationships. Thus in his study of the depth to carbonates in the Thom watershed, Iowa, he found the overall relation between depth to carbonates, D, and percent slope gradient, S, and azimuth of slope direction, A, to be

$$D = 98.06 - 3.1081\,S + 0.9981\,S \sin A + 0.7282\,S \cos A \;(R^2 = 0.659)$$

However, Vreeken (1975a) found that the relationship between depth to carbonates and slope properties depends on the age of landscape elements. On interfluves and upland summits, landscape elements some 14 000 years old, depth to carbonates can be explained mainly in terms of slope gradient, whereas on the valley-side slopes, landscape elements less than 7710 years old, it is explained by slope gradient and slope direction, south-west facing slopes having the shallowest depth to carbonates.

Lithofunctions

That parent material is very important in determining the course of soil development cannot be denied. At a local scale, the pattern of soil types largely reflects the pattern of parent materials. But the relationships between parent material and soil properties are difficult to express in anything but a qualitative way; this is because it is difficult to assign particularly useful numbers to rock types. Most lithofunctions compare just a few rock materials. Thus Barshad (1958) compared the amount of clay formed in soils per unit weight of different parent rock materials for situations in which other soil-forming factors were held more or less constant. Koinov *et al.* (1972) compared the relative migration and residual accumulation of some chemical elements in soils in eluvial portions of landscapes developed in acid rocks (granite), moderately-acid rocks (rhyolite-andesite, trachyandesite, and andesite) and calcareous rocks (marbles and limestones). A problem in establishing lithofunctions is the detection of material added to the soil surface during soil formation. Originally thought to be a localized, episodic process, aeolian deposition has, by recent measurements of dust transport, been shown to be a continuous process of global significance (Yaalon and .Ganor 1973; Jackson *et al.* 1973).

A new departure in the study of lithofunctions has been to relate soil type, and thus for small scale soil units mainly parent material, to landscape position. Bunting (1965) conjectured that soil type and position in a drainage basin are related; this relation is implicit in topofunction analysis but it has not been rigorously tested. An attempt to test it was made by Huggett (1973) and later by Warren and Cowie (1976) using soil maps. The method was to divide the whole soil–landscape under consideration – the 1 : 25 000 sheet of Harold Hill, Essex prepared by Sturdy (1971) in Huggett's case, and the 1 : 63 360 sheet of north-east Wales prepared by Ball (1960) in Warren and Cowie's case, into valley basins of increasing order. Next, the cumulating area of these basins, from first-order headwater valleys, to the largest order valley in the area (100th order in Harold Hill using the Shreve system (Shreve 1966), and 143rd order in the Elwy valley, north-east Wales) is plotted against the cumulating area of each soil type in each valley group, order by order. The regularity of the curves so produced is striking and suggests that they may be evidence of underlying physical principles linking soil type and position in a valley basin. Some examples are shown in Fig. 5.5. The cumulating area of the Windsor Series in Harold Hill, which is formed in London Clay, is zero in headwater valleys, first appears when the drainage area is 0.15 km^2, and rises at an increasing rate with increasing drainage area. The cumulating curve for the Titchfield series, formed mainly in valley gravels, does not start until the drainage area is about 1.5 km^2, and then increases with increasing drainage area at a constant rate. In the Elwy valley study, the cumulating curve for the Denbigh Series (a freely drained brown earth on Silurian shale) shows an initial sharp rise to the second order basins and then onwards a consistent though slightly irregular rise; this is thought to reflect the dominance of Denbigh Series soils on all the middle slopes at intermediate altitudes. The most informative curve in the Elwy valley was that for the Aled Series on alluvium; it does not appear until 5th order basins, climbs slowly to the 24th order, then jumps sharply to the 124th order, after which it climbs very gradually to the 143rd order. Warren and Cowie (1976) suggested this showed that higher order valleys are incised (perhaps by rejuvenation), leaving broader intermediate valleys with wide alluvial flats which tongue out in the 5th order valleys. Work by Arnett and Conacher (1973) on soils and drainage basins in the Rocksberg basin, Queensland, also points to a relation between soil types and stream order (Table 5.3).

Climofunctions

By far the largest part of Jenny's (1941) book concerns the observed relations between soil properties and broad measures of climate such as annual temperature and rainfall means. Since the time Jenny was writing, many more climofunctions have been established; some examples are

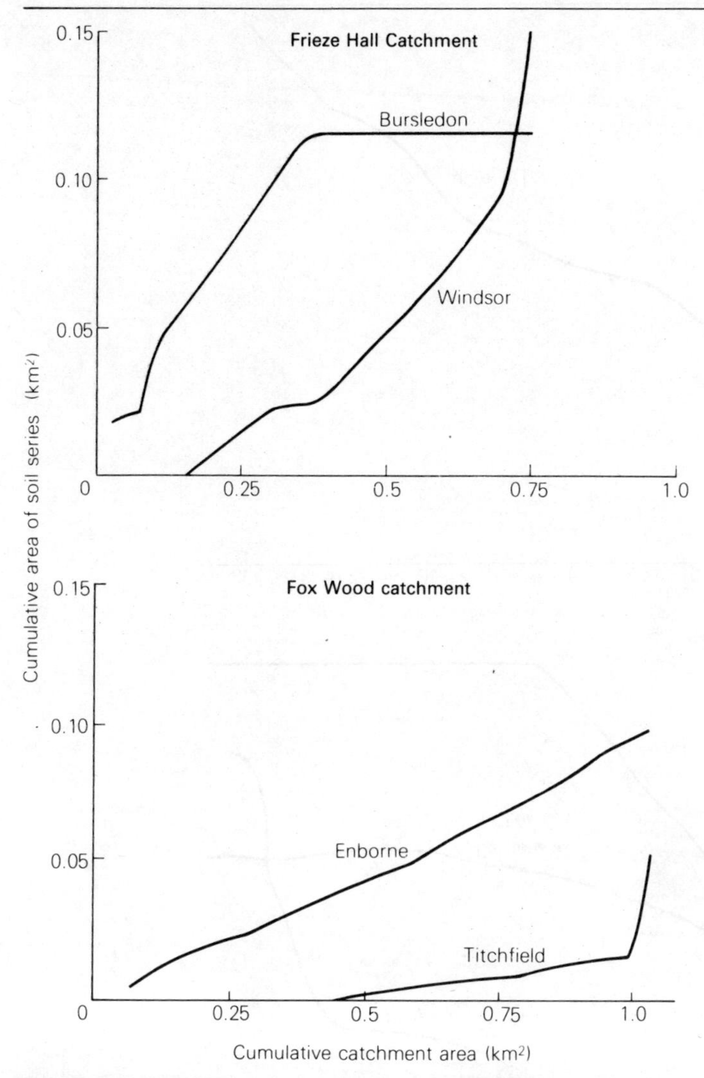

Fig. 5.5(a) Curves showing the relation between cumulative soil series area and cumulative basin area in Harold Hill, Essex. (Burlesdon series: a gleyed brown earth formed in loamy Eocene beds on shoulders; Windsor series: a surface-water gley formed in Eocene clay on backslopes, footslopes and toeslopes; Enborne series: a ground-water gley soil formed in fine loamy, and silty, non-calcareous alluvium along narrow floodplains; Titchfield series: a surface-water gley formed in flinty, loamy, or clayed head on Eocene clay found on footslopes and toeslopes.

given in Table 5.4. Seldom do these studies show exactly what is the connection between the crude climatic measures and the soil property they seem to influence; nor have any general conclusions been drawn

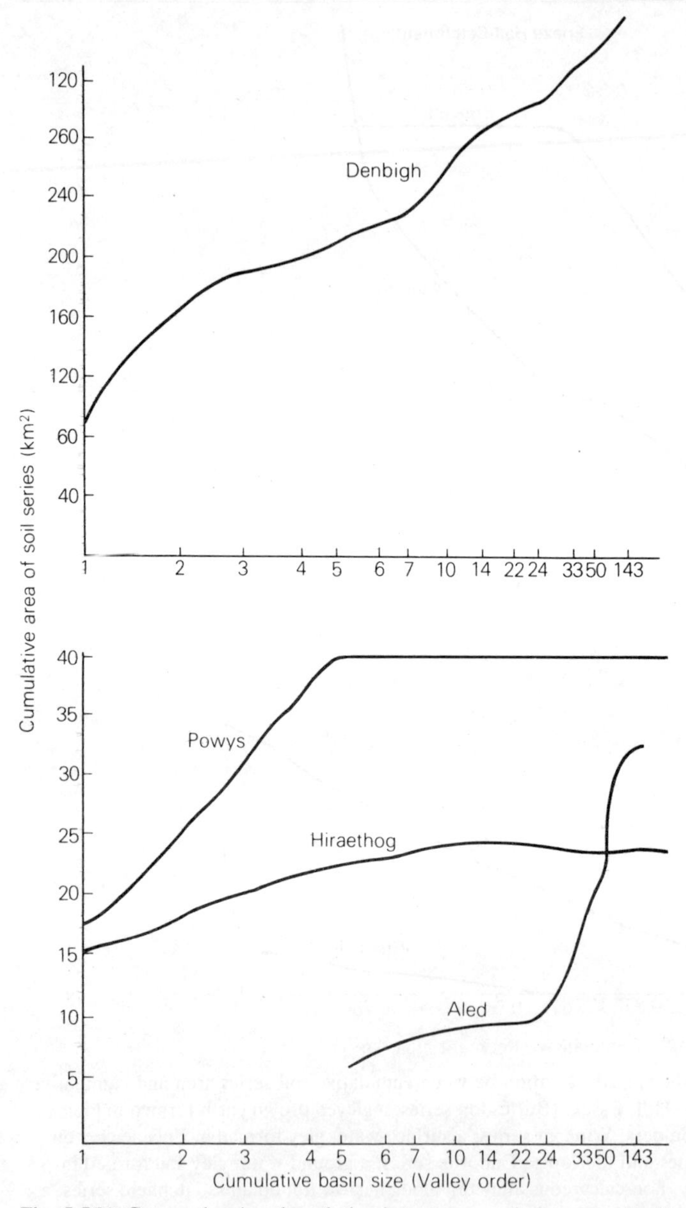

Fig. 5.5(b) Curves showing the relation between cumulative area of soil series and cumulative basin size in the Elwy basin, north-east Wales. (From Warren and Cowie, 1976, Fig. 2).

Table 5.3 Generalized soil and slope characteristics according to stream order in a Queensland site (after Arnett and Conacher 1973)

Stream order	Mean slope angle (degrees)	Basal channel	Soil type		
			Summit (crest)	Backslope (centre slope)	Toeslope
1	9.4	percoline	deep red loam	deep red loam	deep red loam (gleyed)
2	16.9	seepage line	red loam	red loam	acid red loam (gleyed)
3	19.0	poorly defined channel	shallow loam	skeletal loam	deep red podzol
4	20.0	well defined channel in bedrock	skeletal loam	skeletal	deep red podzol
5	21.2	well defined channel in bedrock and alluvium	skeletal loam	skeletal	alluvial
6	16.0	well defined channel in bedrock and alluvium	skeletal loam	skeletal	alluvial

from them. These drawbacks have been overcome in some recent studies which try to get at the process relations between the water and energy balance and soil processes. Scrivner *et al.* (1973) used daily records of temperature and precipitation for a 30-year period in conjunction with available water capacities of various soil horizons, to establish the depths of soil wetting and drying. Their findings included the suggestion that in soils of the Mexico Series, Columbia, Missouri, the A horizon and upper Bt horizon characteristics may be associated with the penetration of summer rains, whereas the thickness of the Bt horizon may be related to the annual depth of drying, and the B2 horizon may coincide with that depth which is subject to one annual wetting and drying cycle. Other work of this kind carried out by Goddard *et al.* (1973) showed that the amount of pedogenic clay (as opposed to that inherited from parent material) in Typic Argiudolls in Illinois seems to be controlled by rainfall frequency; and the distribution of pedogenic clay in the profile is related to the amount and distribution of rainfall, the depth of leaching and the natural drainage class of the soil.

Biofunctions

The relationships between plants (and animals) and the soil have been scrutinized by Crocker (1952) and Perring (1958). The independent biotic

Table 5.4 Climofunctions of specific soil properties (from Yaalon 1975)

Soil environment and region	Soil property tested Y	Climatic determinants X	Function calculated	Reference*
Tropical – Colombia	nitrogen content	mean temperature	exponential	Jenny *et al.* (1948)
Basalt, volcanic ash, etc. – Hawaii	kaolinite, SiO_2/Al_2O_3, SiO_2/Fe_2O_3, SiO_2/TiO_2 in clay	annual rainfall	power	Tanada (1951)
Topsoil – East Africa	organic matter, nitrogen, clay content	altitude, annual rainfall	linear	Birch and Friend (1956)
Igneous rocks – California	nitrogen content	mean temperature	exponential	Harradine and Jenny (1958)
Basalt – Queensland	kaolinite, SiO_2/Al_2O_3, SiO_2/Fe_2O_3, SiO_2/TiO_2, Fe_2O_3/Al_2O_3 of clay	annual rainfall	power	Simonett (1960)
Forest and cultivated soils – India	carbon and nitrogen content	annual rainfall mean temperature	linear exponential	Jenny and Raychaudhuri (1960)
Basalt – Israel	montmorillonite, free Fe_2O_3, free Al_2O_3, CEC of clay fraction	annual rainfall	linear	Singer (1966)
Forest and prairie soils – various	base saturation, carbon, pH, nitrogen content	biofactor (= rainfall/ temperature)	linear, multiple linear	Kohnke *et al.* (1968)
Various – California	nitrogen content	annual rainfall, mean temperature	multiple linear	Jenny *et al.* (1968)
	base saturation	annual rainfall	linear	
Forest and grassland soils – East Africa	carbon, nitrogen	annual rainfall, altitude	multiple exponential	Jagnow, (1971)
Savanna – West Africa	carbon, nitrogen, C/N, clay	annual rainfall, altitude	linear and multiple linear	Jones (1973)
Mediterranean – Israel	non-exchangeable ammonium	annual rainfall	linear	Feigin and Yaalon (in preparation)

* Graphical functions, mainly curvilinear, which have been published but not calculated numerically, are not included. For pre-1941 calculated functions see Jenny (1941).

factor is the potential floristic list of the system – that is, all the available seeds, pollen and spores. The actual growing vegetation is a dependent variable of the other state factors which is intimately linked with the soil. Noy-Meir (1974) has carried out a multi-variate analysis of the inter-relationship of soil and vegetation in arid and semi-arid environments and he found the distribution and composition of vegetation units was mainly determined by those soil and topographical variables which affect the spatial distribution of soil moisture and the depth of the wetting front. Most ecological studies which seek relations between plants and soil adopt the same kind of approach. Thus soil and plants and animals are best considered as interacting components of an ecosystem in which simple causal biofunctions have little meaning.

Chronofunctions

Equations which show how soil properties change with time are potentially an exciting prospect because they provide a means of gauging the pace of soil processes and soil formation. In fact, only a few quantitative solutions of rate functions have been determined, most of which indicate an exponential increase or decrease with time which accords with the laws of chemical kinetics. Examples include Salisbury's (1925) study of sand dune sequences and Crocker and Major's (1955) study of soils in the environment near a retreating glacier. The paucity of useful chronofunctions is attributable to the difficulty of finding datable land surfaces in which the effect of climatic change can be shown to be minimal. Apart from soils developed in a recent sequence of deposits, the best bet is a landscape where the soils have become progressively buried, though there is no guarantee that the soil will not have altered after burial (Vreeken 1975b). Dated surfaces can give some indication of how soil properties alter with time where changing climates can be shown to have had little effect. If the soil-forming process is relatively insensitive to short-term fluctuations of climate, long-term chronofunctions can be obtained; these show a consistent picture of exponential process rates (S-shaped curves), the changes gradually becoming negligible as steady-state is approached (Fig. 5.6).

5.4 MATHEMATICAL ANALYSIS

Using natural laws, it is possible to build mathematical models of a soil system and carry out analysis with them. Mostly, such models consider processes in soil profiles, though recently they have been extended to deal with processes in soil–landscapes and have proved helpful in unravelling three-dimensional soil–landscape relationships.

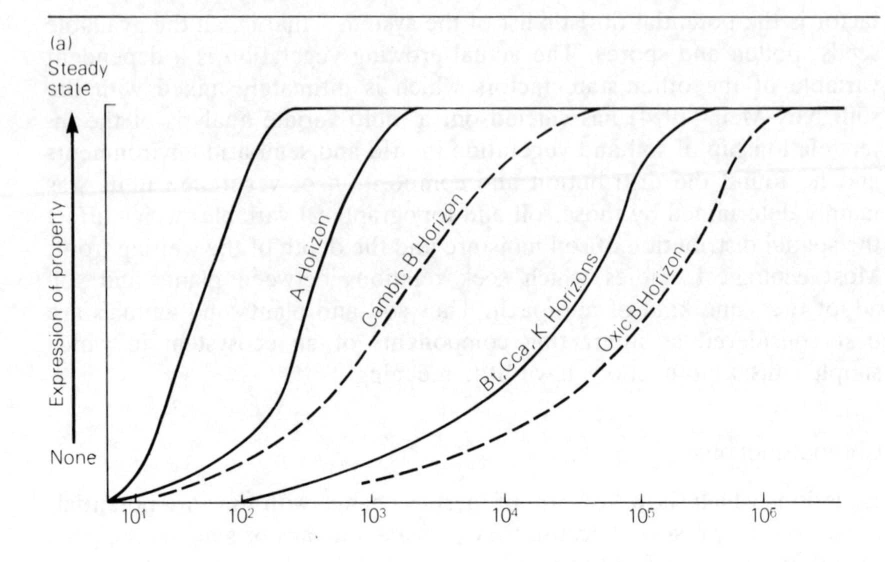

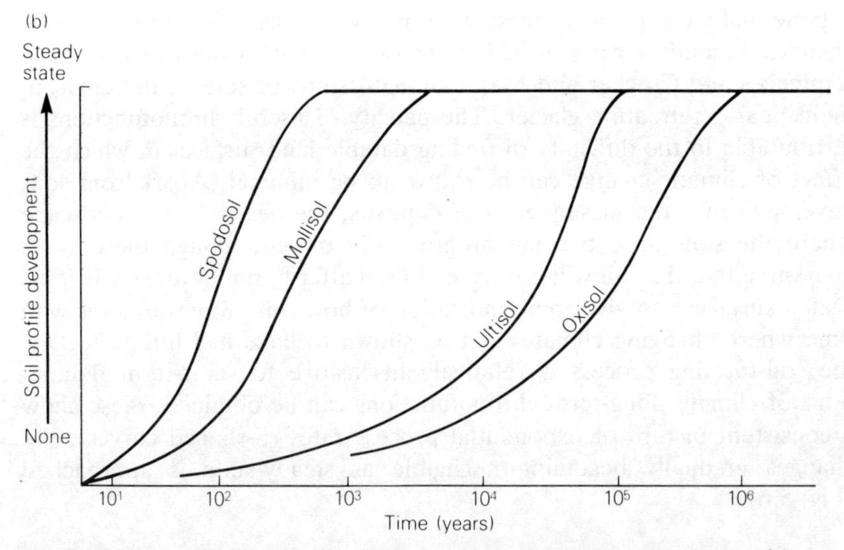

Fig. 5.6 Schematic diagram showing the time needed to attain the steady state for (a) various soil properties, and (b) some soil orders (from Yaalon 1975 Fig. 1, who took them from Birkeland 1974, Figs. 8 to 17)

Modelling processes in soil profiles

Processes in soil profiles have been extensively modelled. To name but a few cases in point, the soil profile water balance has been considered by

Carbon and Galbraith (1975), soil water redistribution by Sasscer *et al.* (1971), and soil redistribution by Burns (1975, 1976, 1977). Boast (1973) summarized the mathematical models used to study the movement and transformation of chemicals in soils and de Wit and van Keulen (1972) gave a detailed coverage of this topic.

One of the simplest models of soil processes is the one built by Yaalon (1965) to establish the relative mobility of a salt. The relative mobility or *R*-value of a salt is defined as

$$R_m = \frac{\text{distance travelled by ion}}{\text{distance travelled by water}} = \text{relative migration coefficient.}$$

For ions moving at the same rate as the wetting front because of little interaction with the soil, the R_m value is close to 1; for ions which do interact with the soil the R_m value is a fraction of 1. From his experiments with soil columns, Yaalon (1965) found that the R_m value of sulphate is about two thirds that of chloride. As Yaalon pointed out, the depth functions for various ions in a soil profile depend on the depth of moisture penetration (because dispersion is proportional to distance travelled), pore-size distribution (dispersion is greater in finer material owing to a greater tortuosity factor), absorption and interaction with the soil (which lead to skewed distributions) and the frequency of wetting–drying cycles (the accumulation peak is more pronounced the greater the frequency of wetting drying). But despite the great variability in the depth function, the position of the peak concentration is independent of all the above-listed factors. Yaalon (1965) defined the depth to peak concentration of an ion as $X = \frac{P_{ef} \cdot R_m}{W_a}$ where X is the depth to maximum concentration (cm), P_{ef} is the effective precipitation (cm), R_m is the relative migration coefficient, and W_a is the fractional volume occupied by available soil water. Despite its gross assumptions, this equation has been successfully used in the loessal semi-arid region of Israel (Yaalon 1963), the Gezira clay soil of the Sudan (Jewitt 1955) and the validity of the model is supported by the findings of Hutton (1968).

More elaborate models of soil profile processes are legion. The steps in their construction and operation go, by and large, like this. The soil profile is divided into several layers. Each layer is regarded as a store of energy or a store of material. The change in the amount of energy or material stored in each layer during a time interval is, by applying the laws of energy and mass conservation, equated with inputs and outputs of energy or mass to and from each layer and sources and sinks of energy or mass within each layer, to produce a set of energy or mass storage equations, one for each layer. The general form of a mass storage equation is

$$\text{change in mass stored in layer} = \left(\text{mass inputs} - \text{mass outputs} + \text{mass sources} - \text{mass sinks}\right) \text{time interval.}$$

Mass inputs and outputs could be affected by processes like translocation between layers; weathering is an example of a mass source, and plant uptake is an example of a mass sink. The energy or mass inputs and outputs are defined by appropriate transport laws, of which Darcy's law for the movement of water through rock or soil, Fick's law for the diffusion of solutes and Fourier's law for the conduction of heat are all examples. Putting the laws defining the rates of input and outputs into the storage equations produces a model of a soil process. This model can be used to study the effect the process has on storage in each soil layer, providing the following conditions are specified: (1) the amount of energy or mass in each layer at the start of the analysis – the initial conditions; and (2) the inputs and outputs of energy or mass at the top and bottom of the profile for the period of time over which the analysis is to run – the boundary conditions. A detailed account of this modelling procedure, as applied to heat transfer in soils, can be found in Thomas and Huggett (1980).

Mansell *et al.* (1977) built a mathematical model to describe the storage, transformation, and movement of orthophosphate in a soil profile. They recognized four phases in which soil phosphorus may be stored – water soluble, physically adsorbed, immobilized and precipitated. The phosphorus may be transformed from one phase to another under the control of six reversible kinetic equations. The transformations are determined by rate constants (these determine the fraction of phosphorus that moves from one phase to another in a unit time interval) and the amounts of phosphorus in each phase. Plant uptake and leaching from the solution phase were regarded as sinks of the chemical. They defined the rates of change of phosphorus in each of the four phases for no flow conditions (when net movement of soil solution is negligible) and for steady flow conditions in which the transport of solution-phase phosphorus was defined. This yielded a set of five storage equations which were solved for specific initial and boundary conditions. The results described phosphorus transformation and transport during steady water flow through the soil. Initially, the five storage equations were applied to a soil column of length 100 cm and devoid of phosphorus. A solution containing 100 μg/ml of soluble phosphorus was applied (mathematically) to the soil surface for two hours and was then followed by water only. A steady water flux of 1.0 cm/hr was maintained throughout the soil for a simulation period of 100 hours. Figure 5.7 shows some results. Figure 5.7(a) is the case where all six rate constants are zero; this means the applied phosphorus does not react chemically with the soil – the rapid transport shown is to be expected from unreactive solutes such as chloride. Figure 5.7(b) is the

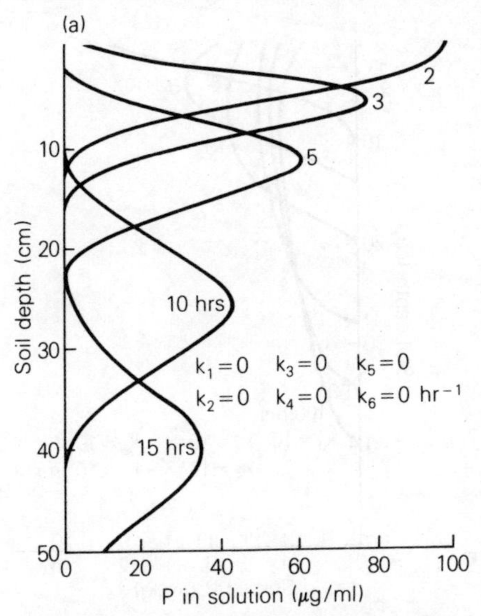

Fig. 5.7(a) Distribution of solution-phase phosphorus is a soil profile for the case in which all six reaction rate coefficients are zero. The *k*'s are the rate constants: k_1 is from solution phase to sorbed phase; k_2 is from sorbed phase to solution phase; k_3 is from sorbed phase to immobilized phase and k_4 is the reverse process; k_5 is from immobilized phase to precipitated phase and k_6 is the reverse process (From Mansell *et al.* 1977, Fig. 2)

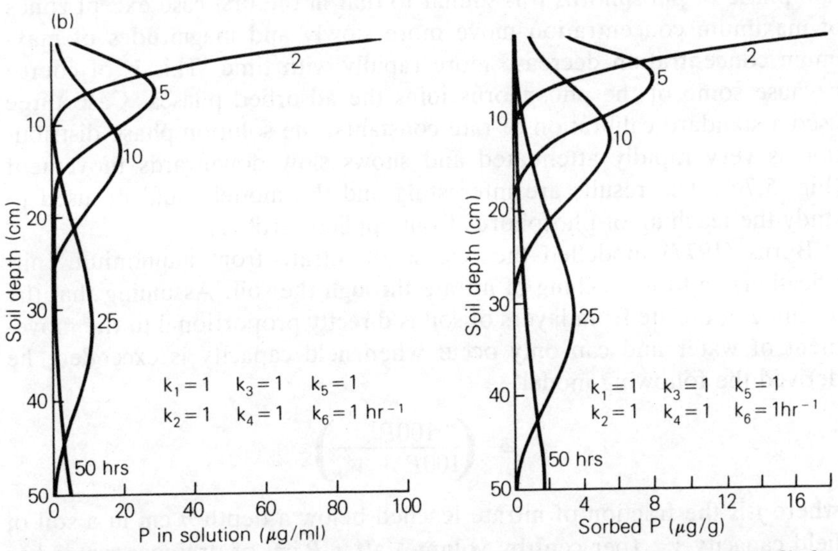

Fig. 5.7(b) Distributions of solution-phase and sorbed-phase phosphorus in a soil profile for the case in which all six reaction rate coefficients are 1 per hour (from Mansell *et al.* 1977, Fig. 3)

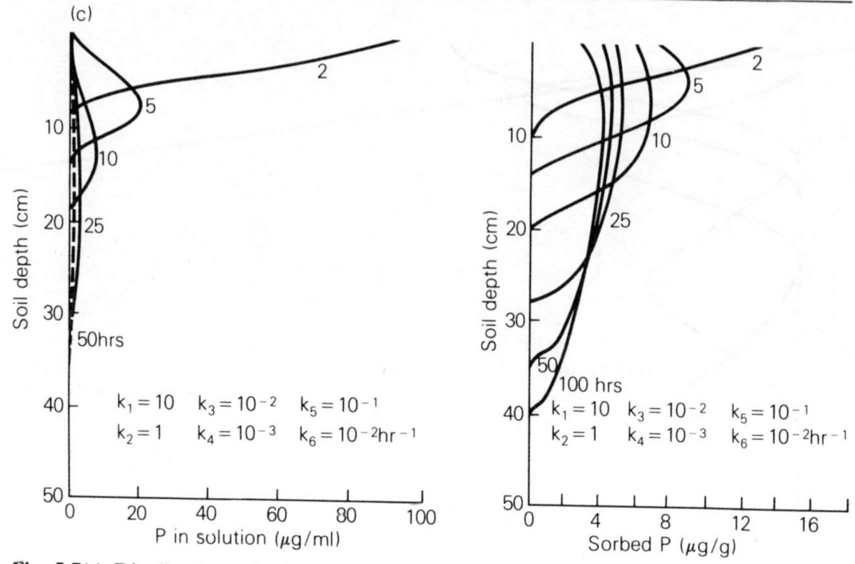

Fig. 5.7(c) Distributions of solution-phase and sorbed-phase phosphorus in a soil profile for the case in which $k_1 = 10$, $k_2 = 1$, $k_3 = 0.01$, $k_4 = 0.001$, $k_5 = 0.1$, and k_6 0.01 per hour: this combination of reaction rate coefficients was deemed typical of many soil types and so referred to as the standard case (from Mansell *et al.* 1977, Fig. 5)

case where all six rate constants are equal to 1/hr. Movement of the solution phase of phosphorus was similar to that in the first case except zones of maximum concentration move more slowly and magnitudes of maximum concentration decrease more rapidly with time. This is of course because some of the phosphorus joins the adsorbed phases. Case three used a standard calibration of rate constants; the solution phase distribution is very rapidly attenuated and shows slow downwards movement (Fig. 5.7c). The results are interesting and the model could be used to study the leaching of phosphorus from applied fertilizer.

Burns (1977) modelled the release of nitrate from ammonium (nitrification) and the leaching of nitrate through the soil. Assuming that the leaching of nitrate from layers of soil is directly proportional to the movement of water and can only occur when field capacity is exceeded, he derived the following model

$$f \simeq \left(\frac{100P}{100P + V_m}\right)^x$$

where f is the fraction of nitrate leached below a depth h cm in a soil of field capacity V_m (per cent by volume) after P cm of drainage water has passed through the profile, and where x is a simple function of h. The form of x depends on the initial distribution of nitrate in the profile: $x =$

h for material added to the surface; $x = h - w/2$ for nitrate incorporated uniformly to a depth *w* cm in the soil; and $x = h/2$ for a uniformly distributed nutrient. Values of *P* can be estimated from a simple water balance equation

$$P = \begin{array}{c}\text{Cumulative}\\\text{rainfall}\end{array} - \begin{array}{c}\text{Cumulative}\\\text{evaporation}\end{array} - \begin{array}{c}\text{Initial}\\\text{deficit of soil}\end{array}$$

The validity of the equation was tested using data from 46 separate nitrate and chloride treatments in several different leaching experiments on soils ranging from a sand to a clay. Agreement between observed and predicted mean displacements was good (Fig. 5.8).

Sasscer *et al.* (1971) modelled the movement of water through an old-field ecosystem in Argonne, Illinois. The system was represented by a layer of vegetation and 49 layers of soil, each 1 cm thick. The behaviour of two types of water was studied: stable water and tritiated water (tritium). The water fluxes or flow rates, all of which were considered to pass vertically through the top and bottom surfaces of the layers, are measured in units of millilitres per square centimetre per hour ($ml/cm^2/hr$) for stable water movement, and in units of degenerations per minute

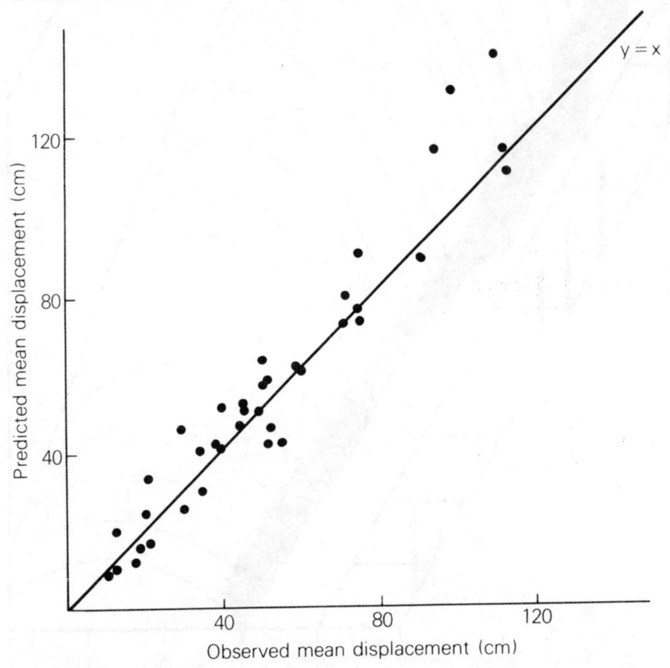

Fig. 5.8 Regression plot for testing the validity of Burn's nitrate leaching equation (from Burns 1977, Fig. 3)

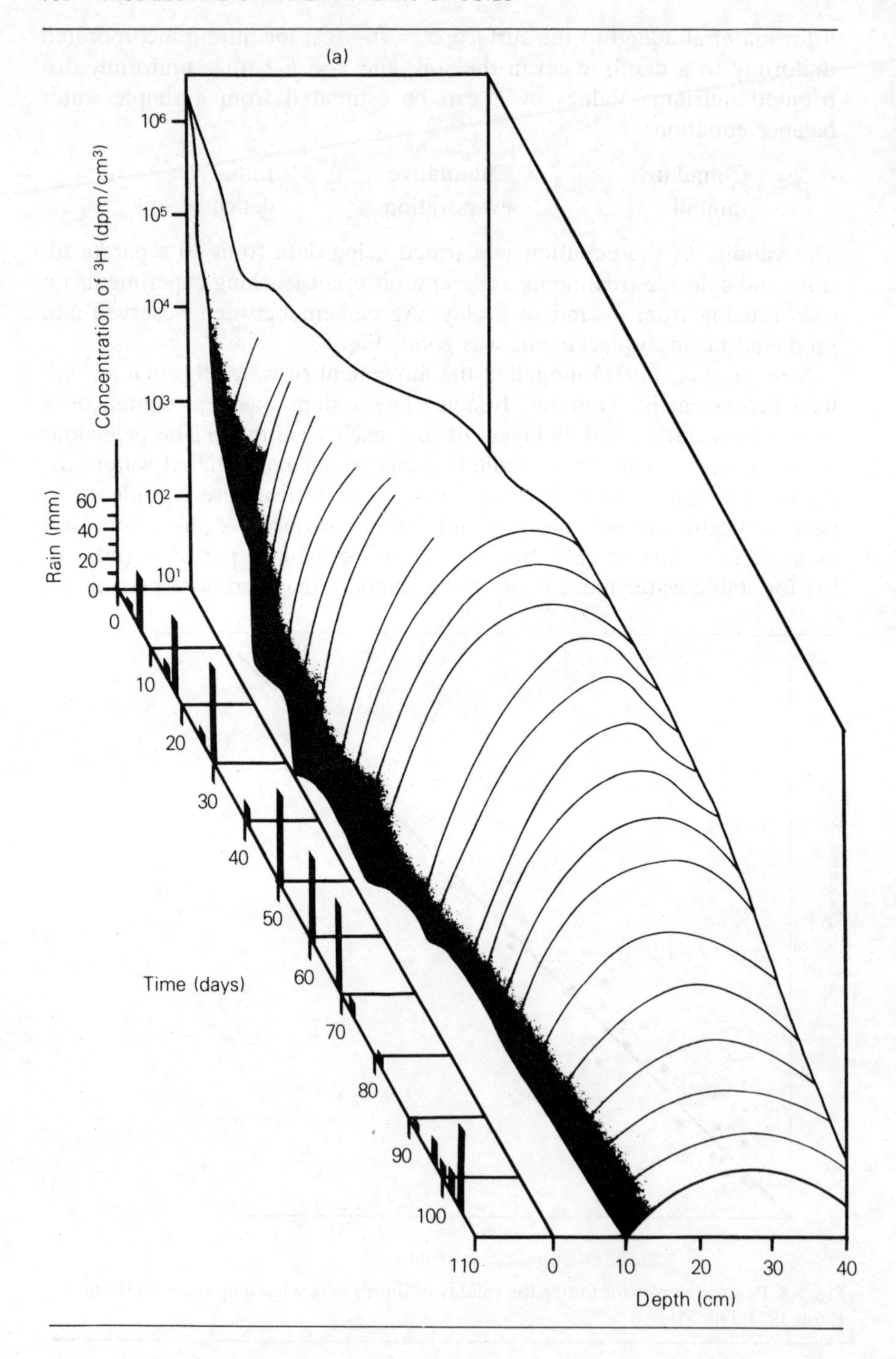
(a)
Concentration of ³H (dpm/cm³)
10⁶
10⁵
10⁴
10³
10²
10¹
Rain (mm)
60
40
20
0
Time (days)
0
10
20
30
40
50
60
70
80
90
100
110
Depth (cm)
0
10
20
30
40

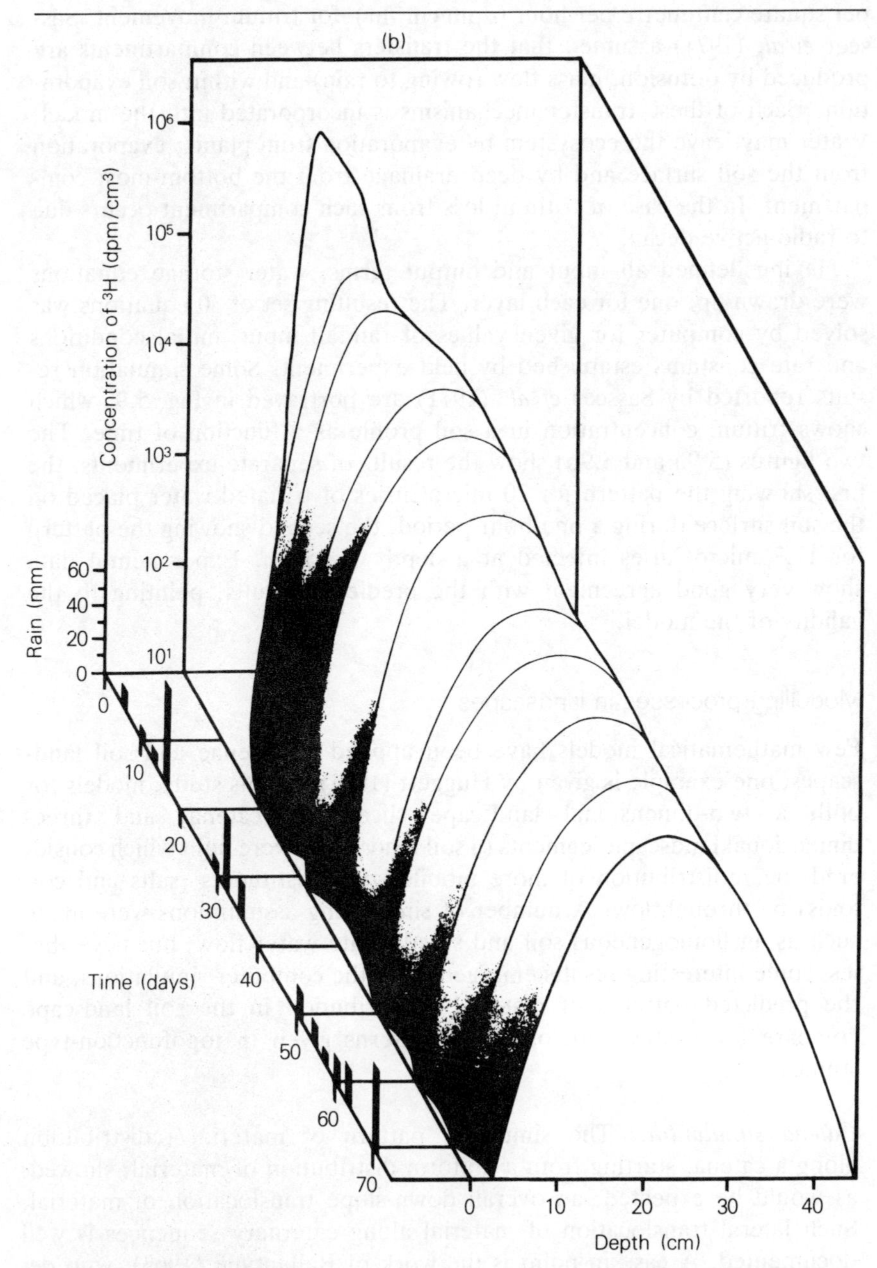

Fig. 5.9 Tritium concentrations in soil as a function of time and depth for: (a) deposition of tritium on the soil surface; and (b) deposition of tritium at 15 cm depth (from Sasscer *et al.* 1971, Figs. 5 and 6)

per square centimetre per hour (dpm/cm^2/hr) for tritium movement. Sasscer *et al.* (1971) assumed that the transfers between compartments are produced by diffusion, mass flow (owing to rain) and within-soil evaporation. Each of these transfer mechanisms is incorporated into the model. Water may leave the ecosystem by evaporation from plants, evaporation from the soil surface and by deep drainage from the bottom-most compartment. In the case of tritium, loss from each compartment occurs due to radio-active decay.

Having defined all input and output terms, water storage equations were drawn up, one for each layer. The resulting set of 50 equations was solved by computer for given values of rainfall input, initial conditions and rate constants established by field experiment. Some simulation results reported by Sasscer *et al.* (1971) are portrayed in Fig. 5.9, which shows tritium concentration in a soil profile as a function of time. The two figures (5.9a and 5.9b) show the results of separate experiments, the first showing the pattern for 20 microCuries of tritiated water placed on the soil surface during a one-hour period, the second showing the pattern for 17.7 microCuries injected at a depth of 15 cm. Experimental data show very good agreement with the predicted results, pointing to the validity of the model.

Modelling processes in landscapes

Few mathematical models have been applied to catenae and soil landscapes; one example is given by Huggett (1973). In this study, models for both a two-dimensional landscape slice (soil catena) and three-dimensional landscape segments (a soil landscape) were built which considered the redistribution of more mobile soil constituents (salts and colloids) by throughflow. A number of simplifying assumptions were made such as an homogeneous soil and steady-state water flow, but nevertheless some interesting results emerged from the computer simulations, and the predicted patterns of material redistribution in the soil landscape compare favourably with observed patterns given in topofunction-type studies.

Catena simulation. The simulated pattern of material redistribution along a catena, starting from a uniform distribution of material, showed, as would be expected, an overall down-slope translocation of material. Such lateral translocation of material along caternary sequences is well documented. A case in point is the work of Ballantyne (1963), who demonstrated that the recent accumulation of salts in toeslope soils of south-eastern Saskatchewan had been derived from summit positions by translocation during periods of abnormally high precipitation. Another example is provided by Holowaychuck *et al.* (1969), who found that strontium-

90 burden levels in very poorly drained soils are higher than in better drained soils in the vicinity of Cape Thompson, Alaska; they accounted for the difference by the lateral redistribution of leachate from high-lying to low-lying landscape positions. One of the few studies which quantifies the down-slope movement of soil material is the one by Smeck and Runge (1971) in a catena in Cass County, Illinois. Smeck and Runge (1971) noticed that in some soils more phosphorus had accumulated in the B horizon than could be attributed to eluviation from suprajacent A horizons, and in other soils more phosphorus had been lost from the A horizons than had accumulated in the B horizon. To try to fathom out the reasons for this disparity in the phosphorus balance, they quantified net gains and losses of phosphorus in each profile. Using zirconium oxide as an index and estimating the area represented in the landscape by each profile, absolute gains and losses from each soil unit were calculated (Table 5.5). In the entire catenary sequence a minimum of 944.9 kg of phosphorus had been laterally translocated within an area of 1.1 hectares. Summit soils had lost 151.34 g/m^2 of phosphorus; footslope soils had gained 493.49 g/m^2 of phosphorus. Adams and Raza (1978) quantified gains and losses of iron along two catenae in mid-Wales; the importance of their findings will be discussed later (p.169).

Table 5.5 Total gains and losses of phosphorus in profiles which are part of a catenary sequence in Illinois (after Smeck and Runge 1971)

Soil	Area	Net change in total phosphorus content to a depth of 230 cm	
	(ha)	(g/m^2)	(kg/soil area)
Site 1	0.0376	+493.49	+188.0
Site 2	0.1596	+284.96	+460.5
Site 3	0.1808	+ 58.86	+107.7
Site 4	0.1708	−103.19	−178.3
Site 5	0.2184	− 99.94	−221.0
Site 6	0.3564	−151.34	−545.6
Landscape balance	1.1236	− 16.60	−188.7

It is hardly surprising that the computer model generates a down-slope movement of mobile soil material. But it also showed more subtle effects, for which some supporting field evidence is available. The down-slope movement in the model led to an initial build up of material at the junction of convex and concave portions of the catena, a situation found in some slope soils by Furley (1971) and Whitfield and Furley (1971). This 'peak' of concentration is not a static feature; as time progresses it moves as a wave down-slope. The notion is of a transient concentration wave along a catena of soil is not unreasonable since ions do progress through a vertical column of soil in this manner (Yaalon 1965) and may well move down a soil catena in the same way (Yaalon *et al.* 1974). Field evidence for the wave-like progress of peak concentration along a catena is meagre. If mobile soil materials do move down a catena in the same way as they move through a soil profile, then the peak of the concentration wave will appear at different down-slope positions for different materials because each material has a different mobility. Huggett (1976a) looked for concentration peaks along a catena in a site in the Northaw Great Wood, south Hertfordshire, England; he found some evidence of this for horizons developed in London Clay where, starting up-slope, the order of 'peaks' of concentration is aluminium, iron and manganese.

Soil-landscape simulation. The redistribution of mobile soil materials in a valley basin was modelled (Fig. 5.10). Not surprisingly, the overall down-slope movement of material was modified by convergences and divergences arising from contour curvature. Convergence of movement into the hollow led to a build up of material there, whereas over the nose divergence of movement led to a removal of material. Field evidence for this very general pattern of material redistribution is plentiful. Early work by Aandahl (1948) and Troeh (1964) gave evidence for a relation between soil properties and landscape position. More recently, research by Vreeken (1973, 1975a) has revealed distributions of clay, organic carbon and carbonates in drainage basins which can in part be explained by topographical influences on three-dimensional material redistribution. For example, Fig. 5.11 depicts the topography and depth to carbonates in the Soucek watershed, Iowa.

The computer simulation also produced more subtle effects superimposed on the overall redistribution in the valley basin. In particular, down-slope movement along a flowline running into the hollow is relatively faster and shifts more material than down-slope movement along a flowline running over the nose. Evidence for this in the field comes from the Northaw Great Wood survey (Huggett 1976a). Figure 5.12 portrays gains and losses of oxalate soluble aluminium, iron, manganese and silicon in the Northaw soil pits. (Gains and losses were calculated using a slightly modified version of the method described by Borchardt *et al.* (1968). In essence, sand was used as a yardstick against which to gauge relative gains and losses of the elements in soils developed in Pebble Gravel, and clay was used as a similar yardstick in soils formed in London Clay. (Details of the method are given in Huggett 1976a). Notice that all horizons formed in lithostratigraphic units II, III, and IV over the nose (Fig. 5.12) show a minor gain of aluminium. Comparable horizons in the hollow show a gain of aluminium, with the exception of the C(g) horizon of profile 5, but in the hollow horizon the largest gain is more peaked, in

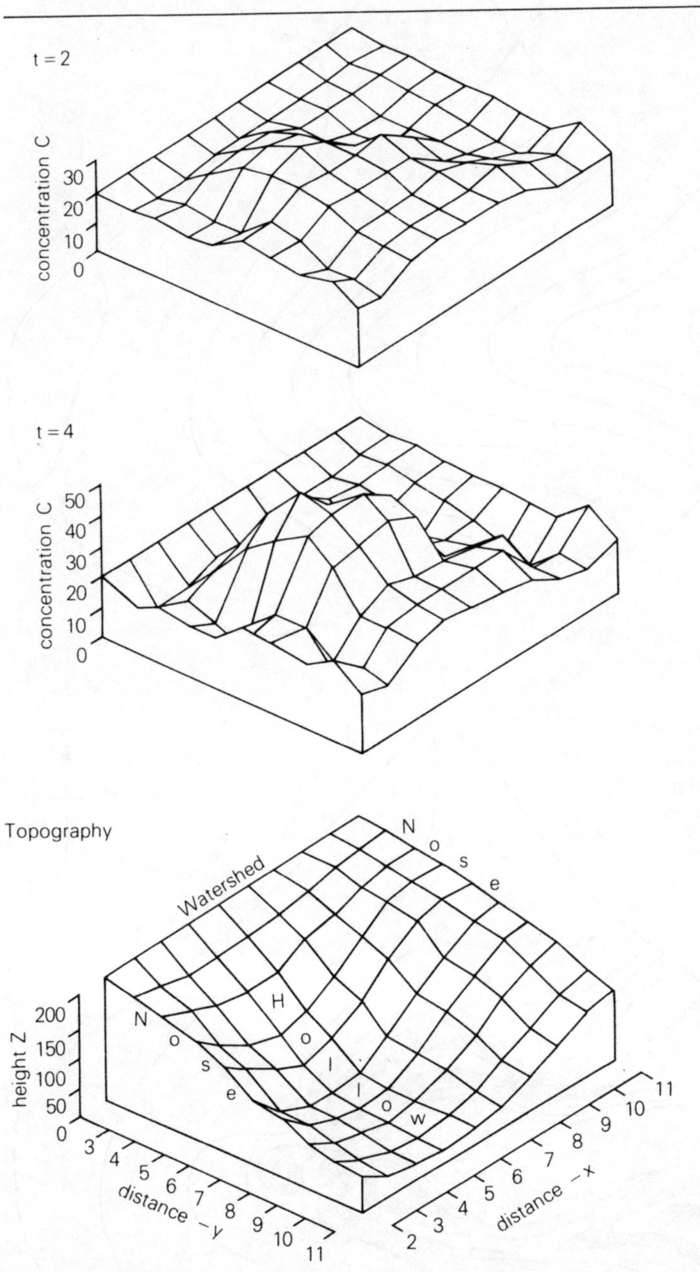

Fig. 5.10 The computer-simulated movement of a mobile soil constituent in a catchment. The surface topography is shown at the bottom of the diagram. As time progresses, the material builds up in the hollow (from Huggett 1973 and 1975)

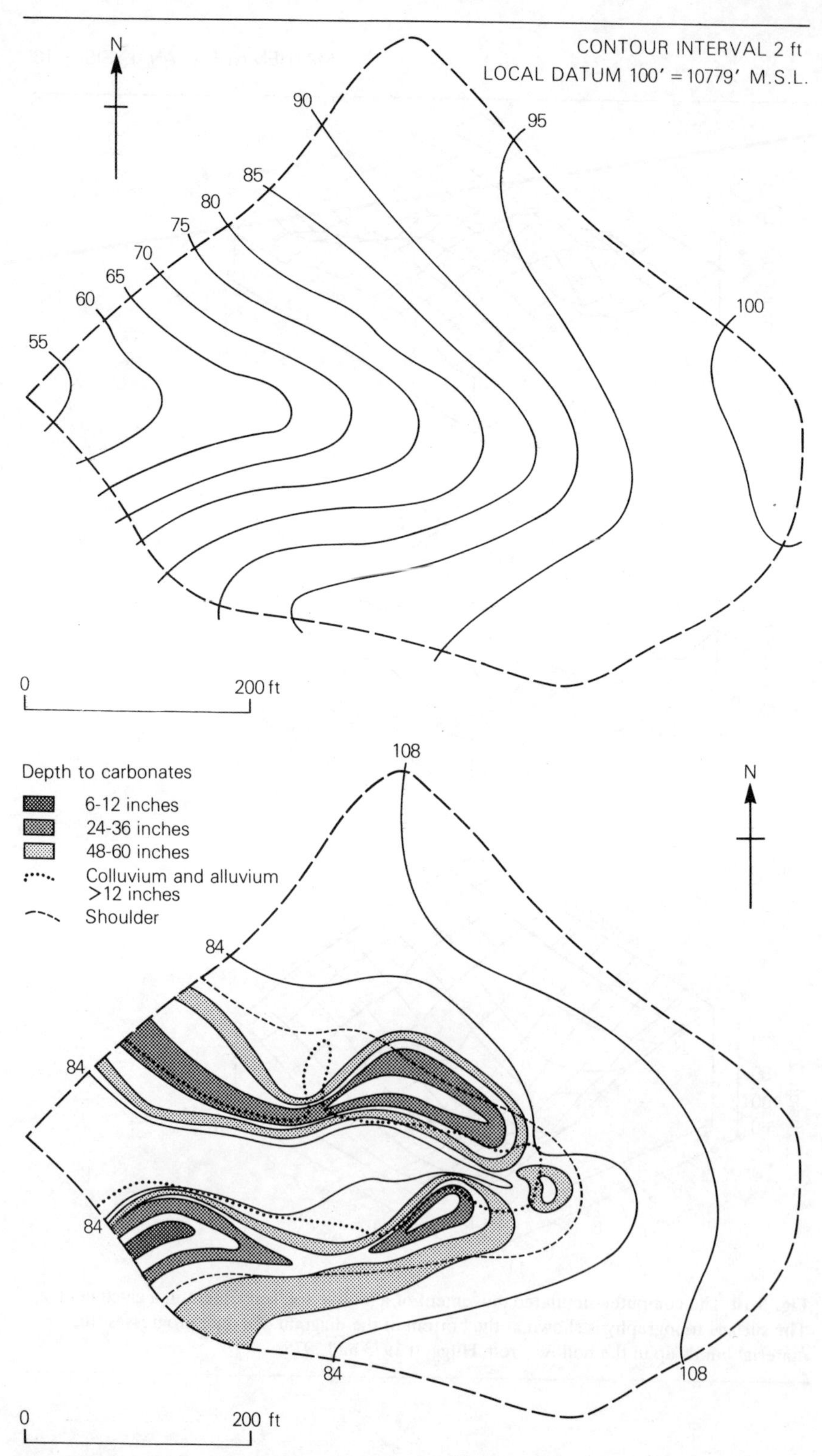

Fig. 5.11 The Soucek watershed: topography and depth to carbonates (from Vreeken 1975a, Fig. 7)

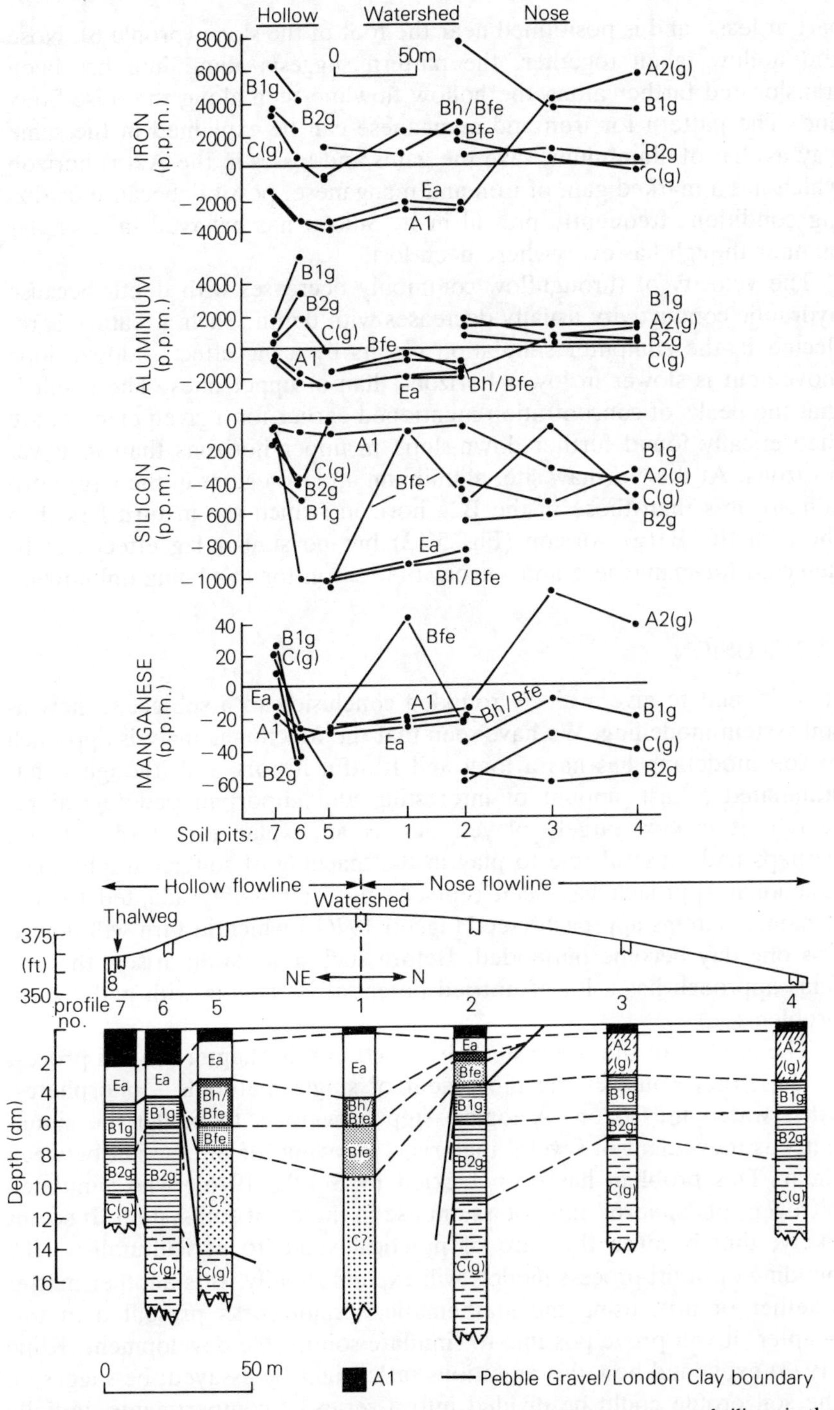

Fig. 5.12 Gains and losses of oxalate-soluble aluminium, iron, manganese and silicon in the Northaw Great Wood pits. N.B. scales differ on the vertical co-ordinate (from Huggett 1976a, Figs 2 and 5)

part at least, and is positioned near the foot of the slope (profile 6). Nose and hollow taken together, the pattern suggests aluminium has been translocated farther along the hollow flowline than along the nose flowline. The pattern for iron and manganese can be explained in the same way as that of aluminium, save the gains and losses in the A2(g) horizon which has a marked gain of iron and manganese, possibly because oxidizing conditions frequently prevail in it. Silicon has behaved in a similar manner though has everywhere been lost.

The velocity of throughflow commonly decreases with depth because hydraulic conductivity usually decreases with depth. Such a feature is reflected in the computer simulation results by a lag effect – down-slope movement is slower in lower horizons than in upper ones. The result is that the peaks of concentration mentioned earlier for a given element are theoretically found further down-slope in upper horizons than in lower horizons. At the Northaw site, aluminium and iron gains in the C(g) horizon are less than those in the B2g horizon, which arc in turn less than those in the B1(g) horizon (Fig. 5.12) but no similar lag effect can be detected for manganese and silicon, the reason for this being unknown.

CONCLUSION

It is difficult to give a nicely rounded conclusion to a subject so new as soil system modelling. We have seen that the functional analysis approach to soil modelling has has a long and fruitful history and, though it has stimulated a vast amount of interesting and important pedological research, it is now largely played out as an explanatory tool but still perhaps had a useful role to play in the teaching of soil geography. The functional approach has been replaced by, or possibly adapted to, the dynamic systems approach (see Huggett 1976b) which in turn will doubtless one day become outmoded. Before such a situation arises, the systems approach has a lot of untried potential in dealing with pedological problems.

The models of the soil system discussed in this chapter are uni-process models; they consider the movement of single materials – phosphates, salts, water and so on. A logical step forward is to model the simultaneous movement of several materials, allowing for interaction between them. This problem has been tackled by Smith (1976), who simulated nitrogen, phosphorus and potassium use in the plant-soil system. It seems likely, that because they are of practical value to agriculturalists, the building of multi-process models will expand rapidly. It is another matter whether or not, using the mathematical frameworks presented in this chapter, it will prove possible to simulate soil profile development. Kline (1973) explained how this ambitious task might be essayed; he suggested the soil profile could be divided into a series of compartments, initially undifferentiated, through which materials pass. Material leaves the profile by plant uptake, erosion and deep drainage; it enters the profile from the atmosphere and in return flows from vegetation (litter and timber fall, and root sloughage). Material in the profile may be moved from one form to another (in much the same way as phosphorus shifted from one phase to another in the model of Mansell *et al.* 1977). The output from the model, Kline (1973) said, would be a set of curves, one for each compartment, showing how storage of a particular material changes with time. The profile could then be reconstructed at fixed times by plotting compartment contents as a function of depth. The major difficulty in running such a model would probably be trying to cope with the huge differences in the magnitude of rate constants – some processes act very quickly (days or even hours), whereas others act over protracted time periods (centuries or even millennia). A way of dealing with this problem might be to ignore the rapidly changing soil properties – solute concentrations for instance, and attempt to model just the longer-acting processes such as clay moment. Calibration of slow processes would be troublesome because their rate of operation is difficult to measure directly in the field; data from chronofunctions would probably have to be used. This might not matter unduly because only the general pattern of soil profile change would be of interest. The same kind of model could in theory be applied to soil landscapes, demands on computer time being a possible limitation.

Kline's (1973) proposed model does not include the effects of geomorphological processes on soil development. Perhaps the biggest untapped area of research in soil geography is the building of mathematical models which link soil and vegetation processes to geomorphological processes on slopes and in landscapes. In the light of work such as that of Adams and Raza (1978), who found that the pattern of gains and losses of iron in the two soil hillslope sequences in mid-Wales was not determined by lateral translocation down-slope but by the truncation of upper soil layers by creep erosion, joint soil–slope–vegetation models would seem worthwhile. Some tentative proposals of a conceptual type have been given an airing by Trudgill (1977) in a book which fosters a novel way of looking at soil–vegetation–geomorphological relationships. A more concrete approach, in mathematical terms, is embodied in the recent attempts of Kirkby (1977) to extend his model of slope development to include the effects of hydrology, soil and vegetation on slopes. Some useful pedological spin-offs may come from this research.

In building a joint slope-soil model, Kirkby came across two main stumbling-blocks. First of all, he could find no soil models which were compatible with the inputs and outputs of slope models. Soil models simulate changes of short duration, a few years at most, whereas slope models generally predict change over hundreds or thousands of years. Kirkby overcame this problem by ignoring daily and seasonal changes in soils and looking just at annual totals of water and solute transport. The other problem he faced was to reduce the number of variables describing the soil profile to those which are most relevant to the slope as a whole, without neglecting important soil-forming processes; he was unable to circumvent this problem very satisfactorily from the point of view of the slope, but the soil model is sound enough.

Kirkby's soil profile model is primarily chemical. Soil inorganic materials are treated as mixtures of elementary oxides. Soil organic matter is considered to consist of three portions: 'carbohydrates', the main source of carbon dioxide in decomposition, including sugars, starches and cellulose; 'amino acids', the principal building blocks of protein, the main source of organically bound nitrogen in the soil; and 'lignins', a varied group of large molecules rather resistant to decomposition. The model works like this. The distribution of carbon dioxide is predicted by simulating the diffusion of carbon dioxide released by decaying organic matter. Soil pH is then calculated with reference to the carbon dioxide distribution. In turn, soil pH allows weathering (solubility) equations to be solved at all depths throughout the inorganic soil profile. To calibrate the model for a particular case, the following items must be specified: climate – mean annual rainfall, actual evaporation and throughflow; the distribution of water uptake by plants at all depths and the distribution of throughflow with depth; the composition of bedrock expressed as oxides; the gross productivity and photosynthesis rate of plants as well as the concentration of elements, expressed as oxides, in various organic substances; and the rate of surface lowering. Kirkby gives a worked example of the model for a site on a steep hillside which receives 700 mm of rain a year, loses 400 mm a year in evaporation and has a surface lowering rate of 0.05 mm a year. The soil profile produced in this case is a rather immature brown earth with a 50 cm 'A' horizon resting on a 'BC' horizon extending to 160 cm. Part of Kirby's model considers the energy changes which take place in weathering reactions (cf. Fig. 5.1). The 'thermochemical' approach to weathering is in fact a fairly new field of research (for example, Curtis 1976a,b) which may provide useful inputs to models of soil and slope development.

Clearly, the modelling of the soil spatial system is only just beginning but the work done so far looks promising. The would-be modeller will find no shortage of problems to crack in this relatively new field of soil geography.

6

AGRICULTURAL USES OF SOIL SURVEY DATA

E. M. Bridges and D. A. Davidson

INTRODUCTION

In all countries of the world, the most fundamental resource of the agricultural industry is the soil. As soil is the medium used to produce food and industrial crops essential for life, every effort should be made to safeguard its continued existence and well-being. Historically, this has been embodied in the concept of 'keeping the land in good heart' and has resulted in farmers passing to their heirs fertile land in good physical condition. Good soil of high fertility is a limited, scarce resource which is in great demand for many purposes other than crop growing. Whether it is for agriculture, horticulture, housing, factories, airports, motorways, reservoirs or sewage works, the land needs to be level with loamy soils having free drainage and few physical limitations. This results in extreme pressure being placed upon the best agricultural land; alternative schemes with high short-term economic advantages compete with traditional users. Once urbanized, the flexibility of land use is lost and if needed for food production at a later date a return to agriculture or horticulture is virtually impossible. There is in England and Wales approximately 10 000 000 ha of improved agricultural land of which only slightly more than 2 000 000 ha fall into grades I and II on the agricultural land classification maps (Agricultural Land Service, 1966) but grade I land amounts to only 2.3 per cent of the total (Fig. 6.1). It is necessary to have a clear idea of the geographical extent and distribution of this limited amount of good soil and land when evaluating any change in use from agriculture.

The growth in importance of national and county planning during the last 40 years in Britain makes it extremely important that a knowledge of soil distributions is known both at local and national levels. Decisions concerning land use are made by professional planners who are not familiar with matters of soil science and agriculture. It is therefore vital that these people are provided with information so that decisions are at least made with the best advice available. Soil survey and soil geography pro-

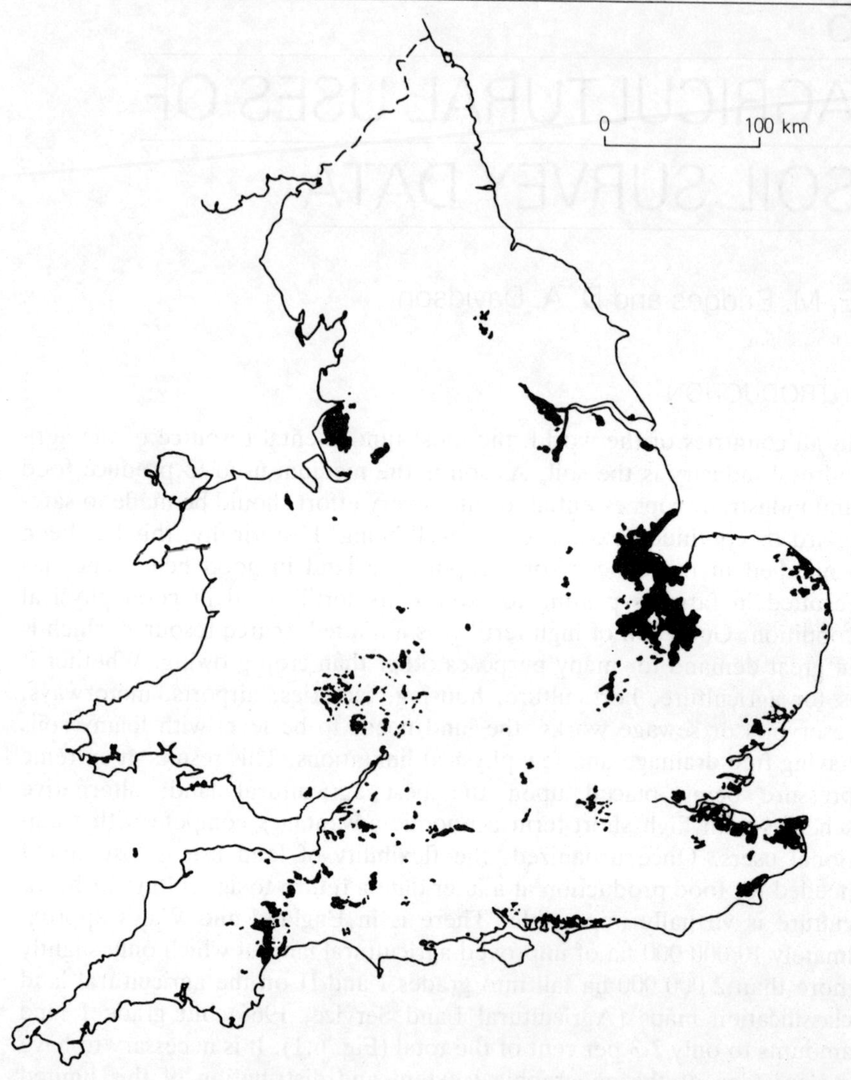

Fig. 6.1 Grade I farmland in England and Wales (after Coleman 1976)

vide an inventory and a framework within which decisions concerning land use can be reasonably made.

Soil surveys are financed in many cases from government funds from the Ministry of Agriculture and the work is done by qualified surveyors. Alternatively, in the private sector, soil surveyors are employed on a contract basis to produce soil surveys for specific development projects. The fact that both government agencies and financially accountable companies support soil survey projects is a strong indication that beneficial results accrue from the knowledge obtained. Mapping units shown on soil survey maps are 'agriculturally significant because they are defined by properties that directly or indirectly affect crop production and management. Their actual behaviour can be predicted by assembling the results of agronomic research and experience' (Mackney 1974). It follows that soil survey maps interpreted for specific properties of the soil can be used extensively in agricultural advisory work, farm planning and planning in a wider sense.

Interpretations of soil survey maps began many years ago and Aandahl (1958) has outlined the purposes of this exercise in the following words: 'soil survey interpretation comprises the organization and presentation of knowledge about characteristic qualities and behaviour of soils as they are classified and outlined on maps'. The art of soil survey interpretation is to identify and present both broad and specific alternatives or opportunities in the use of soils. However, in the final analysis, the farmer's choice does not depend upon soil alone, but on the economics of supply and demand as well as the personal satisfaction which results from farming in a particular manner.

In all interpretations of soils survey data, it is necessary first to understand what the user of the information requires. Close collaboration is essential between interpreter and client to establish which properties of soil and land are relevant and which affect specific uses. Once these properties have been established, the next step is to relate them to soil behaviour and performance. This may not be a simple task, for the information is not readily available, being scattered throughout published and unpublished reports of crop trials, scientific journals and surveys of fertilizer usage. Much information is only available as unwritten knowledge in the minds of experienced farmers and agricultural advisers. Once assembled, the information can be shown on maps or be tabulated according to the requirements of the client. The relationship of soil properties and crop production is always complicated by the weather, which is generally outside man's control, and by management factors. Consequently, most interpretations of soil behaviour have to be presented assuming average weather conditions and in terms of a moderately high level of managerial competency.

There are several kinds of interpretation which can be made from basic soil surveys. The most obvious is to take a soil property which is known to be linked with a particular problem, and its distribution can then be traced on a *single-factor* or parametric map. Unfortunately, it is rare that only a single factor is responsible for a particular problem, and so other data, such as climatological information, have to be considered as well. When this occurs, the result is to produce land or soil quality maps. The land use capability systems of the USA (Klingebiel and Montgomery 1961), Canada (Canada Land Inventory 1965) and Britain (Bibby and

Table 6.1 Land use capability classes and subclasses (Bibby and Mackney 1969)

Land use capability classes	Class 1	Land with very minor or no physical limitations to use.
	Class 2	Land with minor limitations that reduce the choice of crops and interfere with cultivations.
	Class 3	Land with moderate limitations that restrict the choice of crops and/or demand careful management.
	Class 4	Land with moderately severe limitations that restrict the choice of crops and/or require very careful management practices.
	Class 5	Land with severe limitations that restrict its use to pasture, forestry and recreation.
	Class 6	Land with very severe limitations that restrict use to rough grazing, forestry and recreation.
	Class 7	Land with extremely severe limitations that cannot be rectified.
Land use capability sub-classes	*Capability sub-classes are divisions within classes based on the kinds of limitations affecting land use.*	
	w	wetness limitations
	s	soil limitations
	g	gradient and soil pattern limitations
	c	climatic limitations
	e	liability to erosion

Mackney 1969) are designed to disseminate information to non-soil scientists such as agricultural advisers, farmers, planners and other people who are concerned with decisions about land use (Table 6.1). Interpretations of a more complex kind including many soil properties and associated data are considered as land evaluation and are couched in terms of the suitability of the land for growing specific crops. Evaluations such as these can be carried out for farm-sized areas or for continental areas such as Africa (Dudal 1978).

Although the interpretation of soil survey maps for agricultural purposes is still of great importance, interpretation for a wide range of other purposes is a growing aspect of the soil scientists' work. Kellogg (1974) indicated that only 25 per cent of the benefits from soil survey in the USA now arise from agricultural applications; the other 75 per cent relate to non-agricultural uses, as described in the next chapter.

6.1 SOIL PROPERTIES OF IMPORTANCE TO AGRICULTURE

It was demonstrated in Chapter 2 that soil surveys indicate the distribution of soil classes or mapping units. These, at a detailed or very detailed scale of mapping, are soil phases, series or complexes, and the more intensive the survey, the greater purity in each mapping unit. Once the properties of a soil are established and it is known which characteristics affect crop production, it is possible to extrapolate from areas already mapped to other areas where the same mapping units are recognized.

This predictive use of soil maps and accompanying information is particularly important in advisory work and interpretations for agricultural purposes. There are many soils series (about 600 in England and Wales) and information about their productivity under different crops and management is far from comprehensive.

Early surveys in Britain partly defined soil series by the stratigraphical age of the presumed parent material, but present-day approaches rest upon the intrinsic soil properties whenever possible. The main criteria used to differentiate one soil series from another within sub-group of mineral soils are particle-size class, and presence and nature of contrasting horizons or layers impenetrable to roots, origin of the soil material and mineralogical or related characteristics of the soil material (Avery 1973). As a result, each soil series or mapping unit has a combination of a typical colour, organic matter content, texture, structure, consistence and moisture regime, which, although not absolutely unique, are sufficient to distinguish between soils at a local level. Moreover, these soil properties have been shown to relate to productivity by numerous studies and practical experience (Mackney 1969). Brief comments follow on each of these soil properties and their significance to agriculture.

Soil colour

Colour figured very largely in former attempts to describe and classify soils as it had an immediate impact upon the observer. The colour of soil is determined by the minerals of the parent material and by the presence of organic matter. Ferric iron compounds give the soil a brownish or reddish colour and are usually associated with aerated conditions in freely drained soils. Reduced iron compounds develop in poorly drained conditions by the process of gleying and have a grey or blueish-grey colour, often associated with structure faces and pore linings, but in continually saturated circumstances, are seen throughout the whole soil. In temperate environments organic matter darkens the soil colour at the surface and in certain subsoil horizons of the podzolic soils. In the absence of these masking colours of iron compounds and organic matter, soil minerals have a grey or even white colour, as can be seen in the eluvial horizon of podzol soils.

In terms of practical agriculture, certain colours of soils do confer an advantage. Observable temperature differences occur between soils with dark surface horizons and those with light coloured topsoils. In horticultural terms it is possible to alter the surface albedo by spreading chalk or soot to lighten or darken the surface, increasing or decreasing the amount of reflection of incoming radiation. A soil which warms rapidly in the spring has an advantage over a soil which has a lower temperature regime; this can be used to advantage in the case of potatoes where the first crop of early potatoes can command premium prices. However, the water content of soils greatly influences the soil temperature regime and considerable improvements can be made by draining which outweigh the small advantages conferred by a different colour. Although a dark colour is indicative of higher humus content in the soils of temperate regions, it is not necessarily true for soils in the tropical regions of the world.

In the interpretation of drainage in wet soils, colours are a helpful diagnostic feature. The grey colours of reduced iron compounds on structure faces in the E and B horizons are a good indication of the presence of standing water in the soil pores and fissures and the necessity for drainage operations. The appearance of subsoil colours at the surface, particularly at the crests of slopes, is indicative of the action of soil erosion, accelerated by agricultural activity. Management should be vigilant for these signs and take appropriate action.

Soil texture

The particle-size distribution of a soil is one of its fundamental characteristics; moreover, it is a characteristic which is difficult to change by agricultural practices. It may be possible to modify soil texture to a limited extent, but on a field scale it is virtually out of the question. Most definitions of soils suitable for arable agriculture stress they should be deep, well-drained loams, sandy loams, silt loams or peat, lying on level sites. The range of particle sizes included in loams ensures that the soils will be able to retain adequate supplies of 'available' moisture but not be saturated for long periods and it follows that these soils can be used for a wide range of field crops. Soils which are dominated by any one size range can pose difficulties for farming operations but adverse effects can be mitigated by the incorporation of organic matter and liming.

Coarse textured soils with large stone contents normally have severely limited available water capacities and they are prone to drought. The extreme stoniness can interfere with cultivation and root development, consequently such soils are usually left under rather poor permanent pasture. Sandy soils (sand and loamy sand) are normally well drained but continued cultivation lowers the amount of organic matter so that structural problems may develop and low moisture-supplying powers result. Certain root crops, carrots in particular, require an open sandy soil without stones for a high quality crop. Retention of lime and fertilizers in sandy soils is poor, deficiencies of micro-nutrients occur and erosion by wind and water is always a danger. Some fine sands pack tightly, set hard when dry and when wet become almost fluid with consequent difficulties in cultivation. Thus, sands generally provide ease of cultivation over a fairly wide range of moisture contents with low inputs of fuel energy; they are not without their difficulties.

Silty soils, with silt loam and silty clay loam textures, are common where soils have formed from alluvial deposits. Fairly extensive areas in eastern England, especially the Fenland, are characterized by soils with these textures, where they are used for intensive farming and horticultural enterprises. The silty soils possess a better water-holding capacity than the sandy soils but structural difficulties may occur with surface capping developing on intensively used arable soils. This results from the impact of raindrops which disrupts the structure of the surface soil, thus compacting it, and so preventing seedling emergence. Mole drainage (see p. 193) is difficult in silty soils as the channels collapse, a problem which is experienced in Wales.

Clayey soils (clay, sandy clay and silty clay) are mainly composed of small particles of silicate clay minerals. The layered structure of the silicate minerals and their extremely small particle size gives rise to a large surface area within the soil (1 ha topsoil containing 20 per cent clay would have a surface area of 2.5 million ha). This large surface area is reactive, having small electrical charges present as a result of ionic substitution in the lattice structure of the clay minerals. Cations are attracted to the surface of the clay particles and the degree to which this occurs is the cation exchange capacity of the clay or soil. The amount of cations and the manner in which they are held also affects the colloidal properties of flocculation and dispersion, and this in turn determines many of the mechanical and engineering properties of soils (Brown 1974).

In agriculture, clay soils are referred to as heavy land, and considerable care is necessary to obtain the best results in terms of yield. Practically, the addition of lime and organic matter can assist structural stability, and drainage removes excess moisture. With good aggregation, excess surface water is lost but sufficient remains in clay soils for adequate plant growth. In a drier than normal year higher yields may be obtained from clay soils through their ability to supply greater amounts of moisture. As a result of this water uptake the volume of certain clays expands greatly; conversely, as water is removed the clay shrinks. Where the mineral montmorillonite is present cracking of dry clay soils is very marked and can result in self-mulching and pedoturbation.

As a source of plant nutrients the clay fraction of soils plays a very important role. Minerals of the mica group slowly release potassium over a longer period of time to provide a natural low-level supply of this major plant nutrient. The cation exchange capacity of clays helps to retain potassium (K^+) and ammonium (NH^+) against leaching in humid climates. Phosphate is also held by clays, but formation of insoluble aluminium and iron phosphates complicates the picture. Clay soils are strongly buffered chemically, so that as lime is added further hydrogen (H^+) ions are released from the clay. Thus greater quantities of lime are required to raise the pH of clay soils compared with sandy or loamy soils.

The exchange properties of soils also affect the pesticides and herbicides. When a herbicide such as paraquat is used to remove plant growth, any that falls upon the soil is immediately immobilized and made unavailable to plants, so it can be used just before planting or before emergence of the crop to control weeds.

Soil structure

Soil structure is a term which refers to the shape, size and degree of aggregation, if any, of the primary soil particles into naturally or artificially formed structural units (peds) and the spatial arrangement of these units including the description of pores and fissures (voids) between and within the aggregates (Hodgson 1974). The physical condition of these soil aggregates is of particular importance to agriculture and soils with defects or extra good structural properties can usually be attributed to particular soil series. Consequently, it is possible to indicate the distribution of soils with or without structural problems under cultivation and predict where problems might occur.

Soil structure is the product of the natural aggregation of soil mineral material aided by organic matter and soil fauna. Some smaller soil structures originate as faunal excreta such as worm casts, but larger subsurface peds result from wetting and drying cycles which subsequently become coated with organic matter and/or oriented clay skins (cutans). On the soil surface the terms clod or fragment are preferred where soils are cultivated and artificially formed structures are created. Good soil structure is essential for adequate root penetration and for seedling emergence.

Although many soils are reasonably tolerant of misuse, it has been a matter of concern that under modern intensive systems of agriculture soil structures show signs of breaking down (Strutt 1970). Certainly, modern agricultural techniques require more from soils than traditional methods. Peas grown for freezing are grown under a strict schedule of sowing and harvesting with intervening cultivations and spraying for pests quite regardless of weather and soil conditions. Damage to soil structure can occur through the passage of heavy harvesting machinery. Similarly, cultivations undertaken when soil moisture content is unsuitably high result in smearing below the plough sole to give a plough-pan. In South Wales, on the coastal lowlands of the Gower peninsula, winter ploughing-in of brassica debris in preparation for spring barley can result, in certain years, in a poor barley crop as the brassica debris putrefies in the saturated conditions above the plough-pan. The benefit of grasses to soil structure has been known for a long time in that the dense root system helps crumb formation, but the activities of earthworms casting at the surface and lower in the soil are equally important. Additional benefits arise from nitrogen fixed by clovers which usually form part of the grass mixture.

Organic matter content

Organic matter in the soil may come from plant or animal sources and it forms an important constituent of the surface horizons of any soil. Experience has shown that when the soil organic matter content falls below 3 per cent in fine sand and silt soils, maintenance of structure becomes a problem. Levels of 2 to 3 per cent are common in soils under arable cultivation in England. The benefits which the presence of organic matter brings are that the soil has a much better physical condition, enabling better root development and improved drainage. At the same time, organic matter increases the capacity of the soil to retain moisture, particularly in coarse-textured soils. In sandy soils, the organic matter helps to bind the particles together and restrict loss from blowing. Breakdown of organic matter provides a small amount of plant nutrients to a growing crop, but insufficient for its total requirements; it also provides food for soil-living organisms.

Accumulation of organic matter takes place naturally in soils from annual leaf fall and root decay. Breakdown and incorporation of organic matter is achieved by the soil fauna, most of which live in the immediate surface layer. Deeper incorporation occurs when earthworms carry organic debris to their burrows, by ingestion and subsequent casting. The earthworm burrows also encourage drainage and aeration of the soil. On agricultural land organic supply is derived from a number of sources. Root decay *in situ* is an important source of fibrous material which breaks down slowly, providing stability for soil structures. Crop debris and farmyard manure are spread and ploughed-in when available, but often farmyard manure is in short supply. Intensive dairy farms, often, have large amounts of slurry which can be spread safely upon certain soils. Where groundwater is at a high level, pollution of water courses may result, but elsewhere if the slurry is applied at a rate suitable for the soil to accept without deleterious effects it can be a valuable source of nutrients. Winter applications may be difficult on soils too wet to support traffic. In such cases, storage may cause added expense for the farmer.

Consistence

The term consistence is used by the soil surveyor to describe the cohesion and adhesion of soil particles. It can be conveniently described as the way in which the soil responds to disturbance. Consistence is described under the headings of strength, failure characteristics, cementation, maximum stickiness and plasticity. The soil property of consistence is closely related to the structure of the soil, but instead of describing these structures it attempts to measure the forces which hold structures together. Consistence varies greatly according to the moisture content and surveyors take this into consideration. For the practical agriculturalist, consistence is an approach towards assessing the tilth of a soil. However, tilth is applied normally to topsoils, whereas the soil surveyor's assessment of consistence can be applied to any soil horizon. In subsoil horizons, this can be an important guide to the advisability or otherwise of different forms of under-drainage. At the soil surface, the production of a satisfactory tilth can be critical for many crops as small seedlings emerge. A coarse cloddy surface or a sealed surface where structure has broken down can inhibit or even prevent seedling emergence and establishment. As consistence is one of the criteria used in soil series description, the soil map may be interpreted for the likelihood of such problems and their distribution. For example, on the Freckenham series, compaction has been shown to reduce the yield of sugar beet (Harrod 1975).

Moisture regime

The presence of water in soil pores and fissures for longer than a few days can inhibit growth of roots and even cause their death. Soils where this condition occurs frequently are known as gley soils; their distribution can be appreciated from any soil survey. The surveyor attempts to describe two other features in addition to the distribution of gley soils; these are the morphological effects of the gleying process and the soil moisture regime. Throughout lowland England most gley soils have been drained at some stage of their history, increasing their rooting depth, extending the period when the soils can be cultivated or grazed and raising their yields.

The moisture regime and the supply of moisture to the growing crops is critical and strongly influences yields. Surveys conducted from Rothamsted Experimental Station (1968) show that wheat yields in England and Wales are significantly related to soil moisture regimes and Jarvis (1973) cites soil moisture regime and texture as influencing the yield of crops in the Wantage district. One of the first attempts to relate productivity to soil properties was by Clarke (1951). Taking the depth of soil, texture and a rating for gleying, he was able to account for 90 per cent of the variation in yield in a 'good' harvest year and about 70 per cent in a poor year.

Many other features are described and recorded in a soil survey as well as the actual distribution of soils over the landscape. From the previous brief comments it will be apparent that the information revealed by a soil survey can provide a great deal of useful assistance for the agriculturalist either as a practising farmer or as an advisory officer. The usefulness of a soil survey may be summed up in terms of the bringing together of environmental information which can be used to assess the suitability of soils for specific projects, such as drainage schemes, irrigation schemes or special forms of cultivation, or the suitability of specific crops and of appropriate fertilizer and cultivation techniques. Finally, a soil survey can form the basis of land use capability classification which can be used in the evaluation of different plans for reorganization of farming activities.

6.2 USE OF SOIL MAPS IN AGRONOMY

Progress in relating the yield of crops to soil series or soil properties has been retarded by the limited extent of the country covered by detailed soil surveys, a prevalent association in people's minds of soils and geological formations and an over-simplified chemical approach. It has earlier been stated that soil series are defined by properties which are significant for crop production and much work remains to be done in the search for correlations between soil properties and yield. When more of these relationships are established, soil maps will be more widely used by farmers and advisers in decisions about fertilizer usage. The problem is complicated by interactions of weather, site and management which can affect yields greatly, and although soils may have the same morphology and genesis this does not always ensure similar cropping potential. Many experiments have been carried out in Britain since 1945 but they have taken place on sites where the farmer happens to be co-operative and competent rather than on a typical example of a particular soil series. Although experiments began on specific soil series in the 1950s even the Rothamsted annual report of 1965 perpetuated the grossly over-simplified linkage of soils with Clay with Flints, Oxford Clay and Chalky Boulder Clay in an account of some experiments on potatoes (R.E.S. 1965; Coulter 1974).

Field observation on the growth and recording of the ultimate yield of crops on different soils series is highly desirable in many ways. The individual farmer would have a better yardstick with which to measure his own production, advisers would have greater confidence in their recommendations for specific crops and nationally, a greater production and more efficient agricultural industry would result. Unfortunately, the exercise of measuring crop yields is not as simple as it seems. If one looks to past records, individual farmers do not keep consistent records and often, only crops sold off the farm are weighed. Production of fodder crops and grass are rarely measured; also, where crops are measured, it is usually by the better than average farmer. Yields depend very much on the two elements of weather and management and to include variations of these it is necessary to continue measurements of yields over a period of several years which raises problems of staff and consistency of methods.

In practice, experimental farms of the Ministry of Agriculture are located in typical farming districts. These farms have reference plots upon which a range of fertilizer treatments is given to the crops commonly grown. The site and weather of these should not be too different from the surrounding farms and the problems of management variability are under the control of the agronomist. Providing these farms are sited on typical examples of a soil series, they are ideal sites for 'benchmark' studies where a number of nutrient levels and cultivation techniques are standardized for a range of crops over a period of years. A second technique for soil and crop evaluation is to collect yield data from a number of sites under strictly controlled management. This data is then subject to statistical analysis. Alternatively, Wilkinson (1968, 1974) has used micro-plots to study the relative productive potential of a group of six soil series in Berkshire and in Nottinghamshire/Leicestershire. The study collected data on the management requirements and performance of crops on six soil series in each area. These data were compared with yield and management data recorded on six fields on the same soil series under practical farming conditions.

Other approaches include finding what the maximum yield of a soil series would be after nutrient and water needs for the crop have been satisfied. It has also been suggested that large blocks of soil could be transported to one site to overcome the problem of microclimate, but this has not been attempted on any scale. In practice, one simple method whereby management and climatic differences can be overcome and the influence of past history minimized, is by the use of two different soils in the same field for experimental purpose. Selection of possible sites becomes feasible when a detailed soil map of the study area exists.

Some of the earliest experiments on soils series, beginning in 1955, were carried out by Boyd and Dermott (1964), who found soil texture to be a significant factor and that responses of N and K dressings on a main crop potatoes reflected the different textures of soils. The series used in the experiments were the sandy and sandy loam Bridgnorth and Newport series and the Evesham series silt loam and clay loams. A knowledge of the geographical distribution of these soils series would indicate where there would be a lack of response to N and K (on the Evesham series) and where it would repay the farmer to apply smaller and more frequent applications as on the coarse textured soils.

Burnham and Dermott (1964) report an interesting comparison of the Bromyard and Munslow series, two of the most widespread soils in the Church Stretton district. Although the yields recorded on farmers' fields were not too dissimilar when the average figures were compared, these average figures disguised some real differences between the two series.

Table 6.2 Yield response potatoes (after Coulter 1974)

Soil series	Mean yield t/ha
Newport and Bridgnorth	28.42
Evesham	19.41
Saltmarsh	25.28
Blacktoft	20.82

Table 6.3 Potassium released to ryegrass (after Mackney 1974)

Soil series	Released non-exchangeable K mg/100 g soil
Gaerwen	2.8
Bromyard	47.7
Hanslope	59.2
Denchworth	70.3
Evesham	102.3

Compared with results of experiments on similar crops at the Rosemaund Experimental Husbandry Farm the farmers' yields were falling consistently short of the reference plot results.

Links between field and laboratory studies are greatly strengthened if laboratory experiments are designed to investigate properties of soil series. The findings of the laboratory investigations are more easily transferred into the field situation. An example of the applications of this type of investigation is given by Arnold and Closes' (1961) determination of the K release properties of topsoils from a number of different soil series. By means of a pot-culture technique, they showed that the non-exchangeable potassium of topsoils from different soil series varied greatly in their ability to release potassium to ryegrass (Table 6.3).

Non-exchangeable K is released mainly from the fine clay fraction and it was found that a strong correlation existed between these two characteristics ($r = + 0.72$). Other studies have shown that percentage clay and its clay content reflect the ability to release K well with a correlation of $r = + 0.79$. Some soil series have large K reserves: Evesham, Worcester, Hanslope, Charlton Bank, Long Load and Rowsham series and are unlikely to respond to potassium fertilizers but other series such as the Banbury series are naturally deficient. Once again it must be stressed that knowledge of the soil geography aids sensible use and economical fertilizer utilization.

The amount of phosphorus used by plants depends very much on the soil moisture, soil structure, rooting volume and amount of P in the soil, and it is well appreciated that wet clayey soils tend to fix added P in less readily available forms. Using P_{32}, Caldwell and Jones (1965) have shown that the Hanslope series, according to conventional methods of analysis, is lacking P. However, it is unresponsive to applications of P and can yield well for many years without further additions of P fertilizers. By contrast, peaty soils such as the Adventurers' and Prickwillow series have a good response to amendments of P even though they are classed as having moderate to high levels of soluble P according to laboratory analyses.

Cooke (1967) states that there is no satisfactory basis for recommending how to use about four-fifths of the P and K fertilizers bought by farmers, but in other publications (1965 and 1972) he acknowledges that knowing the soil type has often been more successful than knowing soil analyses in the interpretation of tests of K fertilizers. Expanding this idea, Cooke suggests that soil maps combined with agronomic experience could, in turn, help the soil chemist arrive at methods of analysis suitable for particular groups of soil types.

Geographical patterns of trace element deficiency have emerged from studies of crop and animal disorders. These patterns in turn have been associated with certain soil series. Avery (1955) identified molybdenum-induced copper deficiency with the Evesham series in Somerset but on the same series in the East Midlands the problem is sub-clinical or absent (Thomasson 1971). In Wales, similar problems exist with soils developed from a 'black shale' facies of Silurian strata. Molybdenum-induced copper deficiency has been identified west of Carmarthen and sub-clinical conditions extend eastwards between Carmarthen and Llandeilo. Straightforward copper deficiencies have been identified in the Scottish counties of Berwick, Roxburgh and Fife. The Hobkirk, Eckford and Sourhope series are deficient in this element and farmers have resorted to adding copper salts to the land, thus reducing the problem. In England, nitrogen-induced copper deficiencies have been reported in spring barley and wheat on the shallow, chalky Icknield series.

Sandy soils often lack trace elements, or contain very low levels, which can easily be rendered unavailable by poor farming practice such as over-liming. Manganese deficiency in oats and barley on Flint, Cottam and Newport series, boron deficiency in sugar beet on Newport series and boron deficiency in swedes on the Bodavel and Arvon series are common occurrences. Deficiencies in magnesium have been observed on humus-iron podzols in southern Scotland which farmers have countered by the addition of the element to animals' diet and to the soil through liming with magnesium limestone. Cobalt deficiencies have been found to be less clearly related to particular soil series but even so, a soil map does help to identify areas where potential problems may be found (Speirs 1976). In all these examples, soil maps have been useful in advisory work, particularly where conditions are subclinical in animals or plants. The characteristics of the soils themselves, as identified in the mapping units, can assist in decisions about the best methods of rectification of deficiencies.

Finally, the presence of enhanced levels of heavy metals, e.g. zinc, cadmium in soils can be demonstrated to have an influence upon human health. Information upon this interesting area of study has only recently become apparent, and further studies are proceeding.

The ability of the soil to retain and release moisture is of particular interest to the agronomist as moisture is frequently a limiting factor in crop production. The moisture in the soil available to plants, referred to as the available water capacity, is strongly influenced by texture (Salter and Williams 1965). As texture is one of the main criteria used to define soil mapping units, the distribution of different available water capacities can be interpreted from soil maps (Fig. 6.2). Soils which are liable to moisture deficits can be indicated generally throughout the country, or at a farm scale, on maps of appropriate scale and plans drawn up for irrigation purposes and water supply requirements. It has been discussed earlier how the moisture status of soil influences many other soil properties such as structure, thermal properties, trafficability and cultivability.

Certain agronomic pests have been related to particular soils. It has been demonstrated that larval phases of liver flukes prefer certain types of gley soils and wireworms are known to be more prevalent on freely drained soils of clayey textures.

Suitability of soils for tillage

Surveys of land use indicate that the soils of certain areas are used preferentially for arable farming. In Britain this is reflected by the dominance of arable farming in the south and east and pastoral farming in the west and north. This distribution broadly reflects climatic and relief conditions in the country, but it also suggests the soils of the south and east are better suited to tillage. Ease of cultivation is dependent upon the soil moisture content which in turn results from an interaction of basic soil characteristics, such as particle-size distribution, organic matter content and permeability, with climate. Soil physical conditions are equally important as chemical (nutrient) supplies, for without satisfactory physical conditions the growing crop cannot take advantage of the nutrients even when supplied at optimum rates.

Cultivation of the land is a skilled occupation as can be seen at a ploughing match anywhere in the country. It is more than just a technical skill with implements, as it involves an understanding, scientific or otherwise, of the interaction of soil properties, rainfall, drainage and evapotranspiration. Ploughing and cultivations attempted at the wrong time, with less than optimal moisture conditions, can result in structural breakdown

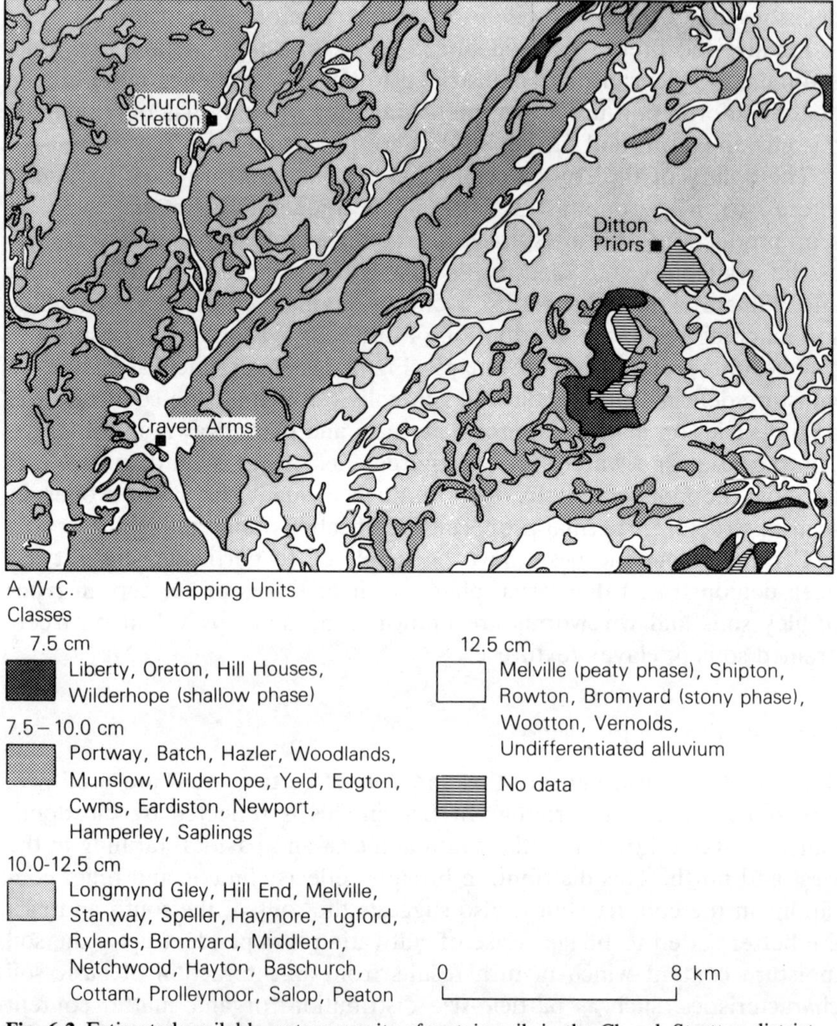

Fig. 6.2 Estimated available water capacity of certain soils in the Church Stretton district, Shropshire (after Mackney and Burnham 1966)

rather than an improvement in the tilth. Soils range in their suitability for tillage and it has been shown that ease of cultivation can be related to particle-size, organic matter content, drainage status and permeability (Jones 1979). After allocating the main soil series of the West Midlands to six groups according to particle size and their behaviour it was found that significant differences emerged when clay contents (C), packing density (L), air capacity (C) and retained water capacity (θ_{vt}) were statistically evaluated. The limits for cultivation groups thought to be critical are given in Fig. 6.3.

CULTIVATION GROUP	GROUP I Easy				GROUP II Moderately easy				GROUP III Moderate				GROUP IV Moderately difficult				GROUP V Difficult				GROUP VI Not normally cultivated					
Physical properties	C	L_{dt}	C_{at}	Θ_{vt}	C	L_{dt}	C_{at}	Θ_{vt}	C	L_{dt}	C_{at}	Θ_{vt}	C	L_{dt}	C_{at}	Θ_{vt}	C	L_{dt}	C_{at}	Θ_{vt}	C	L_{dt}	C_{at}	Θ_{vt}	Depth cm	Horizons
0–25	<18	M	>15	25–35	<25	M	>10	35–42	>18 <35	M	>10	40–48	>25 <40	M	<10	45–50	>35	M	<10	47–55	>35	L	<15	50–60	0–25	A (SURFACE)
25–50	<18	M	>20	15–25	>10 <25	M	>15	25–30	>18 <35	M	>10	30–40	>25 <40	M–H	<10	30–40	>35	M–H	<10	40–45	>35	M	5–15	>40	25–50	E,B
50–100	<18	M	>20	10–25	>18 <30	H	<15	25–35	>30 <40	H	<10	30–40	>35	H	<7.5	35–45	>40	H	<7.5	45–50	>30	M	<10	40–45	50–100	B,C (SUBSURFACE)
	I – II				II – IV				II – IV				III – IV				IV – V				V – VI				WETNESS CLASS	

C gravimetric clay content (%)
L_{dt} packing density (g/cm^3)
C_{at} air capacity (%)
Θ_{vt} retained water capacity (%)

Classes of packing density
L low <1.40
M medium 1.40 – 1.75
H high >1.75

Fig. 6.3 Physical properties of soils in the cultivation groups (after Jones 1979)

Brown earths and brown sands with sandy, coarse loamy or coarse silty textures which drain rapidly and which can be cultivated easily at most times throughout the year form Group I. Soils with silty and loamy textures (Group II) are moderately easy to cultivate but as texture becomes finer, and more moisture is retained, increasing care is required in the timing of cultivations (Group III), otherwise structural damage can result. These Group III soils are preferably ploughed in the autumn; other cultivations in the spring must be carefully timed. Soils with fine loamy and clayey textures from the pelosol and stagnogley groups and slow internal drainage are included in Group IV. Early autumn ploughing is essential before moisture contents reach field capacity and the frosts of winter help break clods down to a fine tilth. Group V soils belonging to the pelo-stagnogley group have clayey textures, high packing densities, very slow drainage and are normally above field capacity from early autumn to late spring. There is a high risk of structural damage if these soils are ploughed or cultivated at the incorrect moisture conditions and the ideal land use is a ley grassland system. Soils allocated to Group VI are best left to permanent grassland and cultivation only attempted when the pasture deteriorates to the point where re-seeding is essential.

Another important aspect of the physical capability of the soils is the support it has to give to the machinery moving across the surface. Dry soils are able to support tractors and other implements because the shear forces developed in the soil resist the load and soil deformation is minimal. In moist plastic soils, the load applied often can exceed the soil shear strength and so the tractor wheels sink into the soil until the shear force increases in magnitude and it matches the load imposed. These considerations have many implications for the efficient use of farm machinery, but as far as the soil is concerned the maintenance of structure is as important to successful farming as the correct balance of plant nutrients.

At least some of the problems mentioned previously can be overcome if conventional ploughing and cultivation techniques are minimized. The system of direct drilling developed on easily cultivated soils, but in recent years its applicability to more difficult soils has been explored. One of the main advantages of direct drilling in autumn is that it enables drilling to take place closer to the optimum sowing date. It is recommended that previous stubble is burnt off and weed growth controlled with herbicides. In cases of surface compaction a shallow (5 cm) cultivation has been found beneficial. Soil series thought to be suitable for direct drilling occur widely throughout the cereal growing parts of the British Isles. Series upon which experimental data are available include the Andover, Coombe, Newmarket, Sherborne, Macmerry, Sutton and Downholland, all of which have the most favourable properties for direct drilling. The Evesham, Hanslope, Beccles, Denchworth, Dunkeswick, Ragdale, Salop, Thorne, Batcombe and Tendring come into a second category of soils which experimental evidence indicates that extra care is needed for winter cereals and yields of spring-sown crops are likely to be reduced. Substantial risks of lower yields occur on Bromyard, Charity, Newport and Blackwood series as well as Evesham and Dunkeswick series where field capacity is reached before 1 November (Cannell *et al.* 1979).

Information such as this gives the agricultural adviser much background information about the geographical distribution and behaviour of soils. When a farmer proposes to adopt such techniques it should be possible to indicate where they can best be used and the likelihood of success. The main effects of direct drilling are to retain organic matter in the surface soil where it promotes better structure and tilth. Greater numbers of earthworms are encouraged and plants are seen to root more deeply which is important in an ensuing dry period. Plough-pans are not developed, infiltration is improved and so is trafficability. According to estimates, an appreciable part of the cereal-growing area of England would be suitable for direct drilling at least for winter sown crops.

Soil suitability for specific crops

Statements made earlier in this chapter indicate that the information contained in soil surveys may be interpreted in terms of specific crop requirements. An exercise such as this is useful both nationally and locally to identify areas of suitable land. In those areas of the British Isles where soil maps have been in existence for many years, advisers have found them helpful in a wide range of agricultural work including education. Some of the first work on soil suitability for specific crops was concerned with fruit trees. Long-term soil conditions are probably even more important for tree crops than annual field crops as the trees remain for a period of several years. Surveys by Osmond *et al.* (1949) demonstrated that poor growth and yields were related to shallow rooting depth, waterlogging and sandy texture. More recently, Dermott *et al.* (1965) proposed soil/site suitability ratings for fruit trees. Over the years, experience has shown the usefulness of relating specific crops to soils.

Speirs (1976) describes a project in which the possibilities of growing horticultural crops in southern Scotland were examined. A development agency was desirous of bringing processing plant and employment into the area and envisaged the cultivation of brussel sprouts, cauliflowers, peas, beetroot, carrots and soft fruit for the factory. Horticultural crops such as these place heavy demands upon the soil. For example, contract pea growing for quick freezing necessitates sowing as early as February and harvesting by heavy machinery at a predetermined date regardless of soil conditions. Brussel sprouts are harvested in winter when soil moisture conditions are high and structural damage is likely; carrots are similarly harvested in winter but additionally require a soil with low clay content. At first it was suggested that there was over 32 000 ha of potential land for the project, but when information concerning the soils was assembled it was found that the possible area was only about 8000 ha. Knowledge of the soil geography in this case enabled a realistic estimation of the scale of the project, indicated the distribution of suitable soils and the location of the processing plant. Hooper and Crampton (1964) attempted a similar assessment for the former county of Glamorganshire by taking into consideration climate, relief and soil information in a search for potential horticultural soils.

The suitability of soils for grassland has been reviewed recently (Harrod 1979). Maximum grass production is only possible where growth is not limited by soil moisture deficits. From a climatic point of view the milder, wetter areas of western Britain are most favourable as soil moisture is regularly recharged and extended droughts are rare. The available moisture held in the soil occupies the pore spaces including intra-ped pores and inter-ped pores. By experiment it has been found that the water content correlates well with soil texture, depth, stoniness and organic matter content, all of which are more conveniently measured than the pore space. These properties are all recorded in a soil survey and the resultant maps can be interpreted for the suitability of the soil to support a good grass crop.

The ability of the soil to stand up to the passage over it of tractors and other farm machinery, as well as the impact of animal hooves, is equally important. Grasslands with a low bearing capacity soil will be limited in their productive capability because early spring and late autumn grazing and/or forage harvesting will be restricted. Grass, like any other crop, requires fertilizer and it is usually necessary to chain harrow a pasture to spread dung pats; in both cases the soils must allow these operations to be carried out successfully. Grassland soils which poach under the feet of cattle suffer from loss of structure, part of the crop is wasted and future productivity is lowered by compaction.

The suitability of soils for grassland can be subdivided into four categories. Class I are deep, mainly well-drained loamy or silty brown soils, brown podzolic soils or drained lowland peat or loamy ground-water gley soils. Such soils are sufficiently resilient to stand up to moderate amounts of winter traffic and grazing. These include the Moretonhampstead series in Devon, the Denbigh series in Wales and the Adventurers (drained) series in East Anglia. Large areas of grassland occur on soils which come into a class II where minor soil limitations exist, restricting productivity. These limitations include poor drainage, shallow or stony soils, soils which are occasionally droughty and textures which give trafficability difficulties. The risk of fresh-water flooding and a short growing season are other limitations. Examples of soils placed in Class II include the Powys series of Wales, Dunnington Heath series of the Midlands and the Waveney (drained) series of East Anglia.

Where there are limitations which prohibit the intensive use of grassland, the soils are placed in Class III. Many upland soils with wetness limitations and surface-water gley soils in the lowland zone with large summer moisture deficits are included in this group. These soils are only satisfactory for seasonal pasture and include the Hexworthy series of Devon, the Cegin series of Wales and the Fladbury series of the Midlands. Many upland areas are not suited to intensive grass production as yields are poor and trafficability problems are great. A short growing season with steep slopes down-grades these soils to Class IV. They include the Princetown series of Dartmoor, the Hirwaun series of South Wales and associated lithomorphic and raw soils.

6.3 SOIL MAPS AND LAND IMPROVEMENT

Information about the geography of soils can be used for planning land improvement measures, and two examples are given in the following paragraphs. The first is concerned with land improvement through drainage and the second introduces the problems of land restoration and the role soil studies can play in the process.

Drainage

In soils with impermeable lower horizons which occur where precipitation exceeds evapo-transpiration, or areas where seepage waters are prevalent, it is necessary to remove excess water from the land by a drainage scheme. Although attempts to drain land in Britain can be traced back to Roman times, it was from the seventeenth century onwards that field drainage became a widespread agricultural practice. Tile drains were introduced in the eighteenth century and in the early nineteenth century the government encouraged drainage measures by tax exemption and low interest rates on improvement loans. In the period 1940–80 just over 1 million ha have received grants of public money for land drainage in Britain.

A soil map indicates the distribution of poorly drained soils, as the presence of gleying is one of the morphological features soil surveyors use to distinguish different mapping units. The major soil groups of direct interest for land drainage specialists are the surface-water gley and the ground-water gley soils, both of which have gleying beginning within 40 cm of the soil surface. The surface-water gley soils are characterized by impermeable subsurface horizons and their wetness results from the slow movement through the soil of water received from precipitation.

Groundwater gley soils, by contrast, are relatively permeable and are affected by the presence of groundwater within normal soil depth. Other soil groups which may require drainage before satisfactory conditions for crop growth are obtained include the peat soils. The gleyic or stagnogleyic subgroups of otherwise freely drained soils may also require measures to remove excess water from their lower horizons.

Experimental data obtained by the Soil Survey of England and Wales has demonstrated the link between gley morphology and the wetness of the soil. By the use of 'dip wells' at numerous sites throughout the country the average number of days the water-table lies within 40 or 70 cm of the surface in many of the more widespread soils is known. When the water-table lies between these limits for between four and seven months of the year, the soils are in danger of being compacted by traffic or poached by stock. It is concluded that where saturation occurs for more than 120 days (± 50 days) drainage measures should be worthwhile in all cases where the land is being used more intensively than for rough grazing (Robson and Thomasson 1977).

The aim of drainage schemes is to bring the soil back to 'field capacity' within 48 hours of being saturated by heavy rain. This can be accomplished by a number of techniques which control the water-table at a certain depth (about 50–60 cm) or which ensure that it falls to a predetermined depth within a short period of time, limiting the period of water-logging experienced in the rooting zone. The planning of a drainage scheme has to take into account the hydrology of the surrounding area and allowance must be made for sufficient hydraulic gradient to remove the excess water from the area. Open drains are excavated around fields to conduct water rapidly into the existing streams. Within the fields, tile drains are laid at depths of between 70 and 120 cm depending upon the depth to the water level in the surrounding open drains. The present practice is to place tile drains at intervals of 20 m (one chain) on the lowlands of England, but on fine-textured soils and in the wetter western areas a narrower spacing would be more effective. In Scotland and Wales a wider spacing of 25 to 40 metres is common (Thomasson 1975). Theoretically, it is possible to space drains at a wider interval if they are placed deeper, but the capacity of water to move laterally through the soil is limited by the pore-size distribution. Although deep cracks may occur in certain soils when dry, these cracks close when the clays expand upon wetting, thus limiting the permeability. Lateral movement is hindered by the absence of continuous channels and by a lack of hydraulic head to force the water into the tile drains. As both pore size and hydraulic conductivity change with depth it is difficult to obtain representative measurements for a scientific assessment of drain placing. In practice, the depth at which many drainage designs are installed is limited by the presence of an impermeable lower soil horizon (defined as an horizon which

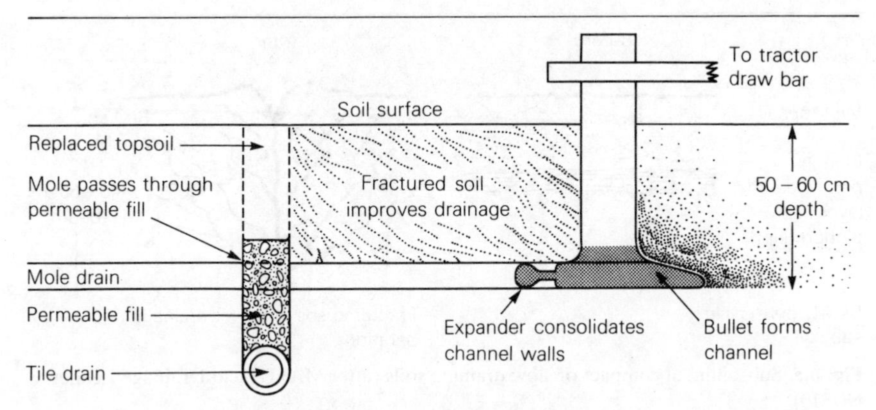

Fig. 6.4 Mole drainage of poorly drained soils (after MAFF Field Drainage Leaflet No. 11)

has a hydraulic conductivity of less than one tenth of the horizon above). Consequently, further measures are necessary to encourage water movement rapidly into the tile drains.

After tile drains have been laid at the bottom of a trench it is normal practice partially to back-fill with porous materials such as gravel or crushed stone, before replacing the topsoil (Fig. 6.4). This has advantages in that it acts as a filter, improves the speed of flow into the tile drain and acts as an effective link between the tile drain and the mole drains. Mole drains are produced by pulling a bullet-shaped object, followed by an expander which smooths the walls of the hole through the soil at a depth of between 50 and 60 cm as well as through the porous fill of the tile drain. This produces a 75–100 mm tunnel through the soil above and at right-angles to the tile drains. This operation can be done satisfactorily in soils with a uniform texture and more than 35 per cent clay. The best conditions occur when the topsoil is dry and the subsoil material is still plastic. Very stony soils and those which contain lenses of sand or silt are not suitable for this treatment as the mole drains collapse to produce local patches of poor drainage where water is retained.

A second practice which is used extensively is subsoiling (Fig. 6.5). Like mole draining, it has to be carried out during correct soil moisture conditions, and when both surface and subsoil conditions are sufficiently dry. A subsoiler consists of a tapered, square share mounted on a blade like the mole-plough, but its purpose is to shatter the impermeable subsoil and to produce many fissures. This can be done over tile drains and the results are similar to that of moling as the permeability of the subsoil is increased. On the Tedburn series in Devon widely spaced tile drains with subsoiling have proved an inexpensive and effective means of draining pasture land. Both moling and subsoiling eventually lose their effectiveness and the procedure has to be repeated. Although it is customary

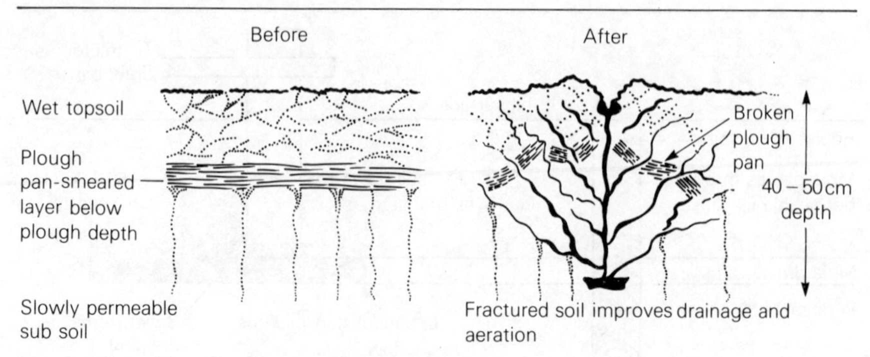

Fig. 6.5 Subsoiling of compact or slow draining soils (after MAFF Field Drainage Leaflet No. 10)

to write off the cost of drainage in ten years, its benefits often last very much longer. Some drainage systems of 150 years ago are still functioning but others suffer from silting and may require renewal in less than ten years.

The beneficial effects of land drainage are demonstrated in larger crop yields where root growth is not inhibited by saturated conditions. A rapid return to field capacity enables the passage of farm machinery and lessens the damage done by stock grazing the land. An effective drainage scheme gives the farmer greater flexibility for cultivation, extending the time in autumn and spring when the soil can be worked. Maps which show the distribution of grants for moling and subsoiling in England and Wales indirectly indicate a concentration on the soils formed from glacial till parent materials, particularly those derived from chalky boulder clay. Taking all drainage schemes recorded in 1972–3, there is a clear concentration in areas where higher value crops are grown, particularly Essex, Cambridgeshire and Lincolnshire and east Yorkshire. There is a scatter of schemes elsewhere throughout the country, with the marked exception of the areas underlain by chalk in south-east England, limestone districts of the High Peak and Northern Pennines, most of Cornwall and the South Wales Coalfield uplands (Fig. 6.6). Individual soil series are too numerous for mention but lists are given in *Soils and Field Drainage* (Thomasson 1975).

Restoration of derelict land for agriculture

Disturbance of the soil by man has increased in intensity throughout historical time as demonstrated in Chapter 1. During the last 50 years machinery has been developed which can literally 'move mountains'. Mining operations for coal, ironstone, gravels and a variety of other minerals leaves in its wake highly disturbed ground. The restoration of

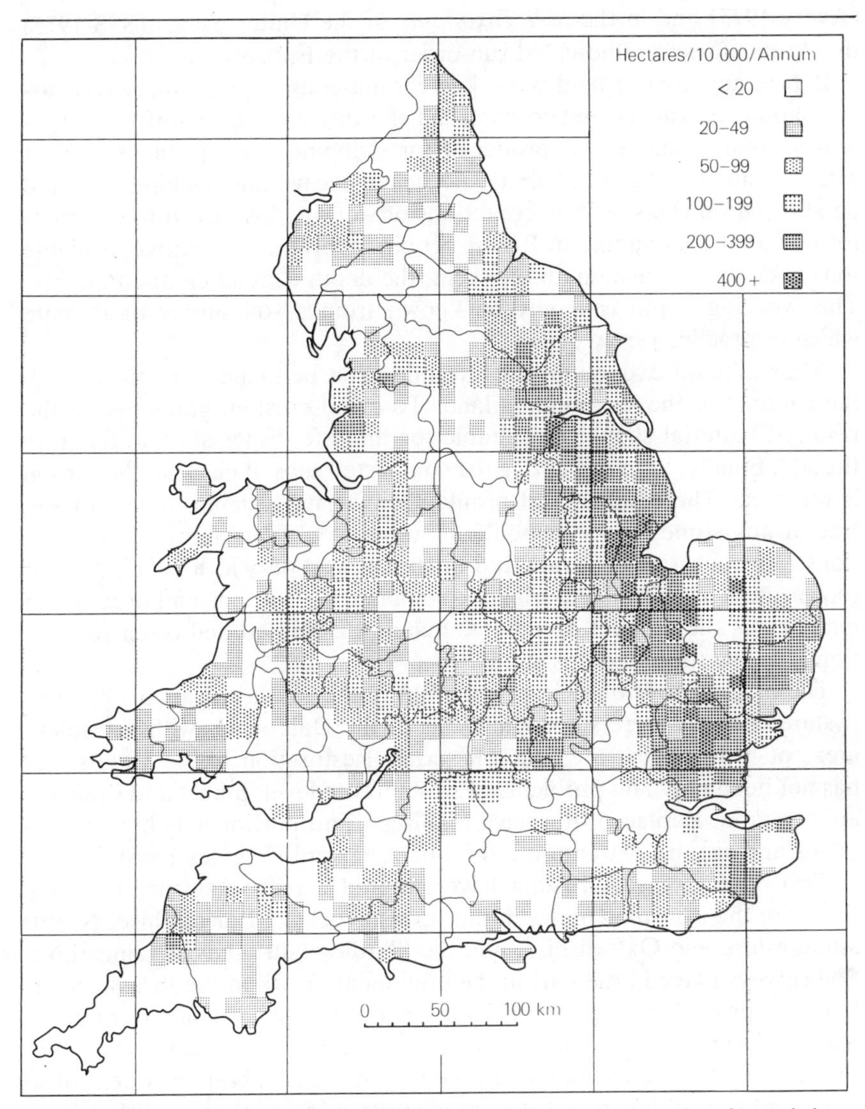

Fig. 6.6 Drainage activity as hectares/1000 hectares, 1972–3; computer plot of recorded schemes (after Thomasson 1975)

this derelict land can be greatly assisted if there is an understanding of the processes of soil formation which have to begin again on the newly formed 'parent material' remaining after disturbance. As areas of disturbed ground do not have a normal soil profile, they are usually distinguished on soil maps by a special category in the legend and recent soil classifications have included categories for disturbed land. In Britain 'man-made raw soils' is one of the groups of the Terrestrial Raw Soils

(Avery 1973) and in the *Soil Taxonomy* of the United States (SSS 1975) the Arents form an undivided sub-order of the Entisols.

Before any area of land is worked for minerals, a preliminary soil survey should be carried out to ascertain the distribution of soil types present so that plans can be produced for stripping the soil and storing it safely. Until this has been done, no heavy excavating machinery should be allowed on the site. The conditions of working inserted into contracts for opencast coal mining in Britain require operators to remove available topsoil down to a maximum of 30 cm, the depth depending upon quality. This working depth can only be known from a soil survey of the site which in practice rarely occurs.

After mineral extraction, the site must first be made to conform with the contour of the surrounding land. To avoid erosion, gradients on the restored material should be suitable for the safe dispersal of water from the site. Finally, after disrupting the compacted subsoil material the topsoil is replaced. The restored soil profile should have a depth of 1 m of soil free of any stone, rocks or boulders which would interfere with cultivations. Any substrate encountered in the excavation which is suitable for replacement as subsoil material can be placed on one side and used but in many cases the amount of suitable material is limited and often restoration is below the standards stipulated.

The original soils of widespread areas in the southern part of the Derbyshire coalfield were stagnogley soils of the Dale series with subsidiary areas of brown soils, the Seacroft series. Restoration of the Dale series has not been easy and problems of structural redevelopment and drainage are common. Replaced sites have required disproportionately high investments and require extremely careful management (Bridges 1966).

Two other areas of Britain have suffered extensive opencast mining these are the ironstone mines in the Jurassic rocks of Lincolnshire, Northamptonshire and Oxfordshire, and the Shirdley Hill sands of Lancashire. Soil surveys played little part in the preliminary work in the Jurassic areas but were proved to be useful in Lancashire where sand required for glass-making formed the eluvial horizon of gley podzol soils developed over the Shirdley Hill sands (Hall and Folland 1967). As operations were not so great and subsoil layers not disturbed, restoration of these soils has been relatively successful.

Given adequate information regarding distribution and behaviour of soils, planners can make soundly based proposals for redevelopment of land. In upland areas the availability of soil and land use capability maps can help to resolve conflicts of use. In many upland districts, water supply, farming, forestry and leisure activities compete for the same ground area. Greater use and understanding of soil maps by planners would be advantageous.

Use of soil survey data in farm planning

It has been demonstrated that information gathered in soil surveys is of considerable value in many branches of agriculture. The soil maps produced by the soil surveyor are most useful in planning the cropping and stocking programme of a farm and can be used to show whether the land can sustain the required production with minimal risks. In a wider perspective it is possible to base planning decisions for a whole community or indeed the whole country on the factual basis provided by soil geography.

The system of land use capability currently favoured in Britain has been developed from the system of Klingebiel and Montgomery (1961) and by Bibby and Mackney (1969) (Table 6.1). The system relies upon the physical characteristics of the land which have to be interpreted in the light of the current economic situation. It links features of soil, site and climate in a workable land classification for farming, avoiding the complications introduced by a rapidly changing economic situation. Assessments of land capability are based upon a knowledge of the behaviour of soil types assuming a moderately high standard of management. In the system there are seven Classes, of which Classes 1 to 4 include land suitable for arable crops; Classes 5 and 6 are most appropriately used for grazing and forestry and Class 7 is not suitable for cropping but is best left as amenity or wildlife reserves. Land use capability subclasses identify the particular factor or factors which impose limitations on land use, these are wetness (w), soil (s), gradient and pattern (g), climate (c) and liability to erosion (e). These subclasses embrace landscape units which require specific management and improvement techniques.

An example of the use of this approach to farm planning is given by Hooper (1974) for a farm of 498 ha with additional 34 ha of rough grazings. Under a revised management plan it was estimated the farm would carry 275 dairy cows and followers, 67 beef cattle and 200 sheep. The map of land use capability (Fig. 6.7), derived from the soil map, was modified to give each field its dominant class; the detailed map of classes and subclasses may be used for re-location of field boundaries if necessary. After consideration of the limitations and potentialities of the farm, cropping was arranged in four blocks (Fig. 6.8) and the rotations proposed were as shown in Table 6.4.

In this example, land use capability classification is seen to be a valuable aid to farm planning and reorganization. Should economic conditions change then a reassessment must take place and a new farm plan evolved using the same basic physical constraints but placing them in a new economic framework. The land use capability system has the advantage of presenting inter-connected soil, site and management knowledge in a logically organized framework which makes possible a quantitative evaluation of proposed changes in agricultural systems. The use of land

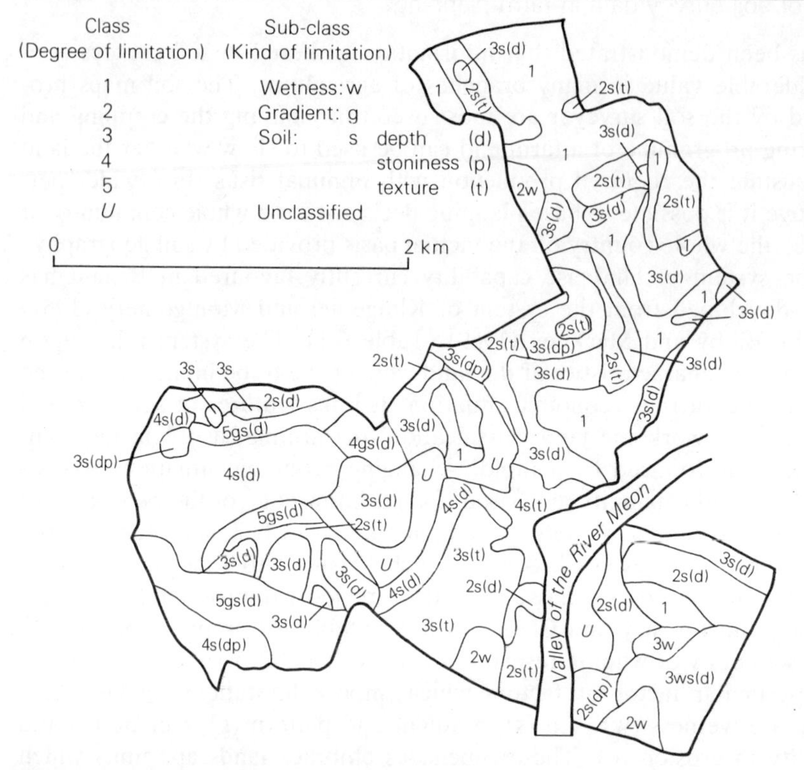

Fig. 6.7 Land-use capability classes and sub-classes

evaluation techniques in a wider planning situation will be considered in the following section.

6.4 METHODOLOGY OF LAND EVALUATION FOR AGRICULTURE

From the preceding sections of this chapter it should be clear that agriculture both in a planning and a management sense can benefit from some knowledge of soil conditions. This has been realized since the earliest days of the Neolithic period, but the modern agricultural planner has to deal not only with a tremendous amount of information about plant–soil relationships, but also with sophisticated economic analysis. In the late 1960s the world Food and Agriculture Organization (FAO) realized the need for some kind of methodological framework to aid agricultural planning. Another problem was that many countries had or were in the process of developing their own land evaluation methods and FAO feared that major problems of information exchange would result. In 1970 a working group was established to develop a Framework for Land Evaluation and the result is a publication with this title. For simplification this

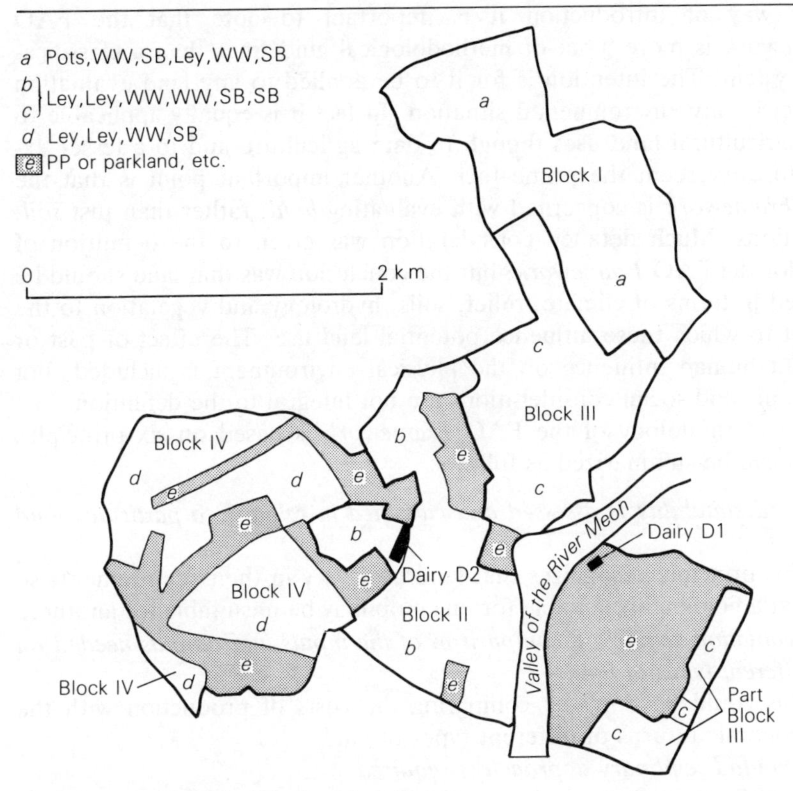

Fig. 6.8 Reallocated farm blocks and rotations

Table 6.4 Reallocation of farm units, rotations and areas (Hooper 1974)

Block	Rotations	Area (ha)
I	Potatoes, winter wheat, spring barley, ley winter wheat, spring barley.	123
II	Ley, ley, winter wheat, winter wheat, spring barley, spring barley undersown.	96
III	Ley, ley, winter wheat, winter wheat, spring barley, spring barley undersown.	116
IV	Ley, ley, winter wheat, spring barley undersown	106
	Permanent pasture	57
	Rough grazing	34

document will be referred to as the FAO *Framework*. Readers are encouraged to examine the full document (FAO 1976) as well as some of the preliminary publications (Brinkman and Smyth 1973; FAO 1975). Outlines of the FAO *Framework* are given by Vink (1975) and Davidson (1980).

By way of introduction it is important to note that the FAO *Framework* is more a set of methodological guidelines than a classification system. The intention is for it to be applied to any land evaluation project in any environmental situation. In fact it is equally applicable to non-agricultural land uses though to date agriculture and to a lesser extent forestry seem the prime foci. Another important point is that the FAO*Framework* is concerned with evaluating *land*, rather than just *soil*, conditions. Much detailed consideration was given to the definition of land for the FAO *Framework*, but the conclusion was that land should be defined in terms of climate, relief, soils, hydrology and vegetation to the extent to which these influence potential land use. The effect of past or present human influence on the physical environment is included, but economic and social considerations are not integral to the definition.

The methodology of the FAO *Framework* is based on six principles which can be summarized as follows:

1 *Land suitability is assessed and classified in relation to particular land uses.*
  This principle recognizes that land uses vary in their requirements so that a field highly suitable for one crop may be unsuitable for another.
2 *Evaluation requires a comparison of the inputs and outputs needed on different types of land.*
  This could be done by comparing the costs of production with the economic returns of different types of land.
3 *A multidisciplinary approach is required.*
  Contributions from such specialists as crop ecologists, agronomists, pedologists, climatologists, economists and sociologists are necessary in order to make a comprehensive and sound assessment of land suitability for a specified use.
4 *The evaluation is made with careful reference to the physical, economic and social context of the area under investigation.*
  It is fairly obvious that any land use proposals have to be realistic for an area. It is important to take into account such factors as cost of available labour and skills of the labour force.
5 *Suitability refers to use on a sustained basis.*
  The proposed use of land must not result in its degradation through processes such as wind erosion, water erosion or salinization.
6 *Different kinds of land use are compared on a simple economic basis.*
  This means that the suitability for each use is assessed by comparing the value of the goods produced to the cost of production.

It should be evident that there is a certain degree of overlap between these principles. Fundamental to the approach is the identification of relevant land uses in the first instance in order that the evaluation procedure is executed with specific reference to these land uses. This means that the land use requirements of the various land uses have to be established and then the actual characteristics of land mapping units have to be assessed in terms of how well they can provide optimum conditions. This comparison of land mapping units with land use requirements is known as matching. A land evaluation procedure is represented in Fig. 6.9. A brief explanation of land qualities is necessary. According to the FAO

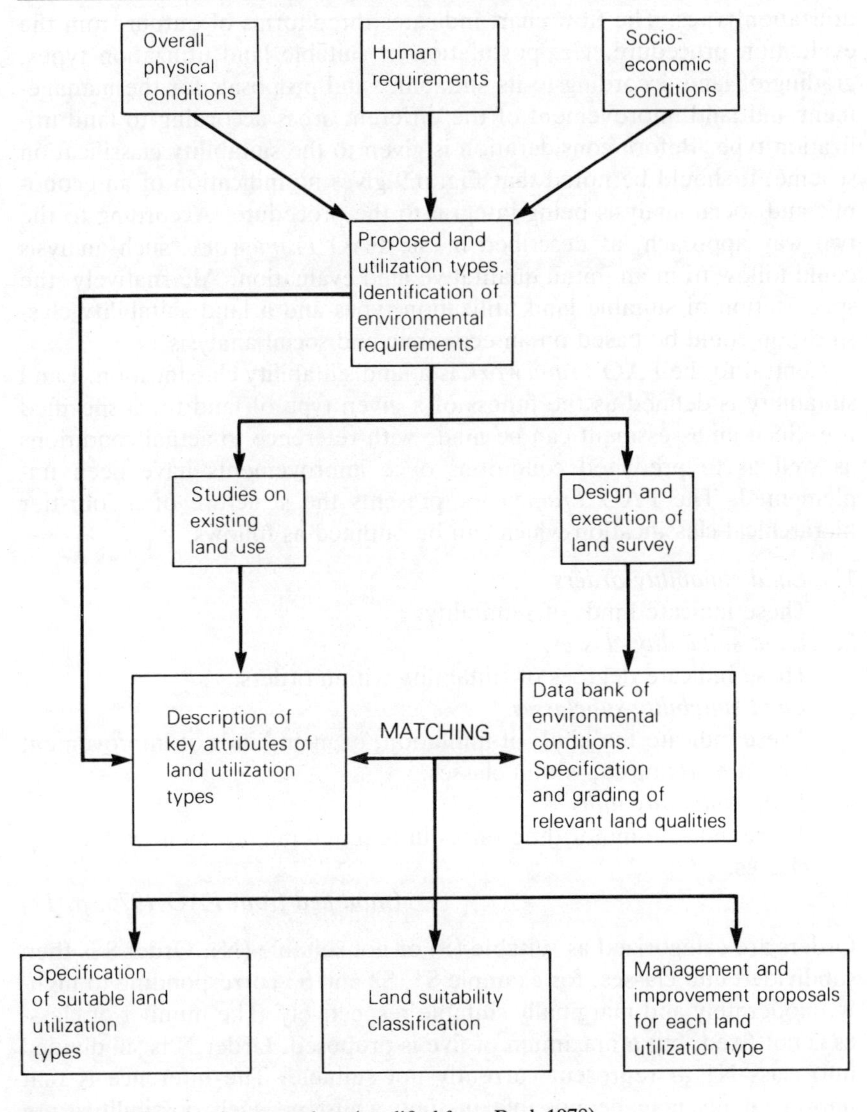

Fig. 6.9 Land evaluation procedure (modified from Beek 1978)

Framework (1976, p. 12): 'a land quality is a complex attribute of land which acts in a distinct manner in its influence on the suitability of land for a specific kind of use'. Examples of land qualities are moisture availability, and resistance to erosion. These qualities result from a number of specific properties (called land characteristics) such as soil texture, angle of slope, rainfall, etc. Thus Fig. 6.9 shows how the process of matching involves comparison between the land qualities of specific areas (land mapping units) and the environmental requirements of particular land utilization types. The flow chart indicates three forms of output from the evaluation procedure, *viz.* postulation of suitable land utilization types, grading of land according to its suitability and proposals for the management and land improvement of the different areas according to land utilization type. Before consideration is given to the suitability classification scheme, it should be noted that Fig. 6.9 gives no indication of an economic and social analysis being integral to the procedure. According to the two way approach, as described in the FAO *Framework*, such analysis could follow from an initial qualitative land evaluation. Alternatively, the specification of suitable land utilization types and a land suitability classification could be based on an economic and social analysis.

Central to the FAO *Framework* is a land suitability classification. Land suitability is defined as the fitness of a given type of land for a specified use. Such an assessment can be made with reference to actual conditions as well as to predicted conditions once improvements have been implemented. The FAO *Framework* presents the structure of a four-tier hierarchical classification which can be outlined as follows:

1. *Land suitability orders.*
   These indicate kinds of suitability.
2. *Land suitability classes.*
   These indicate degrees of suitability within orders.
3. *Land suitability subclasses.*
   These indicate the kinds of limitation, or main kinds of improvement measures required, within classes.
4. *Land suitability units.*
   These indicate minor differences in required management within subclasses.

(*modified from FAO 1976, p. 17*)

Orders are categorized as suitable (S) or not suitable (N). Order S is then subdivided into classes, for example S1, S2 and S3 corresponding to highly, moderately and marginally suitable respectively. The number of classes is not fixed, but a maximum of five is proposed. Order N is subdivided into class N1 to represent currently not suitable. The inference is that improvements may be possible in time whilst no such possibilities are envisaged with class N2 which represents land permanently not suitable.

Sub-classes indicate specific limitations so that a land unit labelled S2m is assessed as being moderately suitable, but possessing a moisture deficiency. Other illustrative codes are o for poor oxygen availability in root zone, n for low nutrient availability and e for poor resistance to erosion. The lowest tier in the classification structure is land suitability units. These are indicated by numbers such as S2m-1, S2m-2, etc. and they differ in production characteristics or in minor aspects of their management requirements. They are applicable to land evaluation at the individual farm level.

As already stated, reference to the complete document is necessary to obtain full comprehension of the FAO *Framework*. Its ambitious aim is to be applicable to any area of the world at any scale. It will be a decade before sufficient studies have been executed in order to decide whether the FAO *Framework* has proved to be useful. Nevertheless, there are a few studies which already indicate the validity of the approach and the remaining part of this chapter focusses attention on the applicability of the FAO *Framework* at three different scales.

Evaluation at the local scale

The selected example designed to illustrate land evaluation for agriculture at the local scale is based on the work of Veldkamp (1979). He carried out research in south-western Nigeria with the specific aim of determining the agricultural land use potential of hydromorphic soils. He notes that such areas are currently avoided for agriculture and he poses the question of the agricultural potential of such lands. Rice is the obvious crop for hydromorphic land, but other relevant crops are maize, yam, cassava, cocoyam, sweet potato, grain, legumes, vegetables and plantain/banana. Veldkamp states that the major constraints on the use of hydromorphic soils are the difficulty of clearing the vegetation, the lack of knowledge about water management and conservation and the problems posed by local farmers lacking the knowledge of rice cultivation.

The study area is a large one covering the forest zone of Oyo, Ogun and Ondo States in south-western Nigeria, but the project is based on the detailed examination of land conditions at the individual valley scale. Emphasis is placed on determining the performance of major food crops on different parts of soil toposequences. The results are then extrapolated on a three-dimensional basis using the location of land units on the soil toposequences. Before this can be demonstrated, consideration must be given to the important land qualities as described by Veldkamp.

The concept of land quality as used in the FAO *Framework* has already been introduced. In summary, a land quality is an attribute of land resultant upon a number of land characteristics and such an attribute influences the suitability of an area for a particular land use. The land qualities as used by Veldkamp can be summarized as follows:

A General land qualities
- 1 Availability of water
- 2 Availability of oxygen
- 3 Availability of nutrients (fertility)

B Additional land qualities
- 1 Probability of occurrence of iron-toxicity (in rice)
- 2 Probability of occurrence of soil erosion
- 3 Difficulty of land preparation and harvesting of root crops
- 4 Impediment to root development

(*from Veldkamp 1979*)

As an example, the availability of water was graded on the basis of four properties, *viz.* moisture situation of the root zone throughout a specific season, ground-water class in that season, available water-holding capacity of the soil in the root zone, and the availability of water in the root zone by capillary rise from the ground-water. Reference must be made to the original publication (Veldkamp 1979) to obtain detailed descriptions of how these properties were measured. Suffice it to say that field observations combined with laboratory analysis of samples from sites along his transect lines permitted the grading of sites and thus soil units according to availability of water.

Any land evaluation project has to be conducted with reference to defined management and technology levels. Veldkamp (1979), in his Nigerian study, considers management at a traditional as well as at an improved level. The latter implies the better care of crops and use of modern technology as well as the implementation of various minor land improvements. He excludes capital intensive farming systems and major land improvements which would involve substantial investment. The only exception is land improvement for the cultivation of rice which requires bunding and the partial levelling of fields. Certain management practices can have an effect on land qualities. Fairly obviously the application of fertilizers will influence the availability of nutrients. Thus Veldkamp (1979) has to grade land qualities according to management levels of particular crops. As another example, the techniques of water management influence three land qualities, *viz.* the availability of water, the availability of oxygen and the probability of occurrence of iron-toxicity.

Consideration of land qualities, crops and management levels then allows Veldkamp to assess ecological crop suitability. This is achieved by predicting individual crop yields on a seasonal basis according to ground-water class and assessing suitability classes according to crop requirements. In the first instance land qualities and data about seasonal ground-water levels are matched with crop requirements in order to produce a 'calculated ecological suitability of a crop'. This is done on a seasonal basis and for a specified management level. The results are then compared with the expected yield figures and modifications are made as required. Data on expected yields were obtained by Veldkamp (1979) through growing and harvesting crops on an experimental basis for five years on eight toposequences. He grew such crops as cowpea, pigeon pea, soybean, rice, maize, cassava, yam, sweet potato, banana, sweet pepper, okra, plantain, tomato and celosia. The results were incorporated into the qualitative process of determining ecological suitability classes. Crops were grouped according to ecological requirements and then graded into one of four suitability classes (high, moderate, restricted and low) for each of the land qualities. This permitted the overall assessment of ecological crop suitability by taking into account the nature of land qualities, the suitabilities of land qualities for different crops and the expected yield data.

Such a crop ecological assessment was followed by consideration of economic and social factors. To aid such an analysis, Veldkamp calculated an ecological crop suitability index which is conditioned by the suitability and importance of a crop, the number of possible cultivation seasons, the dietary significance of the crop, the length of the growth cycle of the crop and how the crop relates to other suitable crops. The actual method of calculation is rather involved and reference must be made to Veldkamp (1979, ch. 9) should the reader wish details. In addition to crop suitability, consideration has to be given to the accessibility of the proposed area for development. Vegetation clearance is essential and particular problems can be posed by the use of machinery on poorly drained land. The consequence is also a major change in land use management from a traditional shifting system to the more permanent cultivation of a specific area. The ultimate crop choice will also be influenced by such factors as the dependability of food supply, crop preferences for food, traditions, the cash return on cropping and the labour requirements of crops.

The nature of Veldkamp's results can be illustrated by considering one of his toposequences. A typical hillslope transect is represented in Figure 6.10 for his Westbank area which is located on the land belonging to the International Institute of Tropical Agriculture near Ibadan. The slope is divided into management groups labelled I, IV, VI, VIII and X on the basis of topographic position and ground-water regime. Fairly obviously hydromorphic soils are most distinct on the valley floor and the influence of a high water-table decreases upslope. These management groups are further subdivided into management units, *viz.* a, b, k, l, o, p, s, u, v, w, x, and y. Table 6.5 shows which crops can be considered to be highly or moderately suited as a result of the ecological suitability determination. As can be seen, this information is provided for each of the twelve man-

Table 6.5 Recommendations for highly and moderately suited crops for the four seasons according to management units as shown on Fig. 6.10 (after Veldkamp 1979)

primary topographic position	valley bottom				lower slope			upper slope				
secondary topographic position	central		side		lower part		upper part					
	I		IV		VI		VIII	X				
management unit	a	b	k	l	o	p	s	u	v	w	x	y
early first season												
highly suited crops	–	–	–	–	–	–	–	–	–	–	cas	–
moderately suited crops	–	–	–	–	cas	cas	cas pip	cas pip	cas	cas pip	pip	pip
major first season												
highly suited crops	ccy	–	–	–	ric ccy	–	–	–	–	–	mai yam cas spo cop soy	–
moderately suited crops	ric soy	ric	ric mai spo ccy cop soy	ric	mai yam cas spo cop soy	ric mai yam cas spo ccy cop soy	cas spo cop soy pip	cas pip	cas	mai yam cas spo cop soy pip	pip	pip
second season												
highly suited crops	ric ccy	–	–	–	ccy	ric	–	–	–	cop	yam cas cop soy	–

moderately suited crops	–	ric	ric ccy	ric	ric yam cas spo soy	yam cas spo ccy	cas cop soy pip	cas cop soy pip	cas	yam cas spo soy pip	mai spo pip	pip
dry season												
highly suited crops	–	–	–	–	–	–	–	–	–	–	cas	–
moderately suited crops	–	–	–	–	cas	cas	cas pip	cas pip	cas	cas pip	pip	pip

Explanation of abbreviations: ric = rice, mai = maize, yam = yam, cas = cassava, spo = sweet potato, ccy = cocoyam, cop = cowpea, soy = soybean, pip = pigeon pea, tom = tomato, okr = okra, cel = celosia, spe = sweet pepper, pla= plantain, ban = banana.

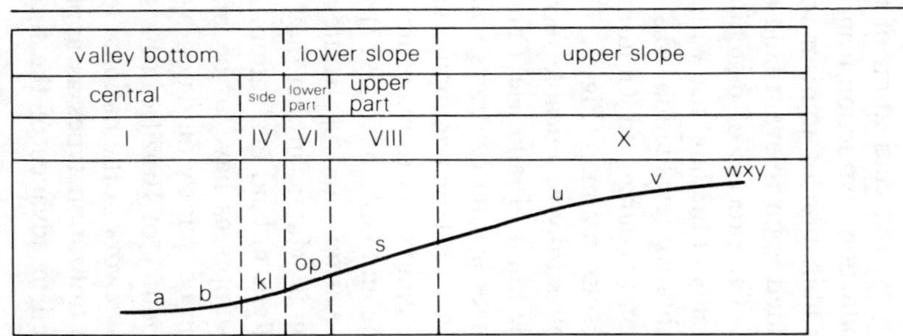

Fig. 6.10 Idealized toposequence for the Westbank area at the International Institute of Tropical Agriculture, near Ibadan, Nigeria. (Lower case letters refer to management units, see Table 6.5) (Veldkamp 1979)

agement units which occur in the toposequence as well as for each of the four seasons. It should also be noted that the highest ecological crop suitability for the two management levels is given in Table 6.5.

This outline of Veldkamp's study demonstrates the applicability of the FAO *Framework* to land evaluation at a local scale in a tropical environment. A strongly ecological approach is adopted whereby the suitabilities of different crops are assessed with reference to local conditions. The work is based on field and laboratory examination of soils coupled with field crop experiments and analysis of local economic and social conditions. As a result Veldkamp's study can be viewed as a model for other detailed land evaluation projects.

Evaluation at the regional and national scale

A study by Young and Goldsmith (1977) demonstrates the applicability of the FAO *Framework* at the regional level. They carried out a survey of the Dedza area of central Malawi, a region of 1935 km^2. A conventional soil survey, supported by extensive use of aerial photographs, permitted the identification of seven soil landscapes. These landscapes were evaluated in terms of six major kinds of land use, *viz.* arable farming for annual crops, the cultivation of perennial (tree and shrub) crops, livestock production, extraction from natural woodlands, forestry plantations and a combination of tourism with production. Young and Goldsmith then decided upon relevant land qualities for each of the major kinds of land use. The process of matching was then possible by comparing the requirements of the land use with the land qualities of the mapping units. This qualitative evaluation was followed by a quantitative economic analysis and Young and Goldsmith (1977) were able to present their overall results in a map (Fig. 6.11). Thus they are able to state that 'these are the physically practicable, economically viable, and environmentally acceptable kinds of land use' (Young and Goldsmith 1977).

This investigation contrasts with the Veldkamp (1979) one in Nigeria by dealing with larger mapping units. Veldkamp does incorporate some economic analysis in terms of net labour productivity for different crops, but Young and Goldsmith (1977) attempt to quantify the financial returns from the various land use strategies for each soil landscape unit. Their conclusion is salutary in that they recommend caution in the use of economic analysis in land evaluation. The prime reasons for this are that too many assumptions have to be made in the economic analysis and any results can be quickly outdated by price changes.

The country of Brazil can be selected to demonstrate the use of the FAO *Framework* at the national level. In fact the concepts of land quality and land utilization types as applied to land evaluation were in use in Brazil well in advance of the formal FAO publication. This is because

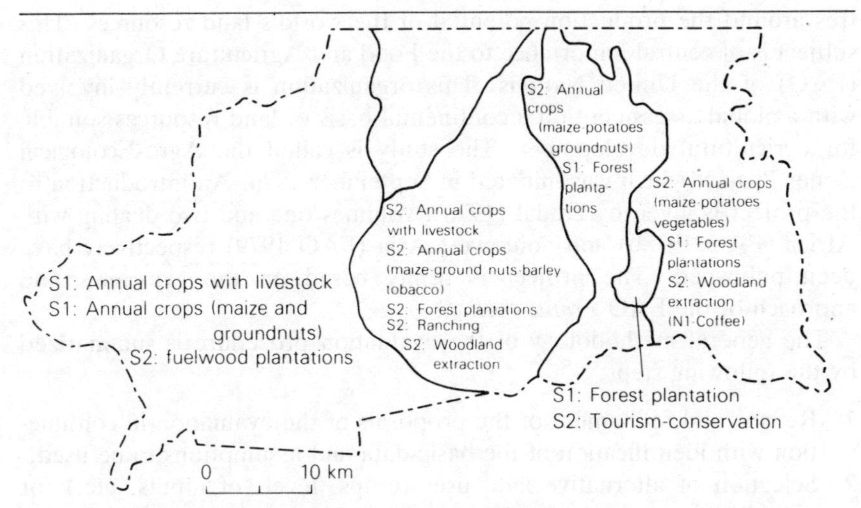

Fig. 6.11 Proposed land use alternatives for the soil landscape regions of the Dedza project area, Malawi (after Young and Goldsmith 1977)

scientists such as J. Bennema and K. J. Beek were involved with land evaluation projects in Brazil and later played key roles in the development of the FAO *Framework*. The research in Brazil is taken as a case study in the FAO *Framework* and a more detailed résumé is provided by Beek (1978). In the Brazilian system, four land suitability classes are distinguished for specific land utilization types. Suitability is determined according to five land qualities, *viz.* susceptibility to erosion, impediments restricting the use of agricultural implements (mechanization), and deficiencies of natural fertility, water and oxygen. Individual land qualities are graded according to the degree of limitation posed to land utilization types and account is also taken of the feasibility of improving land qualities. Conversion tables are used to relate land qualities and their improvement feasibility to land suitability classes. Such tables enable the grading of soils according to their suitability for each land utilization type. Beek (1978) stresses that the main objective in Brazil has been to interpret soil survey information for reconnaissance-type studies for generalized types of land utilization. He visualizes the need for more detailed studies in land evaluation, research of the type demonstrated by Young and Goldsmith (1977) and Veldkamp (1979).

Evaluation at the international scale

Population and land resources for agricultural development is a theme of ever increasing global significance. It is a topic discussed in detail in Chapter 8. As is made clear in that chapter, much scientific debate centres around the production potential of the world's land resources. This subject is of central importance to the Food and Agriculture Organization (FAO) of the United Nations. This organization is currently involved with a global assessment on a continental basis of land resources suitable for agricultural development. The study is called the Agro-Ecological Zones Project which was initiated in September 1976. An introduction to the project is given by Dudal (1978); volumes one and two dealing with Africa (FAO 1978b) and Southwest Asia (FAO 1979) respectively have been published. The project is firmly based on the principles and approach of the FAO *Framework*.

The general methodology of the evaluation procedure is summarized by the following steps:

1 Review and refinement of the proposals of the evaluation in conjunction with identification of the basic data and assumptions to be used;
2 Selection of alternative land uses (crops, levels of inputs, etc.) for consideration;
3 Determination of the climatic and soil requirements of the selected alternative land uses;
4 Compilation of an inventory of the land (climate and soil) and mapping units (agro-ecological zones) with respect to 3;
5 Matching of the requirements 3 with the land inventory 4 and calculation of anticipated production potential in the different agro-ecological zones recognized;
6 Estimation of production costs, and identification of the various suitability classes to be employed and their differentiating parameters;
7 Classification of the land into various suitability classes for the selected alternative land uses, and presentation of results.

(*from FAO 1978b*)

The first step involves an overall assessment of the project with particular regard to basic assumptions and availability of information. As an example the Project excluded irrigation potential due to the lack of consistent water resource data, and instead focusses on rainfed potential only. It was also decided to exclude perennial tree crops such as rubber, tea, coffee, oil-palm and other fruit crops though these crops are of economic significance on a global scale. For the second step, a decision had to be made on the particular crops to be included in the assessment. On the basis of area, the world's 20 most important crops in decreasing order of importance are *wheat, rice, maize, barley, pearl millet, sorghum, soybean, cotton,* oat, *phaseolus bean, white potato*, groundnut, rye, *sweet potato, sugarcane, cassava*, pea, chickpea, grape and rapeseed. Some of these crops only occur in specific regions of the world and thus consideration was limited to the 12 crops in italics which are considered to be widely distributed. Further discussion led to the exclusion of sugarcane since evaluation for this crop ought to include factory management requirements. The resultant 11 crops are wheat, rice, maize, pearl millet, sorghum, white potato, cassava, phaseolus bean, soybean and cotton. These were further considered at two input levels – a low one similar to traditional systems of shifting cultivation or bush fallow rotation, and a high one involving mechanical cultivation under capital intensive management practices.

The third step was probably the most difficult and important in that the climatic and soil requirements of the 11 crops had to be determined. For climatic requirements, particular attention was paid to rainfall, soil moisture and temperature. Importance was given to calculating the growing period when water availability and temperature conditions permit crop growth. In essence this is a season when rainfall is in excess of evapotranspiration and temperatures are above a critical threshold, for example 6.5°C for winter wheat. Consideration was also given to the photosynthetic requirements of the different crops. Five photosynthetic pathways are defined and the crops are assigned to these different groups. In terms of soil requirements, a list was compiled of all properties which were considered to have an effect on the success of the crops. Properties such as soil depth, texture, salinity, etc. were then subdivided in order to define optimum soil conditions.

Following on from the identification of climatic and soil requirements was step 4, whereby data on climate and soil characteristics were collected. For Africa climatic data were obtained from 730 stations; lengths of growing periods were calculated and zones of similar length were delineated by constructing isolines at intervals of 30 days on 1 : 5 000 000 scale maps. The resultant zones were further considered in terms of average values of day and night-time temperature and radiation and the net result was the classification of Africa into eight major climatic divisions. The soil inventory was based on the 1 : 5 000 000 FAO/UNESCO Soil Map of the World using the three sheets covering Africa. Superimposition of the main climatic divisions and the isolines delineating the various growing periods resulted in the identification of 'agro-ecological zones'. These are areas with similar climatic and soil conditions. It must not be thought that these zones have uniform soil conditions since the FAO/UNESCO Soil Map presents the distribution of soil mapping units, each of which may consist of up to eight individual soils. Measurements were made of the areas of different mapping units according to length of growing season. By knowing the proportion of different soils within each soil mapping unit, it was then possible to calculate for Africa the distribution of mapping units according to lengths of growing periods for all major climatic divisions.

The determination of climatic and soil requirements in step 3 and the production of an agro-ecological zones inventory in step 4 then permit the matching process to operate in step 5. By this process the requirements of the specific crops are matched with the actual nature of the agro-ecological zones in order to produce a suitability assessment. Matching of the crop climatic requirements with the climatic inventory is done in a quantitative manner by first determining if the crop temperature requirements can be met; second, ascertaining if the length of growing period is sufficient; and third, calculating potential net biomass and yield of crops. For the matching of soil requirements of crops with the nature of specific soils, a more qualitative approach is necessary because of the lack of models and data relating soils classified according to the FAO/UNESCO legend and crop yields. On the basis of experience, and personal judge-

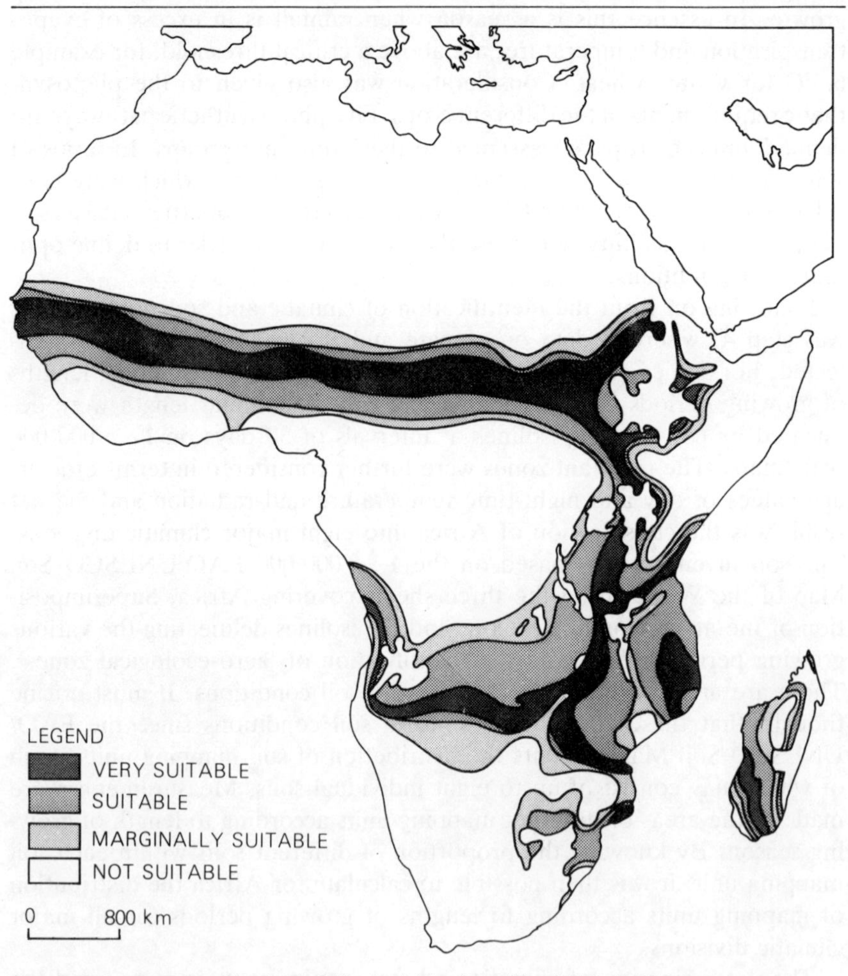

Fig. 6.12 Assessment of land suitability in Africa for millet (FAO 1978b)

ment, the FAO team for Africa were able to grade soils according to suitability for crops. Thus a soil which imposed no constraints on a specific crop would have been classed as S1 whilst a soil which imposed some limitations would have been S2.

The final steps in the methodology of the Agro-Ecological Zones Project are the estimation of production costs and the definition of suitability classes (step 6) and the assignment of agro-ecological zones to these suitability classes for the eleven crops at the two input levels (step 7). In step 7 the anticipated yields from step 5 may be reduced in order to take into account rainfall variability, excess moisture or losses due to pests, diseases and weeds. The results are described as agro-climatically attainable

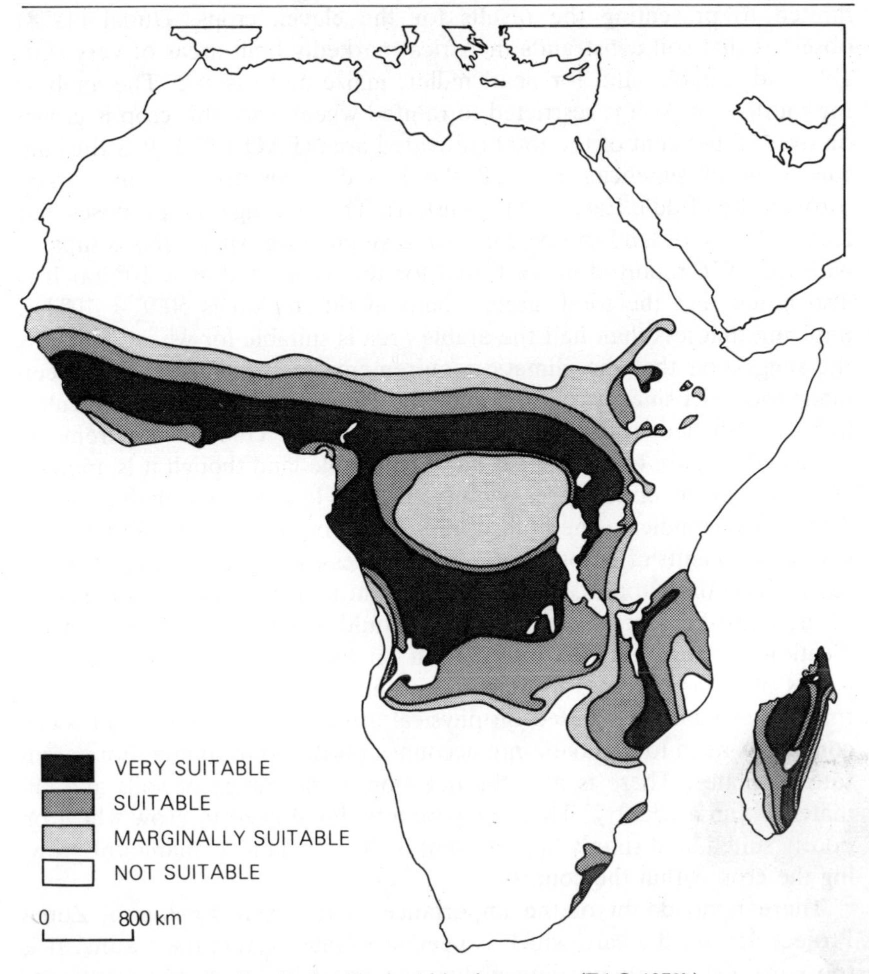

Fig. 6.13 Assessment of land suitability in Africa for cassava (FAO 1978b)

yields. At the end of the land evaluation process, areas are categorized as being very suitable, suitable, marginally suitable or not suitable for each crop at the two input levels.

The FAO (1978b) report for Africa contains a great deal of statistical material illustrating the results of the evaluation exercise for the eleven crops. Estimates of areas in the four suitability classes are made on an individual crop basis for each major climatic division according to length of growing period and input level. Anticipated yields from the suitability classes are also given. Generalized maps of agro-climatic suitability accompany the tabulated results and Figs. 6.12 and 6.13 exemplify contrasting results for millet and cassava respectively.

The FAO (1978b) report for Africa offers no general conclusions and is limited to presenting the results for the eleven crops. Dudal (1978) observes that soil constraints in Africa markedly limit areas of very suitable and suitable land for pearl millet, maize and cassava. The analysis for south-west Asia is restricted to rainfed wheat since this crop is grown on over 75 per cent of the total cultivated area (FAO 1979). It is thus the sole crop of significance in all the included countries (from Turkey through the Middle East to Afghanistan). The investigation proposes that 23.5×10^6 ha of land can be considered suitable for wheat; this compares with the 1976 reported harvest area for that crop of 19.8×10^6 ha. It is also noted that the total 'arable' land in this region is 50.9×10^6 ha, implying that less than half the arable area is suitable for wheat. There is the suggestion that the climatic requirements for wheat may have been made too strict since it is noted that wheat is grown, albeit with difficulty, in areas labelled unsuitable. A relaxation of the climatic requirements produced a figure of 46.0×10^6 ha of cultivable land though it is appreciated that much of this area would produce low and unreliable yields. These results indicate the crucial importance of defining the climatic and soil requirements of crops. Dudal (1978) stresses the need for much more research on these topics. It is thus important to appreciate that the results of the Agro-Ecological Zones Project should be viewed as liable to modification as more detailed information on the climatic and soil requirements of crops becomes available. Another broad reservation to note is that the evaluation is based on physical attributes. Economic and social conditions need to be taken into account in order to postulate more accurate estimates. There is also the question of the range of soils and climates within a country. There may be a national need to grow wheat on poorly suited land simply because that is the only land available for growing the crop within the country.

There is no doubt of the importance of the Agro-Ecological Zones Project despite the care which is needed in interpreting the results. It is too early yet to attempt any global appraisal since only the results for Africa and Southwestern Asia are available at present.

Without doubt this chapter has demonstrated the extent to which soil survey data can aid agricultural planning and management. One focus for current research is the relationship between crop yields and individual soil series. The effect of cultivation on soil properties such as structure and resistance to erosion is also attracting much research effort. The approach to land evaluation as advocated in the FAO *Framework* stresses that soil and climatic characteristics need to be assessed with reference to specific land uses and within the context of defined economic, technological and social conditions. The approach requires the identification of crop growing requirements and the frequent lack of quantitative information on this point can pose difficulties to the use of the FAO *Framework*. More research is needed to determine the optimal growing requirements of crops. The result should be that soils can be evaluated in greater detail in terms of their agricultural use and potential.

7 NON-AGRICULTURAL USES OF SOIL SURVEYS

M. G. Jarvis

INTRODUCTION

Soil surveys are inventories of the soil resources of an area; kinds of soil are identified, their relationships investigated and their extent described on maps and in accompanying reports. Historically, soil surveys have been made primarily to guide agricultural land use and management, but it is now widely recognized that since all uses of land must in some way manipulate or be related to soil, knowledge of the soils in an area of interest, their location and properties is a fundamental prerequisite for all kinds of land use planning and management. Consequently the non-agricultural use of soil surveys and data has increased markedly in recent decades and this trend is likely to continue. Kellogg reports that in 1972 the total benefits derived from soil surveys in the USA were divided as follows: 50 per cent from planning towns and suburban extensions, 25 per cent from locating highways, airports, pipelines and other structures and 25 per cent from guiding use of soils in agriculture, forestry and recreation (Kellogg 1974). The balance differs elsewhere, especially in developing countries where agricultural improvements continue to be a principal use for soil surveys; but even in such areas soil data are frequently used to guide installation of related services such as roads and waste disposal systems. This chapter reviews and illustrates the use of soil surveys for non-agricultural purposes by means of selected examples in forestry, hydrology, engineering, recreation and planning.

Because of the way in which soil data are commonly presented on maps and in reports, users need help and guidance in assessing land for their particular purpose. It is useful here to distinguish between special purpose and general purpose soil surveys. Special purpose surveys are usually made with a specific use of land in mind or to answer specific questions. The relationship between land use and specific soil properties is known either from prior investigation or by experience and it is only then necessary to determine the distribution and variation of the relevant properties in the area of interest. The data collected are of direct relevance to the user and he can produce designs or identify suitable locations from them. This type of survey is especially favoured by engineers (Chapters 2 and 3).

Most surveys produced by national soil survey organizations and others are general purpose surveys intended to guide a wide range of land use and evaluation. They are based on local or international classifications designed to facilitate the orderly presentation of soil data and correlation of soils in different areas or countries. Such classifications are usually based primarily on soil morphology and are not specific to any purpose, though selection of class criteria is guided in part by the needs of potential users. Map legends use the terms of the classification which are commonly unfamiliar to users not versed in soil science terminology, and thus for effective communication and practical use the data need to be interpreted and presented in a suitable, meaningful format. This involves extracting relevant soil data, expressing it in the terminology of the user and at a level of simplification appropriate to his needs. Interpretation is thus essential if general purpose soil surveys are to realize their full potential, and should be regarded as an integral part of a soil survey project.

7.1 PRINCIPLES AND FORM OF INTERPRETATIONS

Soil survey interpretation is an attempt to predict soil performance under defined conditions of management or manipulation (Kellogg 1962). It is in essence a process of synthesizing basic soil, site and performance data and experience with the user's requirements of land. Two main stages can be identified: 1, understanding what the user needs; and 2, relating the synthesis of data to these needs.

For successful interpretation it is essential to have a clear statement of the objectives and critical questions to be answered. This requires close and patient teamwork between soil surveyor, land user and specialists in related disciplines. Each needs to learn the others' terminology and seek to understand mutual problems. In particular it is necessary to identify properties of land and the critical values affecting a specific use. Relationships between soil properties measured or observed by the surveyor and others of interest to the user may need investigation. The significance of non-soil attributes of land, such as slope and vegetation, must be evaluated. The soil surveyor should make good use of his intimate knowledge of a district gained during mapping but not recorded in the soil report. He should also be aware of the limitations of his map, for confidence will be eroded if interpretations purport to be of greater precision than the amount of available information warrants.

Once soil properties of significance have been identified, information must be accumulated linking the properties with soil behaviour and performance. Such data can come from many sources including detailed investigations of soils of interest at trial sites, for example, measured tree growth plots, small stream catchment studies and building sites. If resources allow, studies at benchmark sites, small experimental areas representing important kinds of soil in a district, are valuable. Observations during mapping and especially the judgement of experienced personnel are other guides. It is common experience to find that suitable data are dispersed and not matched with kinds of soil. In these circumstances further work is required to assess the value of performance data. There is almost always a dearth of information and interpretations must extrapolate, to some extent subjectively, from soils about which much is known to others that are similar.

The relative weight of critical factors in soil survey interpretation is frequently of necessity a matter of judgement. Directional effects are often reasonably clear but relative magnitude can only be fully assessed when large bodies of data are available for analysis. Nevertheless an estimate of the relative importance of soil properties is sometimes attempted.

Soil survey interpretations can take many forms, some of which are exemplified below. They can be very specific, e.g. value of land for winter sports fields or much broader, e.g. suitability of land for commercial forestry. At its simplest, the soil map legend can be used as an index of experience and the map for transferring experience between similar soils at different locations. Many users find it difficult to appreciate the differences between the large number of classes listed in map legends and request a grouping of classes with similar *kinds* of limitations or capability. A further step is to order such groupings by their *degree* of limitation or capability. This can be done according to a national scheme when regional comparisons are required, but when such a grouping would obscure locally significant variations a different approach is required. Comparisons can be portrayed positively by arraying land according to its relative potential for specific purposes. As the needs of developers have become more sophisticated and demanding, attempts have been made to compare real costs of land development or provide guidance to alternative methods for overcoming limitations of land for development.

7.2 LIMITATIONS OF INTERPRETATIONS

Though interpretations of soil surveys provide a valuable basis for assessing the relative potential of land, they can be no better than the original soil survey and it is important to have some appreciation of their limitations.

As noted above, general purpose soil surveys are based on a classification unconnected with any specific purpose. Their value as a basis for predictions is thus very dependent on the degree to which properties used to class the soils are the same as, or correlate well with, those of significance for a specific land use. In practice some land uses are well served; but for others where correlation is weaker prediction is poorer, and in some cases the classification is of no value at all. An example of this problem is where available nutrients are a significant factor in the value of land for growing a particular tree crop. Since most soil classifications take little or no account of such properties and the levels of available nutrients can be poorly correlated with physical properties of soils, predictions based on classes in a soil map legend can be very poor.

The degree of correlation can be further weakened by the nature of soil map units. Although each map unit is identified by a named and precisely defined class, the areas of land delineated will also contain other soils. Many such inclusions are sufficiently similar to be managed identically but they can be significantly different. Confidence in interpretations is much improved therefore by knowledge of the spatial variation of critical properties within map units, and users may require a guide to the accuracy and confidence with which they can use recommendations.

Many non-agricultural uses of land, especially in engineering, are concerned with soil conditions at depths below 1 m. Since many soil surveys have been funded by agricultural interests the nature of soil below the crop rooting zone has not been studied in detail, and additional information is essential before interpretations for uses of land that depend on deep soil conditions can be made. In this area the work of the pedologist and geologist are complementary and where co-operation is close very effective predictions about land use can be made.

Reference was made earlier to property variation in soil map units and the confidence with which predictions can be used. It should be stressed that soil survey interpretations are usually intended to be a guide to conditions at any location and it is foolhardy to press the predictions beyond the limits set by data accuracy. Interpretations based on general purpose soil maps provide preliminary evaluations of sites, but especially where intensive use is intended as in the erection of buildings or the development of certain recreational facilities more detailed on-site evaluation including soil surveys is essential.

7.3 SOIL SURVEYS FOR FORESTRY

The broad assessment of land utility for forestry has many parallels with evaluation of land for agriculture. Such an assessment is particularly valuable in under-developed areas where regional decisions about development of land for either agriculture or forestry are needed. Soil data are also useful in assessing specific aspects of forestry such as need for drainage or cultivation, crop establishment and species suitability. The Land

Capability Classification for Forestry, developed as part of the Canada Land Inventory and described below, is a scheme designed for national evaluations. Soil surveys and data are an essential input to this scheme. The appraisal of windthrow hazard is a specific area of forestry where soil data is significant.

The Land Capability Classification for Forestry (McCormack 1972)

The principal aim of the scheme is to rate all land into one of seven classes according to its inherent ability to grow commercial timber. Assessments are based on soil, relief, climate and vegetation data, and the results are published as coloured maps at a scale of 1 : 250 000. Units are broadly defined and commonly complex but are adequate for regional planning. The three categories used in the system are the capability class, the capability sub-class and the indicator species. The Capability Class integrates all environmental factors as they apply to tree growth. The Capability Sub-class identifies specific factors which limit tree growth. The degree of limitation determines the class designation. The Indicator Species is the tree species that can be expected to yield the volume of timber predicted for each class.

Land is classed on the basis of its physical characteristics, many of which can be derived from soil surveys and using all known information about the unit. Land may be of the same capability class but for different reasons; specific limitations are identified by the sub-class. Each class is given a productivity range expressed in volume of merchantable timber and based on the tree species or group of species best adapted to the site at, or near, rotation age and under good management. Economic and social constraints such as distance to sawmill, location and ownership are ignored, as are improvements such as fertilization and drainage, though better management may change the productivity range.

Figure 7.1 is an extract from a published map in central southern Ontario. This area straddles two physiographic regions, the northern part over Pre-cambrian granitic rocks and locally derived till; soils are coarse textured, shallow and infertile with wetter finer-textured soils in valleys. They are of low or moderate capability for forestry (classes 5–7). In contrast, the southern area is underlain by Palaeozoic limestones and shales with deep loamy and calcareous, somewhat dry soils in associated tills, and has a high capability for forestry. The symbols are interpreted as follows: large arabic numerals denote capability classes (see Appendix 1). Small arabic numeral superscripts give the approximate proportion of the class out of a total of 10. Letters placed after class numerals denote the sub-classes, i.e. limitations (Appendix 1). Letters placed below large arabic numerals denote tree name abbreviations.

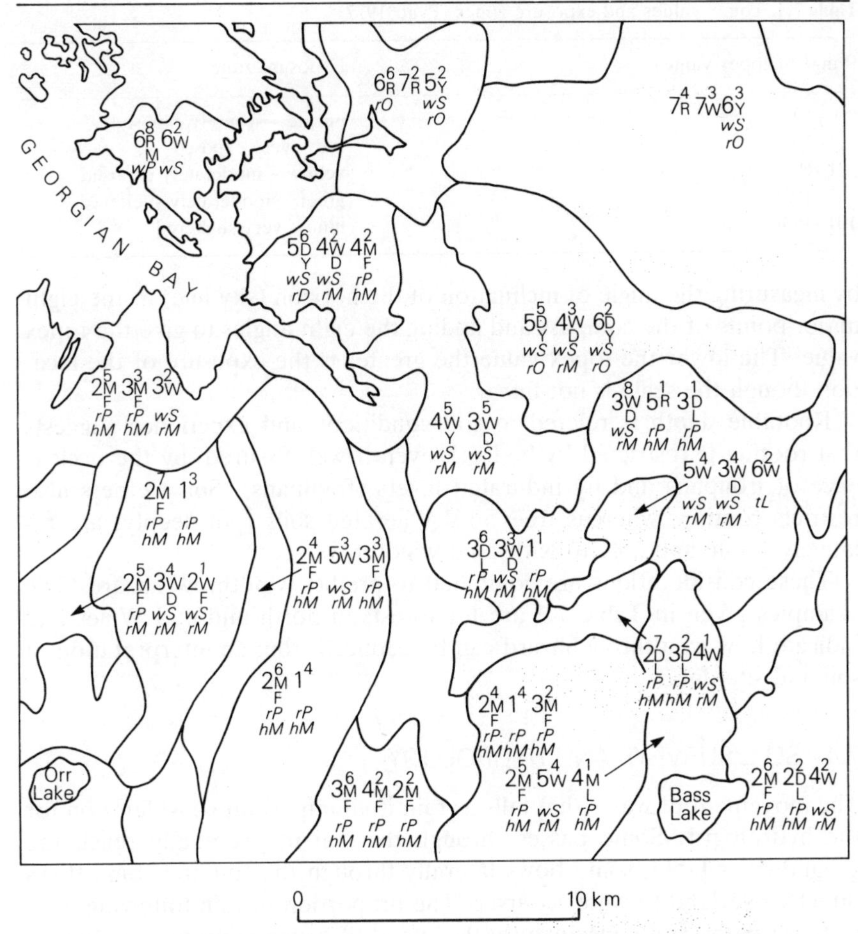

Fig. 7.1 Land capability for Forestry: an example from the Canada Land Inventory

Windthrow hazard

The loss of and damage to tree crops by windthrow is a hazard suffered frequently by forestry enterprises in areas exposed to strong winds. The vulnerability of crops is in part related to soil conditions however and Pyatt (1977) states that 'where rooting is shallow owing to poor subsoil aeration or a mechanical restriction then the crop is almost inevitably disposed to premature windthrow'. He combines an estimate of the degree of exposure with considerations of soil/rooting conditions to predict a 'degree of windthrow hazard' (Table 7.1.). Exposure is gauged by the TOPEX method, in which assessments are made for particular locations

Table 7.1 Topex values and exposure zones (Pyatt 1977)

Range of topex values	Exposure zone
0–10	purple – severely exposed
11–30	red – very exposed
31–60	yellow – moderately exposed
61–100	green – moderately sheltered
101–150+	blue – very sheltered

by measuring the angle of inclination of the horizon (sky line) at the eight major points of the compass and adding the eight angles to give the topex value. The lower the topex value the greater is the exposure of the location though the scale is not linear.

Rootable depth is related to soil conditions and experience suggests that rooting is restricted by bedrock even if well fissured, by the occurrence of ironpans and by indurated layers (fragipans). Soil wetness also restricts rooting, whereas trees in dry aerated soils root deeply, are relatively stable and able to withstand windthrow.

These considerations are combined to predict windthrow hazard. The examples given in Table 7.2 are for forests in north and mid Wales and indicate how windthrow hazard can be deduced from an interpretation of soil and site data.

7.4 SOIL SURVEYS AND HYDROLOGY

The movement of water that falls as rain is an important consideration for the hydrologist. Some passes through the soil to eventually reach the groundwater table, some flows laterally through the soil and some flows directly overland to water courses. The proportion of rain following each of these paths and consequently the speed of water movement to watercourses is closely related to soil properties. Where soils are permeable and able to absorb the amount of water received during average rainfall events there is considerable delay before river levels are affected, but in catchments where soils are dominantly impermeable or quickly saturated, throughflow and overland flow, especially if aided by slope, transmit water to rivers faster, levels rise and, after heavy and prolonged rain, flooding can follow. In order to design flood prevention measures the engineering hydrologist needs to be able to predict river level responses to rainfall events. This is a function of many factors including annual rainfall, slope, the density and form of the drainage network and the infiltration and storage capacity of the soil. The latter could be assessed for individual catchments by many measurements, though there are doubts about the value of such data both as a guide to water movement through or over soil and when extrapolated in space and time. Soils surveys

Table 7.2 Windthrow hazard for forests in North and Mid-Wales (Pyatt, 1977)

Site type* (Soil series)	Rootable depth (cm)†	Causes of shallow rooting	Exposure zone‡ (Topex value range)	Windthrow hazard
Alluvium, poorly drained (Conway, Clwyd)	45–60	Shallow ground-water table	Green and Blue 61–150+	Low or Moderate
Alluvium, freely drained (Teme)	At least 60		Green and Blue 61–150+	Low
Lowland brown earth (Denbigh, Manod)	At least 60		Green and Blue 61–150+	Low
Upland brown earth (Manod)	At least 45, usually over 60	Shallow bedrock occasionally	Yellow and Green 31–100	Low
Upland brown earth, shallow (Manod, Powys)	30–45	Shallow bedrock	Red 11–130	Moderate
Ironpan soil (Hiraethog)	At least 60 except where ironpan not penetrated	Periodic waterlogging in peat and mineral soil above ironpan sometimes restricts rooting to 25–40 cm depth	Purple, Red and Yellow (mainly Red) 0–60 (mainly 11–30)	Low, except where rooting restricted by ironpan when High
Surface water gley (Cegin, some Sannan)	Less than 60, average 40–45	* Inadequate aeration in subsoil † Indurated subsoil in some slope drifts and tills	Red in the Lowland zone 11–30 Green and Blue in the Upland zone 61–150+	High Moderate
Peaty gley (Ynys, some Freni)	Less than 60, average 30–45	* Inadequate aeration † Indurated subsoil in some slope drifts and tills	Red and Yellow 11–60	High, except in least exposed area when Moderate

* Site types are defined in terms of soils, relief and parent material. Correlations with soil series defined by the Soil Survey of England and Wales are as given by Pyatt.
† Rootable depth. Where inadequate aeration is responsible for shallow rooting, the rootable depths quoted are those usually observed in existing pole-stage crops.
‡ Exposure zone in which Site Type usually occurs.

provide an alternative basis for prediction, and interpretations based on an appreciation of the relationships between soil properties and permeability, moisture storage and drainage have been used in the United Kingdom to guide assessments of potential run-off in flooding studies (Farquharson *et al.* 1978). The scheme is summarized below.

Winter Rain Acceptance Potential (WRAP) (source: Farquharson *et al.* 1978)

The requirement was to construct an ordinal classification of run-off potential using not more than five classes and based on readily observable soil properties. The winter situation was considered to eliminate difficulties of assessment introduced by summer soil moisture deficits and assumes that rain falls on soil at field capacity, and therefore cannot be retained within the soil and must be lost either vertically or laterally. A WRAP value is a complex function involving much more than a simple approximation to surface infiltration and the following soil and site properties are considered: soil water regime (or drainage class), depth to an impermeable layer, permeability above impermeable layers and slope.

Soil water regime is an evaluation of the depth to and duration of waterlogging. Three classes are identified, being combinations of six regularly used in soil survey (Hodgson 1976). Where water-tables occur at shallow depth, soils are less able to accept rain since storage capacity is limited, and because water soon reaches the saturated zone subsequent lateral movement is rapid. In soils with impermeable subsoil layers, identified as layers >20 cm thick having less than 5 per cent macropores (>60 μm diam.) and high packing density (bulk density + 0.009 (% clay) g cm^{-3}), vertical movement of water is limited and most is disposed of horizontally. High packing density (Hodgson, 1976). is characteristic of readily identified combinations of structure and particle-size class. The amount of rapid response run-off is considered to be inversely related to the depth to an impermeable layer.

Rapid response run-off is also considered to be related to saturated flow at or near the surface, and thus vertical movement of water to depth in the soil will reduce peak stream discharge following any given rainfall event. The degree to which deep infiltration affects rapid run-off is related to soil permeability which is assessed in three classes for whole profiles or for layers above impermeable horizons from estimates of structure and particle-size class. Since permeability varies and commonly decreases with depth in most soils, assessments are weighted in favour of layers with slower permeability. Slope is of greatest importance where soils are slowly permeable or where the water-table is shallow but has a negligible effect where the soil properties indicate a large WRAP value (classes 1 or 2). WRAP values are estimated using Table 7.3, which is based on experience suggesting that the factors should be given the priority order:

Table 7.3 Winter rain acceptance potential class in relation to soil and site properties (Farquharson *et al.* 1978)

		permeability class (above impermeable horizon)	slope class							
			<2°			2–8°			>8°		
water regime class	depth to impermeable horizon (cm)		rapid	medium	slow	rapid	medium	slow	rapid	medium	slow
1	>80		1	1	2	1	1	2	1	2	3
1	80–40		1	1	2	2	2	3	3	3	4
1	<40		–	–	–	–	–	–	–	–	–
2	>80		2	2	3	3	3	4	–	4	5
2	80–40		2	3	3	3	4	4	4	4	5
2	<40		3	4	4	4	4	4	4	5	5
3	>80		4	4	5	5	5	5	–	5	5
3	80–40		4	5	5	5	5	5	–	5	5
3	<40		5	5	5	5	5	5	5	5	5

Winter Rain Acceptance Class
1 Very high
2 High
3 Moderate
4 Low
5 Very low

Winter Run-off Potential
1 Very low
2 Low
3 Moderate
4 High
5 Very high

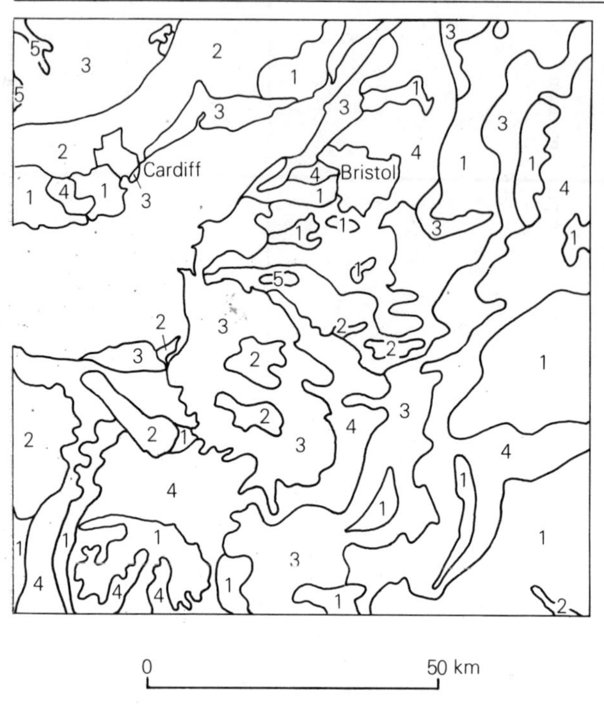

Fig. 7.2 Extract from Winter Rain Acceptance Potential map 1 : 1 000 000 Soil Survey of England and Wales, Harpenden.

soil water regime > depth to impermeable horizon > permeability

though relative magnitude is a matter of judgement. Fig. 7.2 is an extract from a national map of WRAP classes.

The values as displayed on the national map have been used by the Institute of Hydrology in flood studies which aimed to relate recorded flood parameters to physiographic and climatic characteristics of a catchment. The proportions of WRAP classes in each catchment were measured and combined to give a weighted mean. From this a numerical index was evolved using multiple regression of the WRAP class fraction on indices of the catchment's flood potential and morphometric characteristics. Subsequent analysis showed that this index was the principal variable in explaining percentage volume of run-off (35% of the variance) and was also an important element in explaining mean annual flood (an index of a catchment's flood potential – the arithmetic mean in $m^3 sec^{-1}$ of the maximum instantaneous flood recorded in each year of record for the catchment). Further studies have also suggested that the index is a useful tool in predicting low river flows.

7.5. SOIL SURVEYS AND CIVIL ENGINEERING

When a civil engineer is given the task of developing and building a particular kind of structure he must decide where to locate it and what its design should be. These questions are clearly interrelated but to answer either the engineer will need to have information about the characteristics of the land and material on which his structure will stand or over which it may pass. To obtain such information the engineer will probably consult available geological records and also carry out site investigations, whose purpose is to determine the 'suitability and characteristics of sites as they affect the design and construction of civil engineering works and the security of neighbouring structures'. (BSI 1957). Soil surveys provide some, though not all, of the information required by civil engineers about soil horizons in the top 1.5 m of the soil, and are an acknowledged source of background information for site investigations (Dumbleton and West 1971).

Soil to the civil engineer is a material upon which structures and roads are founded and which can be moulded and compacted to form embankments. Although topsoil is differentiated the engineer has little further interest in layers recognized by the pedologist or their significance in soil genesis except insofar as they differ in their mechanical properties. Many properties recorded during soil surveys do however have a close relationship with those of interest to the engineer. Among the latter are the following:

Strength. The effects upon soil of a loading stress or some other applied force.

Consistence. The range of moisture content over which soil behaves plastically or otherwise.

Ease of compaction. When soil is compacted bulk density and strength increase. The engineer needs to ensure that the settlement of an embankment built of soil used as bulk fill is kept to a minimum or that a road sub-grade is as strong as possible by achieving maximum compaction.

Consolidation. Soils beneath foundations are compressed and settlement can result. In some soils, mainly non-cohesive soils, settlement is rapid, in others it can be slow but considerable.

Shrinkage. Clay particles change substantially in volume with changes in moisture content, structures on clayey soils can suffer considerable cracking and distortion as a result.

Susceptibility to frost heave. When water freezes in soil it attracts water from unfrozen zones and ice lenses can build up, causing displacement; when thawing occurs, surface melting over frozen ground below can result in waterlogging and serious loss of strength. Soils with large silt contents are most likely to be affected.

Permeability. Relevant to the installation of soakaways and other drainage works.

Corrosiveness. Wet and acid soils or those containing sulphates or sulphides are potentially aggressive to concrete or steel buried in soil.

For a full explanation and definition of these properties the reader should refer to a standard engineering text.

A number of these behavioural properties can be predicted in general terms from the particle-size distribution and consistence of soils, and the engineer will be able to make his own assessment of them if he is given such data or the engineering soil class. Several engineering soil classifications exist but some of the most widely used are those based on the original scheme of Casagrande (1947); for example, the Unified Soil Classification System (Table 7.4). When using this system all material coarser than 76 mm is excluded and material passing the Standard American Sieve No 200 (0.074 mm) is regarded as fines. A primary division into coarse and fine grained soils is made according to the percentage of fines present and further identification of groups is dependent on either grading of particle sizes or the consistence of fines, the latter being assessed in relation to a plasticity chart (Fig. 7.3.).

Though particle-size data given in soil survey reports may differ from those required by engineers, it is usually possible to place a soil in its engineering class by appropriate conversions and the assessment of consistence or by comparison with similar soils. In addition to particle-size data, soil surveys incorporate information about depth to water-tables, the extent and duration of soil wetness, permeability, run-off potential, occurrence of sulphates and sulphides, soil pH, bulk density and moisture content of soil and indicate where rock is at shallow depth or indurated horizons occur. Soil surveys can also be used to identify the presence of spring lines, swallow holes, solution pipes, areas of potential flooding and peat. Additional data can be keyed into the soil map units either from current experience gained from previous site investigations or surveys on known soils or by carrying out engineering tests on soil from 'benchmark' sites.

The main use of soil maps for planning engineering developments is as a predictive tool in the initial (strategic) planning phase, their advantage being that they provide information over extensive areas. They do suffer the major disadvantage of shallow depth of investigation, but used in

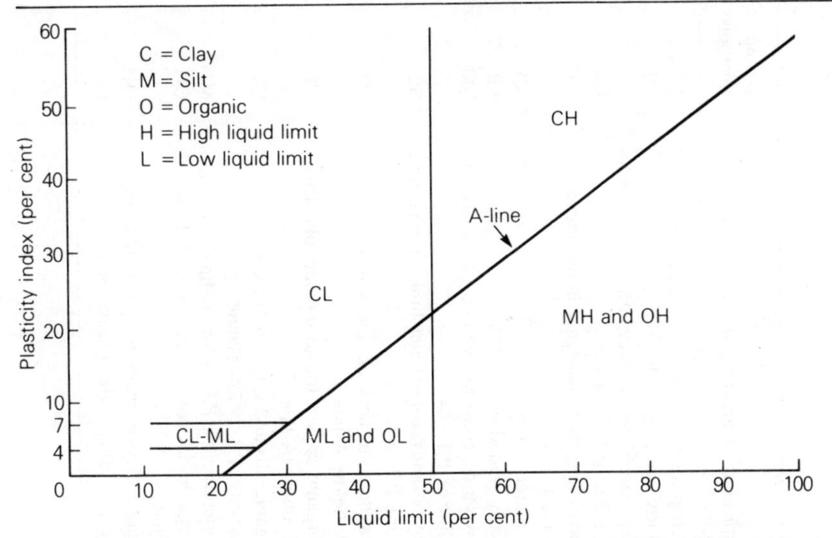

Fig. 7.3 Plasticity chart (Applied Geology for Engineers (1976) HMSO, MOD and Institute of Civil Engineers)

conjunction with geology maps they can be used as a 'first sieve' to identify places worth investigating and eliminate others which are likely to be totally, or relatively, unsuitable for the proposed development.

The value of pedological soil surveys in engineering has been emphasized in recent years with the publication of guidelines for interpreting soil data by such agencies as the United States Department of Agriculture (SSS 1971) and the Food and Agriculture Organization of the United Nations (FAO) (Olson 1973). Each identifies properties of soil relevant to specific engineering uses and provides charts by which land can be rated. Where soil survey data are incompatible with engineering use, tables of conversion are provided. These charts are then used to provide generalized interpretations for kinds of soils in the reports of soil surveys. Table 7.5 is an example of a chart to guide interpretation for shallow excavations and the appropriate rating for any soil is indicated by its most severely limiting property. Tables 7.6 and 7.7 are excerpts from the engineering interpretation summary tables in a USDA soil survey report. These predictions are related to the map units identified and are then applied areally using the accompanying soil maps.

To further illustrate the use of soil surveys in engineering some specific examples are considered below.

Highway construction and design in Michigan USA

Soil maps and classes have been used as a basis for the design and

Table 7.4 Unified soil classification system for engineering (after USDD 1968)

	Major divisions*		Fines (%)	Typical name and classification criteria	Group Symbols†
Coarse grained soils (>50% of material retained on sieve No. 200) ‡	Gravels (> 50% of coarse fraction is >5 mm)	Clean gravels	<5	Well graded gravels	GW
				Poorly graded gravels	GP
		Gravels with fines	>12	Silty gravels: consistency limits of fines below A-line or P.I. § <4	GM
				Clayey gravels: consistency limits above A-line P.I. >7	GC
	Sands (>50% of coarse fraction is <5 mm)	Clean sands	<5	Well graded sands	SW
				Poorly graded sands	SP
		Sands with fines	>12	Silty sands: consistency limits of fines below A-line or P.I. <4	SM
				Clayey sands: consistency limits of fines above A-line P.I. >7	SC
Fine grained soils (>50% passes sieve No. 200)	Silts and clays (liquid limit <50)			Inorganic silts and very fine sands P.I. below A-line	ML
				Inorganic clays of low or medium plasticity, P.I. above A-line	CL
				Organic silts and silty clays of low plasticity, P.I. below A-line	OL
	Silts and clays (liquid limit >50)			Inorganic silts, P.I. below A-line	MH
				Inorganic clays of high plasticity, P.I. above A-line	CH
				Organic clays of medium or high plasticity	OH
	Highly organic soils			Peat and other highly organic soils	Pt

* Material classified is <76 mm (3 in) diameter: gravel is 4.7–76 mm diameter
† Definitions of symbols:
G gravel (4.7–76 mm), S sand (0.074–4.7 mm)
M material <0.074 mm ('silt') with consistency limits below the A-line (Fig. 7.3)
C material <0.074 mm ('clay') with consistency limits above the A-line
H high compressibility, L low compressibility
O organic soils
Pt soils with much fibrous organic matter
W well graded
P poorly graded
Soils near class boundaries are given double, hyphenated symbols eg, GM–GC
‡ Standard American Sieve No. 200 (0.074 mm) has a marginally smaller aperture than BS Sieve No. 200
§ P.I. plasticity index

Table 7.5 Soil limitation ratings for shallow excavations (Soil Survey Staff 1971)

Item affecting use ¶	Degree of soil limitation		
	Slight	Moderate	Severe
Soil drainage class	Excessively drained, somewhat excessively drained, and well drained	Moderately well drained	Somewhat poorly drained, poorly drained, and very poorly drained
Seasonal water table	Below a depth of 150 cm	Between depths of 75 and 150 cm	Above a depth of 75 cm
Flooding	None	Rare	Occasionally or frequent
Slope	0–8 %	8–15 %	More than 15 %
Texture of soil to depth to be excavated*†	fsl, sl, l, sil, sicl, scl	si‡, cl, sc; all gravelly types	c§, sic§, s, ls; organic soils; all very gravelly types
Depth to bedrock ‖	More than 150 cm	100 cm–150 cm	Less than 100 cm
Stoniness class	0 and 1	2	3, 4 and 5
Rockiness class	0	1	2, 3, 4 and 5

* Texture is used here as an index to workability and sidewall stability. Abbreviations for texture classes are as follows:

s	= sand	si	= silt	c	= clay
ls	= loamy sand	sil	= silt loam	cl	= clay loam
sl	= sandy loam	sicl	= silty clay loam	l	= loam
fsl	= fine sandy loam	sic	= silty clay	sc	= sandy clay
				scl	= sandy clay loam

¶ Class definitions are as given in the Soil Survey Manual (SSS 1951).

† If soil contains a thick fragipan, duripan or other material difficult (but not impossible) to excavate with handtools, increase the limitation rating by one step unless it is *severe*.

‡ If soil stands in vertical cuts, like loess, reduce rating to *slight*.

§ If the soil is friable reduce rating to *moderate*.

‖ If bedrock is soft enough so that it can be dug out with ordinary handtools or light equipment, such as back hoes, reduce ratings of *moderate* and *severe* by one step.

construction of roads by the Michigan Department of State Highways for many years (Allemeier 1973). The technique adopted is to index accumulated engineering experience by soil series, classes equivalent to individual or combinations of similar kinds of soil identified in soil surveys made by the United States Department of Agriculture (USDA). The Department of State Highways then uses USDA soil surveys and more commonly strip soil surveys produced by its own staff as a guide to routing and designing new highways. Soil data is used in three stages, first by the Planning Division in its forward planning studies, principally to identify extensive areas of swamp deposits, soft materials and areas with bedrock at shallow depth; distance for hauling fill material is also considered. Secondly, in route location studies similar factors are considered in more detail using data from USDA soil surveys and geological maps supplemented by some detailed work by the Department's own staff. Once the route is fixed a detailed strip map showing soil series is prepared. Soil texture, colour, consistency and depths of horizons are noted and the class identified from abbreviated descriptions given in the Department's Field Manual of Soil Engineering. Designs are then based on recommendations given in Design Charts which embody the results of intensive correlation of experience and performance of each soil series in the state. The proven assumption is that wherever a given soil series occurs, the Design Chart recommendations for that kind of soil are generally valid, though amendment can be needed to meet local conditions.

The detailed soil survey is also used as a basis for planning further investigations for example peat sounding, rock sounding, resistivity and seismic surveys.

Figure 7.4 shows two examples of the soil series profiles and descriptions given in the Field Manual of Soil Engineering and used as the basis for identifying soil classes during mapping. Figure 7.5 is part of a soil strip map showing the extent of recognized classes. Table 7.8 is an extract from the Design Chart showing predicted engineering data and recommendations for series.

Building sites near Utrecht, Netherlands

High ground-water levels and weak soils mean that careful evaluation of land for building sites is essential before construction can begin over much of the western Netherlands (Westerveld and van den Hurk 1973). One building project in an area around Utrecht was the subject of study by a multidisciplinary team whose aim was to compare in monetary terms the cost of improving the land for this purpose. Data available comprised soil maps, water-table levels and fluctuations within 1.5 m, depth to underlying Pleistocene sand, height above sea level and deep ground-water movements. The eastern part of the district consists of relatively high, dry Pleistocene sands but to the west are low and wet peat soils, the peat partly cut for fuels and new lakes; in the south and west are river-clay soils and transitional soils where clay covers sand. Elevation ranges from 15–20 m above sea level in the east to sea level in the peat and back swamp soil areas. Ground-water level varies strongly also, from 5–10 m below surface to < 0.20 m in the peat soil areas.

It was judged that if land were to have sufficient bearing strength for building, ground-water depth should be > 1 m for all soils. For land beneath which shallow ground-water levels could not be lowered, the remedy was to elevate it by dumping sand on the surface, 1 m thick for sand and clay soils and 1.5 m for peat soils which would be compressed when loaded. A standard price for obtaining and dumping sand was adopted.

Table 7.6 Engineering interpretations (Hutton and Rice 1977)

	Suitability as a source of –		
Soil series and map symbols	Topsoil	Sand and gravel	Fill material
*Honeoye:	Fair: contains some gravel; limited volume.	Unsuited: none present.	Good to fair: excessive fines in places, some large stones.
Howard:	Poor: too gravelly.	Good	Good
Kendaia:	Fair to poor: low volume; too gravelly in places.	Unsuited: none present.	Fair below surface layer; excessive fines; wet in places; some large stones.

Soil features affecting –				
Highway location	Embankment foundations	Foundations for low buildings	Farm ponds	
			Reservoir area	Embankment
Seasonal high water table below a depth of 60 cm in places; cut slopes subject to seepage and sloughing; sub-grade in cuts is generally good but is subject to boulder heave in places; trafficability is generally good except on steeper slopes.	Adequate strength for high embank-ments.	Adequate strength; seasonal high water table below a depth of 60 cm in places.	Slowly permeable or very slowly permeable below a depth of about 75 cm; seasonal high water-table below a depth of 60 cm in places; bedrock at a depth of 105 cm in places; most units have sloping to very steep unfavourable slopes.	Low to medium compressibility; low permeability; good to fair compaction characteristics; few to many large stones.
Highway grade location not critical above the water table; local seepage and some sloughing in cuts; sub-grade in cuts subject to differential frost heave; trafficability is generally good; cobbles dislodged in cuts in places because of frost action.	Generally adequate strength for moderately high embank-ments; underlain by wet, compressible soil material in places.	Generally adequate strength; underlain by wet, compressible soil material in places; large settlement possible under heavy or vibratory loads; seasonal high water table at a depth of 90 cm in a few places; subject to caving.	Pervious material; seasonal high water table at a depth of 90 cm in a few places.	Rapidly permeable: low com-pressibility; good to fair compaction characteristics; good for outside shell.
Seasonal high water table at a depth of 15–30 cm; cut slopes subject to seepage and sloughing; sub-grade in cuts generally good, but seasonally wet and subject to boulder heave in places; trafficability is poor when wet.	Adequate strength for high embank-ments.	Adequate strength; seasonal high water table at a depth of 15–30 cm.	Slowly permeable or very slowly permeable below a depth of about 55 cm; seasonal high water table at a depth of 15–30 cm; bedrock below a depth of 105 cm in places.	Low to medium compressibility; low permeability; fair to good compaction characteristics; wet in places.

Table 7.7 Limitations of soils for civil engineering use in community planning (Hutton and Rice 1977)

Soil	Dwellings with basements	Shopping centres and small industrial buildings	Local roads and streets	Septic-tank absorption fields	Underground public utilities	Sanitary land-fill (trenches)
Honeoye silt loam, 8 to 15 per cent slopes	Moderate: slope	Severe: slope	Moderate: slope	Severe: slow permeability	Moderate: slope	Moderate: firm or very firm till; very stony in places
Howard gravelly fine sandy loam, 0 to 3 per cent slopes	Slight	Slight	Slight	Slight	Severe: very gravelly layer; subject to sloughing	Severe: rapid permeability
Kendaia silt loam, 3 to 8 per cent slopes	Severe: somewhat poorly drained	Severe: somewhat poorly drained	Severe: susceptibility to frost action	Severe: somewhat poorly drained	Severe: somewhat poorly drained	Moderate: somewhat poorly drained; firm or very firm till; very stony in places

Litter, leaf mold and humus

Gray to pinkish grey sand.

Dark brown sand, may be weakly cemented.

Pale brown sand, mottled with gray and yellow.

Mottled gray brown and yellow clay or silty clay, low permeability.

Allendale

Series description

The surface texture is usually a loamy sand.

The imperfectly drained Allendale soils represent 20 to 40 inches of sand or loamy sand over clay or silty clay till or lacustrine materials. These soils occur in the central and northern part of the Lower Peninsula and the eastern half of the Upper Peninsula.

Allendale soils are the imperfectly drained member of the catena which includes the well-drained Manistee and the poorly drained Pinconning soils.

Allendale soils differ from Iosco soils in being developed in sands over silty clays or clays while Iosco soils are developed in sands over loam to silty clay loams. Allendale soils differ from Arenac soils which are imperfectly drained but have loams to clays at 40 to 66 inches.

The imperfectly drained Dafter soils consist of 20 to 40 inches of loamy fine sand to fine sandy loams over clay or silty clay. A textural subsoil of loam or sandy clay loam may have developed in the upper part of the profile. In mapping, include Dafter soils with Allendale soils.

Construction Information

The upper 2 feet of this material is generally easily excavated but the high water-table, which makes the sand unstable, may retard hauling operations. Below this depth the underlying clay and seepage water above and in the clay may interfere with hauling.

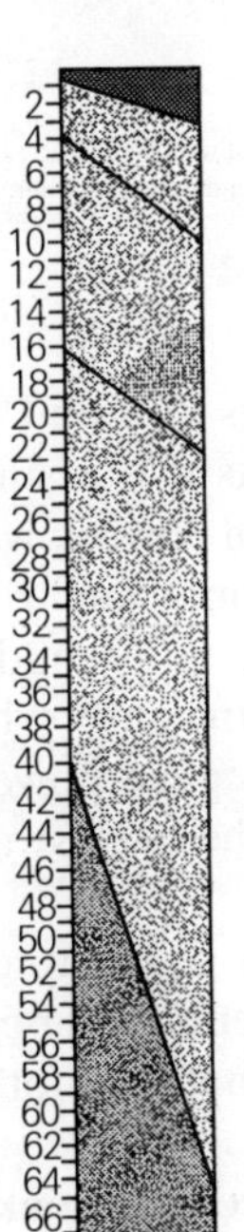

Forest litter, leaf mold and humus.

Pale to pinkish grey sand or loamy sand.

Dark reddish brown to yellowish brown sand becoming lighter with depth. May be weakly to strongly cemented.

Pale brown to brownish yellow sand or loamy sand. Faint yellowish brown mottling is present in some places.

Fine sandy loam to clay.

Melita

Series description

The surface texture is either a sand or a loamy sand.

Melita soils are well to moderately well-drained sands or loamy sands, 40 to 66 inches thick, over fine sandy loams to clays either till or lacustrine materials. The upper part of the profile represents the characteristics of either the Rubicon or Kalkaska soils. They occur on till and lake plains in the central and northern part of the Lower Peninsula and the eastern part of the Upper Peninsula.

Melita soils are the well-drained member of the catena which includes the imperfectly drained Arenac and poorly drained Roscommon soils.

The well-drained to moderately well-drained Bentley soils also have 40 to 66 inches of sands or loamy sands over fine sandy loams to clays. However, Bentley soils have a gravelly sandy loam to light sandy clay loam textural subsoil in the upper part of the profile, usually at a depth below 40 inches. In mapping, include Bentley soils with Melita soils.

Construction Information

This material is easy to excavate to a depth of 2½ to 5½ feet. Below this depth some difficulty may be encountered due to the wet underlying silts, loams, clays and very fine sands. The soggy characteristic of this underlying material will make hauling difficult. The loose and incoherent nature of the sand may interfere with hauling operations.

Fig. 7.4 Profiles of soil series (Michigan Department of State Highways 1970)

Table 7.8 Exerpt from a design chart (Michigan Dept. of State Highways, 1970).

Soil Series	Characteristics			Treatment			
	Brief Description of Typical Profile	Adapted to Winter Grading	Normal Depth to water-table	Grade			
				Recommended location of plan grade with respect to natural ground	Estimated per cent of boulders (rock excavation)	Estimated depth of topsoil (ft)	Sub-base Recommended
Allendae	Imperfectly drained sand over silty clay	No	1′–2′	Fill 4′–5′ (a)	0	0.5–0.8	Yes
Arenac	Imperfectly drained sand over sandy clay	Poor	3′–5′	Fill 1′–2′ (a)	0	0.2–0.4	Yes
Melita	Well drained sand over sandy clay	Poor	3′–5′	5′ above water table	0	0.1–0.4	Yes
Pinconning	Poorly drained loamy sand over silty clay	No	1′–3′	Fill 4′–5′	0	0.4–0.7	Yes
Roscommon	Poorly drained sand	No	0′–1′	Fill 4′–5′ (a)	0	0.5–0.9	No
Rubicon	Well drained sand	Good	10′–20′	Anywhere	0	0.1–0.3	No

General notes
(a) The higher grade heights are a minimum standard for primary trunklines, expressways and interstate routes.
(b) Applies where standards of vertical alignment require cut sections, in variance with recommendations in column 3.

By comparing soil and depth to water-table maps (Figs. 7.6 and 7.7) and using information about deep groundwater movements it was simple to identify areas requiring elevation: for soils with mean lowest water tables below 1.20 m, and the majority of sand, transitional and river-levee soils with water-table classes 1 and 2 (Fig. 7.7) elevation was judged unnecessary since groundwater levels could be lowered sufficiently where required. A few of these wetter soils were affected by seepage however and these needed elevation by the addition of sand as did the back swamp and peat soils.

The costs of foundations, roads and other utilities was similarly estimated. An important factor here was depth to the Pleistocene sand below (Fig. 7.8). Comparing these depths with the soil map, estimates of foundations and utility costs were possible.

Assessments of total cost were made by combining the costs of making the land suitable for building and actual construction costs (Fig. 7.9).

							Resources		
					Embankment				
Estimated lineal feet per 1000 ft of cut below natural ground elevation									
Sub-grade undercutting (per roadbed)	Edge Drains (per roadbed)	*Bank* Use only if cut is deeper than:	Lineal feet	*drains* Granular blanket. For cuts deeper than 6.0 ft (Per cent of slope area)	Suitable borrow for embankment construction	Per cent of shrinkage	Granular material class II	Possible source of gravel	Source of topsoil
300 (b)	600 (b)	3′ (b)	1200	5 (b)	No	20–30	No	No	Fair
300	500	5′ (b)	1200		Ltd.	10–20	Yes	No	Fair
700	800	7′ (b)	800	20 (b)	Ltd.	15–25	Ltd.	No	Poor
400 (b)	700 (b)	5′ (b)	1200	5 (b)	No	15–25	No	No	Poor
100 (b)	1000 (b)	3′ (b)	1200			10–20		No	Poor
100	200	10′	500		Yes	10–20	Yes	No	No

They reflect substantial differences between land and are a valuable guide in planning development. Of the land available in the area, that with dry sand soils offers the widest range of possibilities since the costs of making the wet back-swamp and peat land suitable for building are high, especially where the Pleistocene sand is at depths > 3 m.

7.6 SOIL SURVEYS FOR RECREATION

Soil maps and data can be used readily to assess the suitability of land for recreation. As leisure time increases so does the demand for outdoor recreational facilities. It is important that sites should be carefully located so that initial cost is minimized, subsequent management straightforward and opportunities for maximum use under good conditions provided.

General guidelines for the interpretation of soil maps have been developed in several countries. Table 7.9 is an example for winter playing

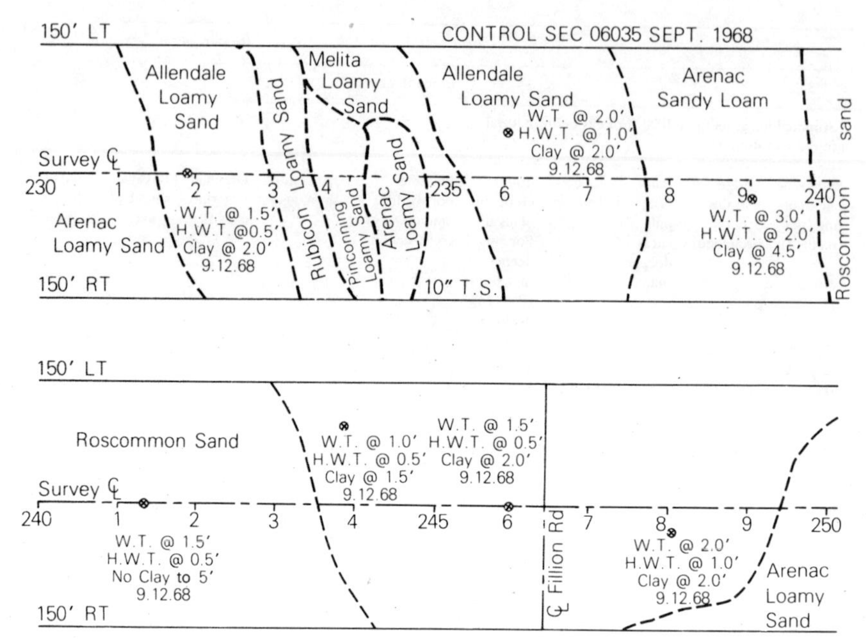

Fig. 7.5 Soil strip map showing extent of recognized classes (Michigan Department of State Highways 1970, *Field Manual of Soil Engineering)*

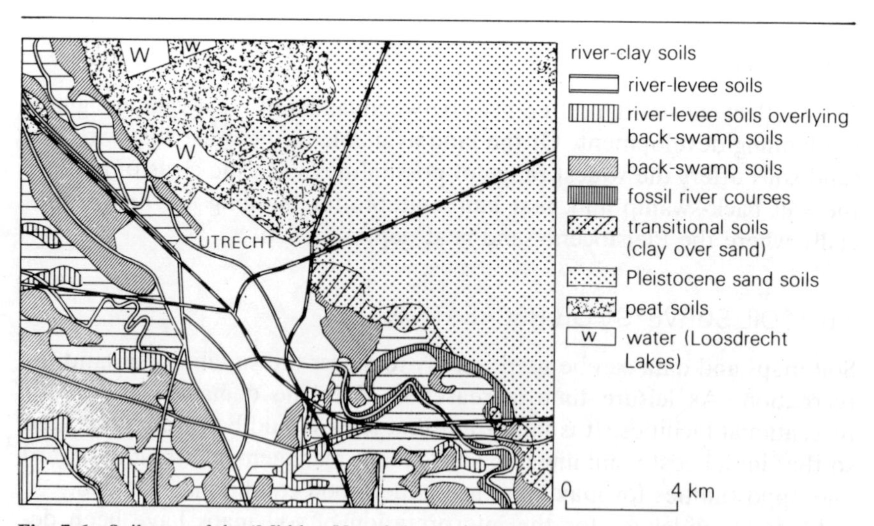

Fig. 7.6 Soil map of the 'Midden-Utrecht' area (Westerveld and van den Hurk 1973)

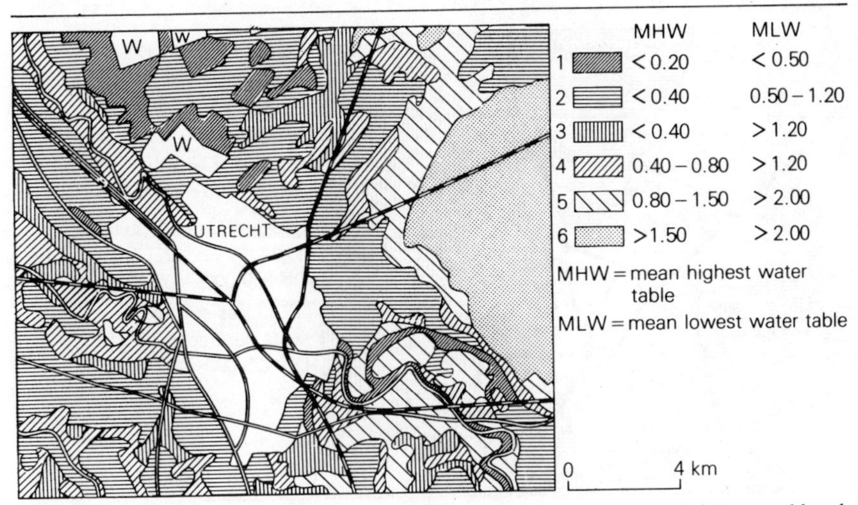

Fig. 7.7 Map of water-table classes (water-tables in metres below surface) (Westerveld and van den Hurk 1973)

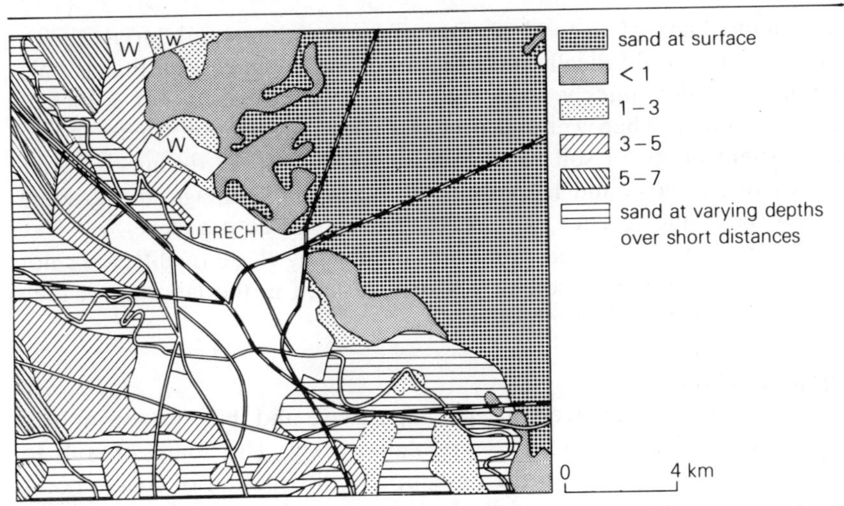

Fig. 7.8 Depth of Pleistocene sand (in metres below the surface) Westerveld and van den Hurk 1973)

fields in England and Wales. Explanations of the criteria are given in some detail to illustrate the reasoning behind their selection.

Land for winter playing fields

The assessment of the suitability of land for winter playing fields should take into account features of soil and site that affect both construction

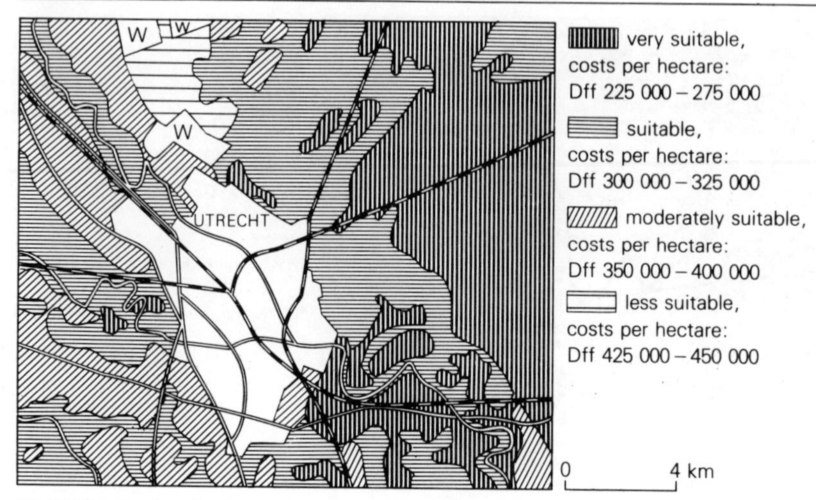

Fig. 7.9 Soil suitability map for building sites, based on the costs of making soils suitable for buildings. (Costs based on the 1968–9 price level, i.e. Dfl 1,000 = approx $280 US)

and maintenance of a good playing surface (Palmer and Jarvis 1979). Suitable land should ideally have the following qualities: soil that is able to transfer water quickly away from the playing surface, has sufficient bearing strength when wet and provides adequate traction; there should be sufficient depth of soil to sustain a good turf cover during the drier periods of the year and to permit any necessary drainage and grading work without interference by rock; the site should be near level.

Though some of the soil properties influencing these qualities are not recorded directly during soil survey or measured in the laboratory, they can usually be related to others that are.

Soil permeability

Sportsfields need a relatively dry playing surface and thus surface infiltration must be adequate to remove rainfall quickly. The ability of soils to drain after rain depends on the amount and distribution of coarse pores (>60 μm) throughout the profile and depth to the ground water-table. These with other factors control the seasonal duration of wetness at a particular depth and hydraulic conductivity.

Soils commonly drain by gravity to a suction of about 0.05 bar; the volume of air-filled pores at this suction, the air capacity (C_a), is therefore a measure of the volume of pores able to conduct water away from the playing surface (drainable pore space), that is the majority of pores >60 μm diameter. Air capacity is measured for individual horizons of many soil profiles or can be estimated from clay content and packing density (Hodgson 1976). Horizons with large or very large C_a (>20%)

Table 7.9 Suitability of land for winter playing fields*

Class	Profile features			Topsoil features			Site features		
	Wetness classes	Permeability†	Depth to rock‡	Retained water capacity	Particle-size classes§	Stoniness	Rock outcrops	Flood hazard	Slope
Well suited	I	C_a 20% or more to 80 cm; K > 1 m/day	>1 m	<30%	SL (with >70% sand) fine and med. LS and S	<1%	None	None	0–1°
Moderately well suited	I–III, IV in drier areas‖	C_a 10% or more to 50 cm, 5% or more below; K >0.1 m/day	>0.75 cm	<45%	As above plus coarse LS and S; SL (with <70% sand) SZL	<6%	<2%	Once in 2 years or less, <3 days on each occasion	0–1.5°
Poorly suited	I–IV	C_a 10% or more in topsoil, can be <10% below; K >0.01 m/day	>0.5 m	<55%	As above plus ZCL; ZC, C, SC each with <45% clay; all humose classes	<16%	<10%	Not more than twice per year <3 days on each occasion	0–3°
Unsuited¶	V or VI	C_a <10% in topsoil; K <0.01 m/day	0.5 m or less	55% or more	Clays with >45% clay, peat	16% or more	10% or more	More than twice per year	> 3°

* To be placed in any class, a site should satisfy the criteria for *all* the features of that class.
† C_a is air capacity, K is hydraulic conductivity.
‡ The importance of the depth to rock limitation varies with the kind of rock present. Sites with soft rocks, such as chalk, at depth will be less limited than other sites over harder rocks.
§ Particle-size classes are abbreviated as follows:

S	sand	SZL	sandy silt loam	ZL	silt loam	CL	clay loam
LS	loamy sand	SCL	sandy clay loam	ZCL	silty clay loam	C	clay
SL	sandy loam	SC	sandy clay	ZC	silty clay		

‖ Class IV soils in areas with <700 mm mean annual rainfall if they have moderate or rapid subsoil permeability.
¶ Land exceeding the limits of the poorly suited class in any feature should be rated unsuited.

will be rapidly permeable, whereas those with small C_a (5–10%) will be only slowly permeable; those with very small C_a (<5%) are effectively impermeable.

The depth to, and duration of waterlogging, the soil water regime, is a function of climate, soil and site conditions and can be subject to improvement or deterioration as a result of drainage measures or intensity of use. The Soil Survey of England and Wales allocates soils to one of six soil wetness classes (I–VI) (Hodgson 1976), each class representing a range of hydrologic conditions. Soils in Class I are rarely wet within 70 cm depth and underdrainage is usually unnecessary. Class V soils are usually wet within 70 cm depth and often to within 40 cm during the winter; drainage of these soils to the standards required for winter sportsfields is unlikely to be economic.

The cause of wetness is also relevant because it affects the design of remedial drainage. Causes can be high groundwater levels caused by river water or seepage from nearby higher land sustaining saturated but usually permeable horizons within the profile, or impermeable soil layers slowing down the dispersal of incident rainfall and so causing surface wetness. Intensive foot traffic during wet conditions can substantially reduce the volume of macropores (>60 μm) at the surface and so reduce permeability. When compacted, topsoil porosity depends largely on particle-size distribution, particularly the sand content. It has long been common practice to improve the topsoil permeability of sportsfields by adding sand. Adams *et al* (1971) suggest that sufficient sand (100–600 μm) should be added to give a topsoil content of about 75 per cent.

Effects of foot traffic

If sportsfields are used when wet, topsoils become compacted, the turf cover can be easily sheared by foot pressure, packing density (compaction) increases and permeability (C_a) of surface horizons decreases so that vertical water movement is slowed and in extreme cases water stands on the surface. The risk of poaching, or trampling by feet of the turf layer, can be assessed by measuring the topsoil retained water capacity, θ_v (0.05), which is approximately the water content of the soil at field capacity. It is an especially important soil property of intensively used winter sportsfields. When moisture content is large the bearing strength is small. Values of retained water capacity for surface horizons are related to the content of clay, organic matter and bulk density.

Surface grip

For all games played on sportsfields it is necessary to have a good grip with studded boots. The best grip is obtained from a thick turf mat especially where it is protected by grass leaf several centimetres in length. Well used patches are often bare or have only sparse grass cover and here the soil surface alone provides the grip. Sands >700 μm diameter give very little traction, whereas there is a considerable increase in potential traction with particles decreasing in size to 60 μm diameter (Adams *et al.* 1971). Soils with large contents of fine sand (60–200 μm diameter) will therefore probably give the best grip. Soils with large amounts of clay, silt or organic matter tend to be slippery.

Surface stoniness

Stones (particles >2 mm) limit the value of land for sportsfields because they interfere with play, are abrasive to players and can make installation of drainage systems difficult. Their size and angularity are almost as important as their quantity.

Slope and depth to rock

Playing fields should be nearly level or have slopes of <1.5° (2.5%) if ball games are to be played without undue fatigue or unfair advantage. Steeper slopes need grading but before construction the soil water regime, presence of impermeable layers and permeability to below the depth of excavation should be assessed. Many soils will show some deterioration when graded even though the topsoil is carefully removed and replaced on the graded surface. Formerly deep-lying impermeable layers will be nearer the surface and pans, which may be difficult to remedy, can be induced in weak-structured layers by passage of implements or foot traffic.

Suitability classes for winter playing fields

Well suited. Land with no or very minor limitations. Soils are deep and permeable to depth with water-tables below 70 cm depth for much of the year. Topsoils are stoneless, have large fine and medium sand content and sufficient bearing strength when wet to resist poaching. Available water is adequate to sustain grass cover during dry months. Sites are level and free from flooding.

Moderately well suited. Land with moderate limitations that requires some modification or improvement. Soils are moderately deep. They can be moderately or slowly permeable and wet within 70 cm depth for much of the winter but rarely within 40 cm. Underdrainage will often be necessary. Topsoils can be very slightly stony and, because of low bearing strength, moderately susceptible to poaching. Sites can be gently sloping or uneven requiring some grading. Floods can occasionally prevent the use of the facilities.

Poorly suited. Land with severe limitations that will usually require major improvement before use. Soils can be slowly permeable or have water-tables above 40 cm depth requiring intensive underdrainage, or they can be shallow, making grading work difficult. Topsoils can be slightly stony, moisture retentive, fine textured or humose and need amelioration. Sites can be gently sloping or uneven and need grading. Floods can restrict the use of the facilities in most years.

Unsuited. Land with such severe limitations that improvement is not feasible or is uneconomic. This includes land with permanently wet soils having a high groundwater table, or with clayey, impermeable soils. Other land can have peaty or very stony topsoils, be moderately sloping, suffer frequent floods or have many rock outcrops.

Guidelines for classes are presented in Table 7.9.

Soil suitability for deciduous trees and grassed playgrounds.

The establishment of a recreational area near Amsterdam, The Netherlands on land comprising a spoil dump of peat and clay material was guided by interpretations of soil surveys (Westerveld and van den Hurk 1973). This particular work has added interest since it illustrates the use of soil survey in reclaimed areas without soil patterns.

The area concerned, the Twiskepolder, is low lying and was originally covered by 3–4 m peat over unripened (soft) clay (0.5–1.5 m) passing to fine sand and with Pleistocene sand at 12–14 m below the surface. Sand has been extracted from part of the area but prior to extraction the unwanted peat and clay layers were dumped as spoil on to part of the neighbouring peat marsh. The whole area is to be developed as a recreational facility and a survey of this spoil area was undertaken to evaluate its suitability and cost of improvements for grassed playgrounds and picnic areas or for deciduous trees.

Figure 7.10a reproduces the soil survey of the spoil dump. There were considerable difficulties in making a survey of the area. The dumped material varied in thickness from 0.8–2.4 m and soil conditions varied substantially. Access was also limited because of the weak nature of the material.

To use the land for the purposes intended, lowering of the water-table to 2.5 m below sea level was essential. This was expected to lead to surface subsidence as the spoil consolidated. Since thicknesses of peaty spoil varies, subsidence would be uneven, greatest for land with most peaty material and least for land with soils developed principally in sandy material. The soil survey was used to predict eventual surface heights and the vertical distance between surface and lowered water tables (Fig. 7.10b).

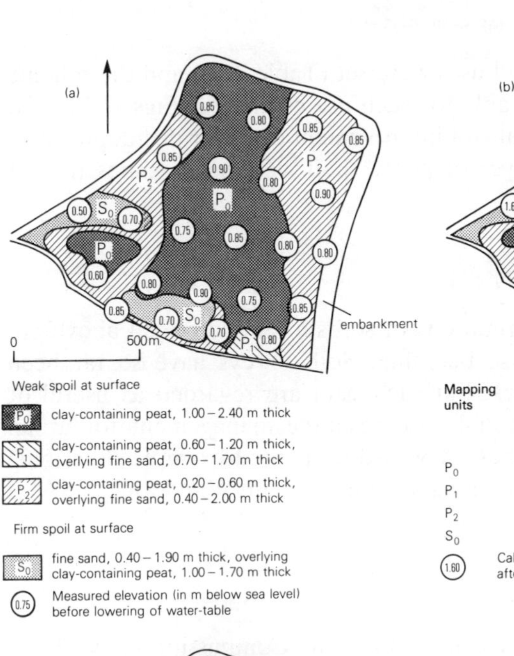

Mapping units	Subsidence of the surface (in m)	Elevation in m below sea level	Distance between land surface and water level in the ditches (in m)
P_0	0.70 – 1.25	1.60 – 1.85	0.65 – 0.90
P_1	0.50 – 0.70	1.40 – 1.50	1.00 – 1.10
P_2	0.50 – 0.95	1.40 – 1.70	0.80 – 1.10
S_0	0.30 – 0.90	1.00 – 1.40	1.10 – 1.50

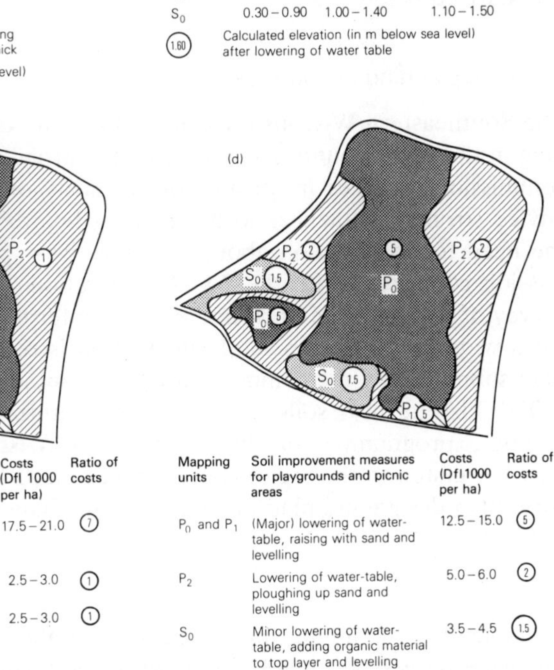

Mapping units	Soil improvement measures for deciduous forests	Costs (Dfl 1000 per ha)	Ratio of costs
P_0 and P_1	Major lowering of water-table and raising with loamy sand or (sandy) loam	17.5 – 21.0	7
P_2	Lowering of water-table and ploughing up sand	2.5 – 3.0	1
S_0	Minor lowering of water-table and adding organic material to top layer	2.5 – 3.0	1

Mapping units	Soil improvement measures for playgrounds and picnic areas	Costs (Dfl 1000 per ha)	Ratio of costs
P_0 and P_1	(Major) lowering of water-table, raising with sand and levelling	12.5 – 15.0	5
P_2	Lowering of water-table, ploughing up sand and levelling	5.0 – 6.0	2
S_0	Minor lowering of water-table, adding organic material to top layer and levelling	3.5 – 4.5	1.5

Fig. 7.10 Soil maps of the northern spoil dump of the Twiskepolder (after Westerveld and van den Hurk 1973)
(a) Elevations measured before lowering the water-table
(b) Elevations (in m below sea level) as calculated after lowering the water table to approx. 2.5 m below sea level.
(c) Investment required per mapping unit to make soils suitable for deciduous trees. (The costs are based on the 1970 price level, i.e. Dfl 1,000 = approx US $280.)
(d) Investments required per mapping unit to make soils suitable for grass-covered playgrounds and picnic areas (costs based on the 1970 price level).

Requirements for the specified uses were set (Table 7.10) and the relative costs of making the land suitable for such use estimated (Figs. 7.10c, d). These estimates indicated that soil improvement costs for deciduous trees on soils with the thickest layer of peaty spoil were seven times that of improving sandy soils.

7.7 SOIL SURVEYS IN PLANNING

Many of the uses and interpretations of soil surveys described above are integrated in orderly land use planning. Soil surveys have so far been little used by British planners, but such data are regarded as useful or even essential guides to rational base resource management for urban development in other countries. A well-documented example describing use of soil surveys by a planning commission in Wisconsin, USA is summarized below.

Regional planning in south-east Wisconsin (source Bauer 1973)

The Southeastern Wisconsin Regional Planning Commission guides planning in the seven counties surrounding the urban area of Milwaukee. The commission operates in an advisory capacity only but its expertise and advice strongly influences local planning boards. Early in its existence, the need for information about resources was appreciated and the commission contracted with the Soil Conservation Service to accelerate soil survey of its area of interest and develop interpretations of the map units for a wide range of uses. Soil information is considered in almost every plan submitted by the commission to local authorities.

To illustrate how soils data have been utilized in the comprehensive planning programme for the Southeastern Wisconsin Region, use can best be related to the regional planning agency's basic functions of inventory, plan design and plan implementation (Bauer 1973).

Soils data and the inventory function

Reliable basic planning and engineering data collected on a uniform, areawide basis is absolutely essential to the formulation of workable development plans. Consequently, inventory becomes the first operational step in any planning process, growing out of programme design. One of the most important of the basic inventories carried out under the commission's work programme concerned the soils of the region. The area has been wholly glaciated and the glacial history has created complex soil relationships and extreme variability and intermingling of soils within even very small areas. The usefulness of generalized soils maps for definitive planning purposes within the region was, therefore, severely limited.

Table 7.10 Technical standards to be met by the soils in the 'Twiskepolder', so as to be suitable for deciduous forests and grass-covered playgrounds and picnic areas (Westerveld and van den Hurk 1973)

Land use	Soil conditions	Depth of water-table	Distance between land surface and water level in ditches	Other requirements
Deciduous forests (including *Acer, Alnus, Fagus, Fraxinus, Populus, Salix* and *Quercus*)	Topsoil of clay–containing mineral material as well as: >0.15 m thick for *Acer, Alnus, Fraxinus, Populus* and *Salix*; >0.50–0.60 m thick for *Fagus* and *Quercus*	>0.20 m for *Alnus, Populus* and *Salix*; >0.30 m for *Acer* and *Fraxinus*; >0.40 m for *Fagus* and *Quercus*	>1.00 m	pH-KCl >4.5–5.0
Grass-covered playgrounds and picnic areas (used intensively)	Topsoil of good permeability and containing: <5% organic matter <5% clay (particles <2 μm) <10% clay + silt (particles <50 μm)	>0.30 m	>1.00 m	Some micro-relief permitted

The widespread occurrence of soils having questionable characteristics for certain types of urban development, coupled with the glacial history of the area, indicated the need for detailed soil surveys as an absolute prerequisite for sound development planning. In order to fulfil the soils data requirements of the regional planning program, a co-operative cost-sharing agreement was negotiated with the Soil Conservation Service, US Department of Agriculture for the completion of modern detailed soil surveys of the entire region, together with the provision of interpretations for comprehensive planning purposes. Each kind of soil within the region was rated in terms of the inherent limitations for specific land uses and engineering applications. These ratings included presentation of the pertinent properties of each soil type relating to:

1. Potential agricultural use, including soil capabilities for common cultivated crops, crop yield estimates, woodland suitability group and crop adaptation.
2. Wildlife–soil relationships, including capability of the different kinds of soil to sustain various food plants and cover for birds and animals common to the region.
3. Non-farm plant material–soil relationships, including suitability of different kinds of soil for lawns, golf courses, playgrounds, parks and open-space reservations.
4. Soil–water relationships by kinds of soil, including identification of areas subject to flooding, stream overflow, ponding, seasonally high water-table, and concentrated runoff.
5. Soil properties influencing engineering uses, including depth to major soil horizons important in construction of engineering works, liquid limit, plastic limit, plasticity index, maximum dry density, optimum moisture content, mechanical analysis, classification, percolation rate, bearing strength, shrink-swell ratio, pH, depth to water-table, and estimated depth to bedrock if within approximately 6 m of the ground surface.

Interpretations of these properties of each soil type for planning purposes were also provided, including:

1. Suitability ratings for potential intensive residential, extensive residential, commercial, industrial, transportational, natural and developed recreational, and agricultural land uses.
2. Suitability ratings for septic tank disposal fields, building foundations for low buildings, trafficability, surface stabilization, road and railway sub-grade and earthwork uses.
3. Suitability ratings for use as a source material for road base, backfill, sand or gravel, topsoil and water reservoir embankments and linings.

4. Rating with respect to flooding potential, watershed characteristics, susceptibility to erosion and susceptibility to frost action.
5. Suitability for wildlife habitat and habitat improvement, lawns, golf courses, playgrounds, and parks and related open areas requiring the maintenance of vegetation.

In order to permit the efficient application of the soils data in graphic form, the Commission staff prepared interpretive maps at a scale of 1 : 24 000 reduced for publication at a scale of 1 : 48 000. The interpretive maps were prepared for various kinds of potential land use: agricultural; large lot residential without public sanitary sewer service; residential with public sanitary sewer service; industrial; transportation route location; and intensely developed recreational. Each interpretive map shows six soil limitation ratings; very slight, slight, moderate, severe, severe to very severe and very severe.

Use of soils data in plan design

One of the most important reasons for undertaking the regional soil survey in south-eastern Wisconsin was to provide data essential to the preparation of regional land use and supporting public works facilities plans. Since the process of plan design is essentially a problem in finding the least costly way to meet stated objectives, in any planning effort it is necessary to link geographic location with development costs. In this way, alternative plans can be explored and the least costly alternative which meets the agreed-upon objectives adopted. Detailed soil surveys provide the means for relating development costs to geographic location since development costs vary with soil type and since the soil types have been geographically mapped. Accordingly, the regional soil survey was utilized in the comprehensive planning programme for the Southeastern Wisconsin Region in the preparation of planning objectives and supporting standards, in the formulation of a regional land use, water resource management and related public works facilities plans.

Planning objectives and standards

Before plans can be prepared objectives must be formulated. Several objectives and standards formulated and adopted by the Commission in its land-use planning programme relate directly to the use of the regional soil survey and its interpretive analyses. For example:

1. Urban development, particularly for residential use, shall be located only in those areas which do not contain significant concentrations of soils rated in the regional detailed operational soil survey as having severe or very severe limitations for such development.

2. Rural development, principally agricultural land uses, shall be allocated primarily to those areas covered by soils rated in the regional soil survey as having only moderate, slight or very slight limitations for such uses.
3. Land developed or proposed to be developed for urban uses without public sanitary sewer service should be located only in areas covered by soils rated in the regional soil survey as having moderate, slight or very slight limitations for such development.
4. All prime agricultural areas, defined as those areas which contain soils rated in the regional soil survey as having only slight or very slight limitations for agricultural uses and which occur in concentrated areas over five square miles in extent that have been designated as exceptionally good for agricultural production by agricultural specialists, should be preserved.

In its comprehensive water-resource management planning programmes, the Commission has also adopted objectives and standards that relate to the regional soil survey and its interpretive analyses.

Comprehensive watershed plans
An important part of each Commission watershed study is the development of a mathematical model, used to simulate the hydrologic and hydraulic performance of the river system under study. Soil data are an important input and detailed soils maps are used to determine the predominant hydrologic soil group in each sub-basin.

Regional sanitary sewerage system plan
This plan has used the regional soil survey data in particular to identify within those areas proposed for urban development in the regional land-use plan land with soils suitable or unsuitable for septic tank sewage disposal systems. In addition, areas of bedrock outcrop, shallow bedrock and high groundwater table have been mapped and analysed as these factors may relate to the planning, design and provision of sanitary sewage facilities.

Soils data and regional plan implementation

Each Commission planning report that recommends for adoption a regional or sub-regional plan element contains specific plan implementation recommendations. Certain of these recommendations relate directly to, and often incorporate, the regional soil survey and its accompanying interpretive analyses. Important among such recommendations are those relating to the incorporation of soils data in local zoning regulations, land sub-division regulations, and health and sanitary regulations.

The soil survey and accompanying interpretations have been directly incorporated into local zoning ordinances in the following ways:

1. The creation of special zones related to certain kinds of soils, with particular emphasis in this respect on exclusive agricultural, conservancy and certain types of residential use.
2. The delineation of zone boundaries and the determination of special hazard areas, such as floodland.

Soils data have similarly been incorporated into local control ordinances, both in the form of general suitability clauses and in the form of special provisions relating to the control of grading operations, the design of drainage facilities, the removal of natural ground cover and control of erosion and sedimentation.

Perhaps most importantly with respect to implementation of the adopted land-use plan has been the incorporation of soils data into local health and sanitary regulations. Such regulations serve in effect to prohibit the installation of septic-tank sewage-disposal systems on soils having high water-tables, low permeability, or excessively high permeability, and on excessively steep slopes.

APPENDIX 1

Land Capability classes for forestry (McCormack 1972)

Class 1. Lands having no important limitations to the growth of commercial forests.

Soils are deep, permeable, of medium texture, moderately well-drained or imperfectly drained, have good water-holding capacity and are naturally high in fertility. Their topographic position is such that they frequently receive seepage and nutrients from adjacent areas. They are not subject to extremes of temperature or evapo-transpiration. Productivity is usually greater than 111 cubic feet per acre per annum.

Class 2. Lands having slight limitations to the growth of commercial forests

Soils are deep, well-drained or moderately well-drained, of medium or fine texture and have good water-holding capacity. The most common limitations are: adverse climate, soil moisture deficiency, restricted rooting depth, somewhat low fertility, and the cumulative effects of several minor soil characteristics. Productivity is usually from 91 to 110 cubic feet per acre per annum.

Class 3. Lands having moderate limitations to the growth of commercial forests

Soils may be deep to somewhat shallow, well-drained to imperfectly drained, of medium or fine texture with moderate or good water-holding capacity. They may be slightly low in fertility or suffer from periodic moisture imbalances. The most common limitations are:

adverse climate, restricted rooting depth, moderate deficiency or excess of soil moisture, somewhat low fertility, impeded soil drainage, exposure (in maritime areas) and occasional inundation. Productivity is usually from 71 to 90 cubic feet per acre per annum.

Class 4. Lands having moderately severe limitations to the growth of commercial forests

Soils may vary from deep to moderately shallow, from excessive through imperfect to poor drainage, from coarse through fine texture, from good to poor water-holding capacity, from good to poor structure and from good to low natural fertility. The most common limitations are: deficiency or excess of soil moisture, adverse climate, restricted rooting depth, poor structure, excessive carbonates, exposure or low fertility. Productivity is usually from 51 to 70 cubic feet per acre per annum.

Class 5. Lands having severe limitations to the growth of commercial forests

Soils are frequently shallow to bedrock, stony, excessively or poorly drained, of coarse or fine texture, can have poor water-holding capacity and be low in natural fertility. The most common limitations, often in combination, are: deficiency or excess of soil moisture, shallowness to bedrock, adverse regional or local climate, low natural fertility, exposure particularly in maritime areas, excessive stoniness and high levels of carbonates. Productivity is usually from 31 to 50 cubic feet per acre per annum.

Class 6. Lands having very severe limitations to the growth of commercial forests.

The mineral soils are frequently shallow, stony, excessively drained, of coarse texture and low in fertility. Most of the land in this class is composed of poorly-drained organic soils. The most common limitations, frequently in combination, are: shallowness to bedrock, deficiency or excess of soil moisture, high levels of soluble salts, low natural fertility, exposure, inundation and stoniness.

Class 7. Lands having severe limitations which preclude the growth of commercial forests

Mineral soils are usually extremely shallow to bedrock, subject to regular flooding, or contain toxic levels of soluble salts. Actively eroding or extremely dry soils may also be placed in this class. Most of the land has very poorly drained organic soils. The most common limitations are: shallowness to bedrock, excessive soil moisture, frequent inundation, active erosion, toxic levels of soluble salts and extremes of climate or exposure. Productivity is usually less than 10 cubic feet per acre per annum.

Capability sub-classes

Limitations affecting the classification of land are identified by capability sub-classes. These fall into four categories and symbols used are listed below.

1. Climate

A drought
H low temperatures
U exposure
C combinations of climatic factors

2. *Soil moisture*
   - M soil moisture deficiency
   - W excess soil moisture
   - X a complex pattern of M and W
   - Z a complex pattern of wet organic soils and bedrock
3. *Permeability and depth of rooting zone*
   - D physical impedance to rooting caused by dense or consolidated layers other than bedrock
   - R shallow bedrock
   - Y a complex pattern of shallowness and compaction or other restricting layers
4. *Other soil factors*
   - E eroding soils
   - F low fertility
   - I soils periodically flooded
   - K presence of perennially frozen material
   - L nutritional problems associated with large carbonate levels
   - N excessive levels of toxic elements
   - P stoniness
   - S some combination of soil factors which together but not separately lower the capability class

8

SOIL RESOURCES AND FOOD: A GLOBAL VIEW

Henry D. Foth

Two of the world's most important problems are rapid population growth during a period of diminishing non-renewable resources and increasing concern for the environment. Consideration of global soil resources is central to these problems. The nature of soil surveys has been examined in Chapter 2 and it is clear that they play a critical role in assessing the potential base for agricultural production and can be used in plans for agricultural development including the amount of investment needed to achieve various levels of soil productivity (Ch. 6). The greatest potential for increased arable land exists where population densities are low and where reconnaissance soil surveys suggest the presence of suitable soils. This final chapter concerns the extent of global soil resources, their adequacy to the end of the century and consideration for maintaining and increasing soil productivity for food production.

8.1 TRENDS IN POPULATION AND FOOD PRODUCTION

Human origins are obscure and date back a few million years. About 1 million years ago *Homo erectus* was using tools and was an efficient hunter in Africa and Asia. By 100 000 years BP *Homo sapiens* had evolved with full brain size and by 10 000 BP humans were on all continents and numbered about 5 million (Deevey 1960). Population growth during this period was in response to both increased land area occupied and improvements in food gathering and hunting techniques. Then, as settled agriculture developed a population explosion occurred that increased world population 16 times between 10 000 and 6000 BP (a doubling every 1000 years) and produced a population of about 86.5 million. World population then entered a period of reduced rate of growth and increased six or seven times to about 545 million over the next 6000 years or by 1650.

The scientific and industrial revolution that began about 1650 ushered in a period of increased resources for food production and created a new dimension to population growth – reduced death rates. As a consequence, recent annual growth rates of 1.9 per cent have produced a doubling of world population in less than 40 years. The world population in 1979 was over 4 billion and growing at an annual rate of 1.7 per cent, slightly less than 1.9 per cent of the 1960–70 period. It is staggering to realize that in another 40 years or so the world population will likely increase as much as in the previous 1 million years. There is some indication that birth control programmes are becoming effective in many countries and the UN estimates annual growth at 1.5 per cent by 2000. However, world population will be likely to increase to six billion or more by 2000 (Kane 1979).

Historically, increases in world food production have paralleled increases in world population. Trends of world food production since 1950 as reported by the United States Department of Agriculture are shown in Fig. 8.1. For the period 1950 to 1972 the increases in total food production were similar in the developed and developing countries. In 1972 total world food production was less than the preceding year. Droughts and other unfavourable weather conditions in 1972 caused a 1 per cent decline in food production and, with about a 2 per cent increase in population, produced a 3 per cent decrease in per capita food production. Total food production in the last decade has been increasing more rapidly in the developing than in the developed countries. In fact, there appears to be a leveling of total food production in the developed countries since 1973. For the 28-year period, 1950 to 1977, total food production in developed countries increased 100 per cent, compared to 140 per cent for the developing countries – increases that are phenomenal compared to historical world population growth rates (Fig. 8.1).

Average annual rates of world and regional food production since 1961 as reported by the FAO are given in Table 8.1. These data also show a slower annual rate of increase in world food production for the period 1970 to 1976 versus 1961 to 1970 and higher annual growth rates of food production for the developing countries compared to the developed countries. In addition, there are considerable regional differences, with Africa showing the lowest rate of increase of 1.2 per cent for 1970 to 1976. In fact, food per capita has declined in Africa in the years 1970 to 1976 because annual population growth has been 2.7 per cent. Food per capita in Africa is now about 90 per cent of the 1961 to 1965 period. Food production failed to match population growth during 1970–76 in no less than 50 (more than half) of the developing countries. Countries with negative annual increases in food production for years 1970–76 are: Mauritania, Benin PDR, Niger, Ethiopia, Chad, Mozambique, Somalia, Mali, Guinea, Ghana and Dem. Kampuchea (Cambodia). Except for Africa, all other major regional areas increased food production more rapidly than population. Since the annual rate of population increase in

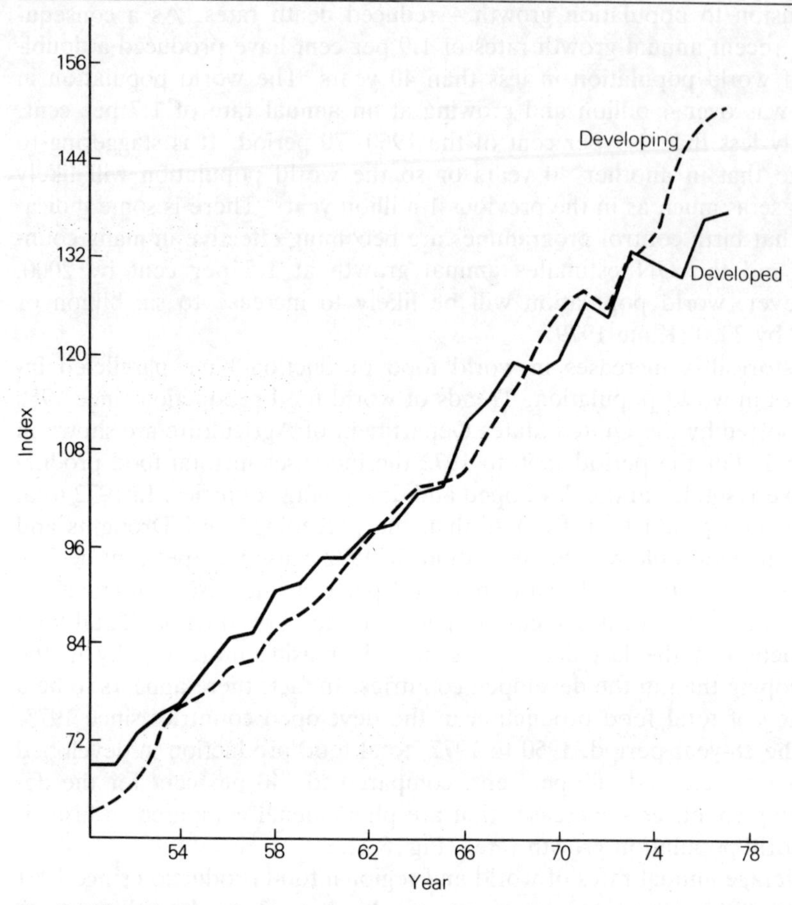

Fig. 8.1 Total food production in developed and developing countries, 1950–1977 with 1961–5 equal to an index of 100 (USDA)

the developing countries is greater than in the developed countries, food production per capita is increasing more slowly in the developing nations, resulting in a widening of the food gap between developed and developing nations.

Total agricultural production includes non-food items such as tea; tobacco and fibre. The agricultural production data for developed and developing countries is similar to that presented for food production. There is a likelihood that FAO data tend to overstate food needs and understate food supplies for developing nations because little is known in detail about the components of the food balance sheets. The trends, however, seem to be real as indicated by the large increases in food grain imports of the developing countries in recent years. Before 1950 many of these developing countries were food grain exporters (Brown 1975).

Table 8.1 Average annual rates of growth of food production in the world and regions, 1961–70 and 1970–76 (FAO 1977)

Region	Total		Per caput	
	1961–70	1970–6	1961–70	1970–6
	Per cent per annum			
Developed market economies	2.2	2.4	1.2	1.5
North America	1.9	3.1	0.7	2.1
Western Europe	2.3	1.6	1.6	1.0
Oceania	2.9	3.1	1.1	1.3
Other developed market economies	3.3	2.1	1.8	0.6
Eastern Europe and the USSR	2.9	1.9	1.9	1.0
All developed countries	2.4	2.3	1.4	1.4
Developing market economies	3.3	2.8	0.7	0.2
Africa	2.7	1.2	0.1	−1.4
Latin America	3.5	3.3	0.8	0.5
Near East	3.0	4.2	0.3	1.4
Far East	3.5	2.8	0.9	0.2
Other developing market economies	2.1	1.5	−0.4	−1.0
Asian centrally planned economies	2.7	2.4	0.9	0.6
All developing countries	3.1	2.7	0.7	0.3
World	2.7	2.4	0.8	0.5

Projected world food and agricultural production for the year 2000 will need to increase about 50 per cent just to maintain the status quo if population increases from 4 to 6 billion. In recent years, 20 per cent of the increased demand for food has been due to increasing affluence. Considering that about 400 million of the world's people suffer from serious malnutrition (Mayer 1976) and that rising incomes are reflected in demands for increased consumption of food, an increase in food production of about 75 per cent or a doubling by the year 2000 would appear necessary (Pimentel *et al.* 1976). Also it appears that some time in the future the world will need to provide for a stabilizing population of 12 billion or more, beyond the year 2100 (Mauldin 1978).

8.2 SOIL RESOURCES

About 65 per cent of the ice-free land of the world has a climate that is suited for some cropping. Only 10 per cent of the land is currently cultivated and this land is that which is generally the most suitable for cropping. To evaluate the potential area of arable land requires knowledge from soil surveys and an estimate of the extent of resource investment that could reasonably be expected to convert non-arable land into arable land. The area covered by soil surveys has greatly increased since 1950 and there have been numerous studies of soil properties and the response of soils to management practices. About 21 per cent of the world's soils have actually been surveyed, ranging from 7.5 per cent in Africa to 76 per cent in Europe (Dudal 1978). The boundaries of soil mapping units for the other 79 per cent are derived from interpretation of general information on land forms, geology, climate, vegetation and from scattered soil studies.

Many estimates of the world's potential arable land have been made. The President's Science Advisory Committee in the United States (1967) made an estimation based on a world soil map at a scale of 1 : 15 million and containing 13 broad geographical soil groups. Climatic boundaries were superimposed on this map, resulting in about 200 soil-climate combinations. Estimates were made of the potentially arable and non-arable land but with grazing potential. Potentially arable land was considered cultivatable and acceptably productive for food crops adapted to the environment. Agricultural technology equivalent to average in the United States was assumed. About 11 per cent of the potential arable land is irrigated or needs irrigation water that is potentially available for at least one crop per year. The study concluded that 24 per cent of the earth's ice-free land is potentially arable (see Table 8.2). This is over twice the area now cultivated and about three times the area of land harvested in any year.

The estimate of potentially arable land is considerably in excess of results of earlier studies. Over half of the potentially arable land is in the tropics, with 30 per cent having humid climate, 36 per cent with alternate wet and dry seasons and 34 per cent with arid and semi-arid climate. The major areas of potentially arable land not now cultivated are in tropical Africa and South America. One prediction was that the removal of vegetation in the tropics would create vast wastelands due to hardening of soft laterite or plinthite into hard laterite or ironstone (McNeil 1964). Plinthite forms on certain topographic–hydrologic positions where ground water seasonally carries reduced iron that is oxidized and precipitated and then accumulates as ferric iron. The extent of such active plinthite formations has been estimated to be less than 2 per cent of tropical lands (Moormann and Van Wambeke 1978). Estimates of the areal occurrence of laterite or ironstone are 2 per cent in tropical America, 5 per cent in central Brazil, 7 per cent for the tropical part of Indian subcontinent and 15 per cent for sub-Saharan West Africa (Sanchez and Buol 1975).

There is little opportunity in Asia and Europe for more cultivated land with 83 and 88 per cent, respectively, of the potential arable land now in

Table 8.2 Arable land compared to potentially arable land by continents (President's Science Advisory Committee)

Continents	Total land area*	Per cent of land area cultivated	Cultivated land*	Potential arable land*	Potential arable land increase*	Ratio of cultivated to potential arable land (%)
Africa	3010	5.2	158	734	576	22
Asia	2740	18.9	519	627	108	83
Australia and New Zealand	820	3.9	32	153	121	21
Europe	480	32.1	154	174	20	88
North America	2110	11.3	239	465	226	51
South America	1750	4.4	77	681	604	11
USSR	2240	10.6	227	356	129	64
TOTAL	13,150	10.6	1406	3190	1784	44

* Millions of hectares.

Adapted from 'The World Food Problem', A report of the President's Science Advisory Committee, White House, Washington, DC, 1967.

cultivation. The United States and USSR are intermediate and a low ratio of cultivated to potentially cultivatable land exists in Australia and New Zealand. North America has recently become the dominant exporting region for cereal foods and feed grains. With the considerable potential for more arable land, it is likely to become even more important in this regard in the future.

In 1961 the FAO and UNESCO in co-operation with the International Soil Science Society embarked on the production of a world soil map to scale 1 : 5 million with a common legend for all countries. Intensive soil correlation was needed and publication was completed with presentation of the entire world coverage on 19 maps at the 11th Congress of ISSS at Edmonton in 1978 (Ch. 3). This map was used by Buringh (1978) to group the world's soils into 222 broad soil regions. For each soil region details of soils, vegetation, topography and irrigation possibilities were studied. Taking these factors into consideration the following were determined: the area of potential arable land, the average soil productivity, the average water availability, the possible areas to be irrigated, the maximum production per hectare and the total maximum production. This was part of the Model of International Relations in Agriculture (MOIRA) study. The conclusion was that 3419 million hectares are potentially arable or 25 per cent of the world's ice-free land. The similar results of these two studies seem to affirm that the supply of arable land for the year 2000 and for some years beyond could be more than doubled and land is likely to be sufficient to meet world food needs when allowance is also made for increased yields in the years ahead.

The areas of major soils in the world which are potentially arable are given in Table 8.3. The three orders with the largest extent of potentially arable land are Alfisols, Mollisols and Oxisols. The first two, Alfisols and Mollisols, tend to be naturally fertile soils located mainly in the temperate and to less extent in subtropical zones, and are the basis for much of the agriculture in Europe, USSR, North America, northern China and Argentina. One of the largest areas of Alfisols, however, occurs south of the Sahara in Africa (see Fig. 8.2). Oxisols are naturally infertile soils of

Table 8.3 Potentially arable soils of the world by soil order (USDA 1971)

Order and generalized description	Total area*	Potentially arable		Per cent total land area	Approximate FAO equivalents
		area*	per cent		
Alfisols – argillic horizon and medium to high base status	1.73	0.64	37	13.2	Luvisols, Planosols, Podzoluvisols
Aridisols – with aridic moisture regime	2.47	0.08	3	18.8	Yermosols, Xerosols
Entisols – little or no pedogenic horizon development	1.09	0.15	14	8.3	Regosols, Lithosols, Rankers, Arenosols, Fluvisols
Histosols – organic soils	0.12	0.001	1	0.9	Histosols
Inceptisols – weakly developed without diagnostic illuvial horizons	1.17	0.23	20	8.9	Cambisols, Andosols, Gleysols
Mollisols – base rich soils of steppes with mollic epipedons	1.13	0.63	56	8.6	Chernozems, Greyzems, Phaeozems, Rendzinas
Oxisols – oxide rich soil of tropics, with oxic horizon	1.12	0.65	58	8.5	Ferralsols
Spodosols – with spodic horizon, illuvial horizon enriched with amorphous material	0.56	0.10	17	4.3	Podzols
Ultisols – argillic horizon and low base status	0.73	0.27	37	5.6	Acrisols, Nitosols
Vertisols – clayey, cracking soils	0.24	0.14	59	1.8	Vertisols
Soils in mountainous areas	2.59	0.23	9	19.7	

* Billions of hectares.

Adapted from 'Estimates of Areas of Potentially Arable Land', compiled by Soil Conservation Service USDA, mimeograph, 1971.

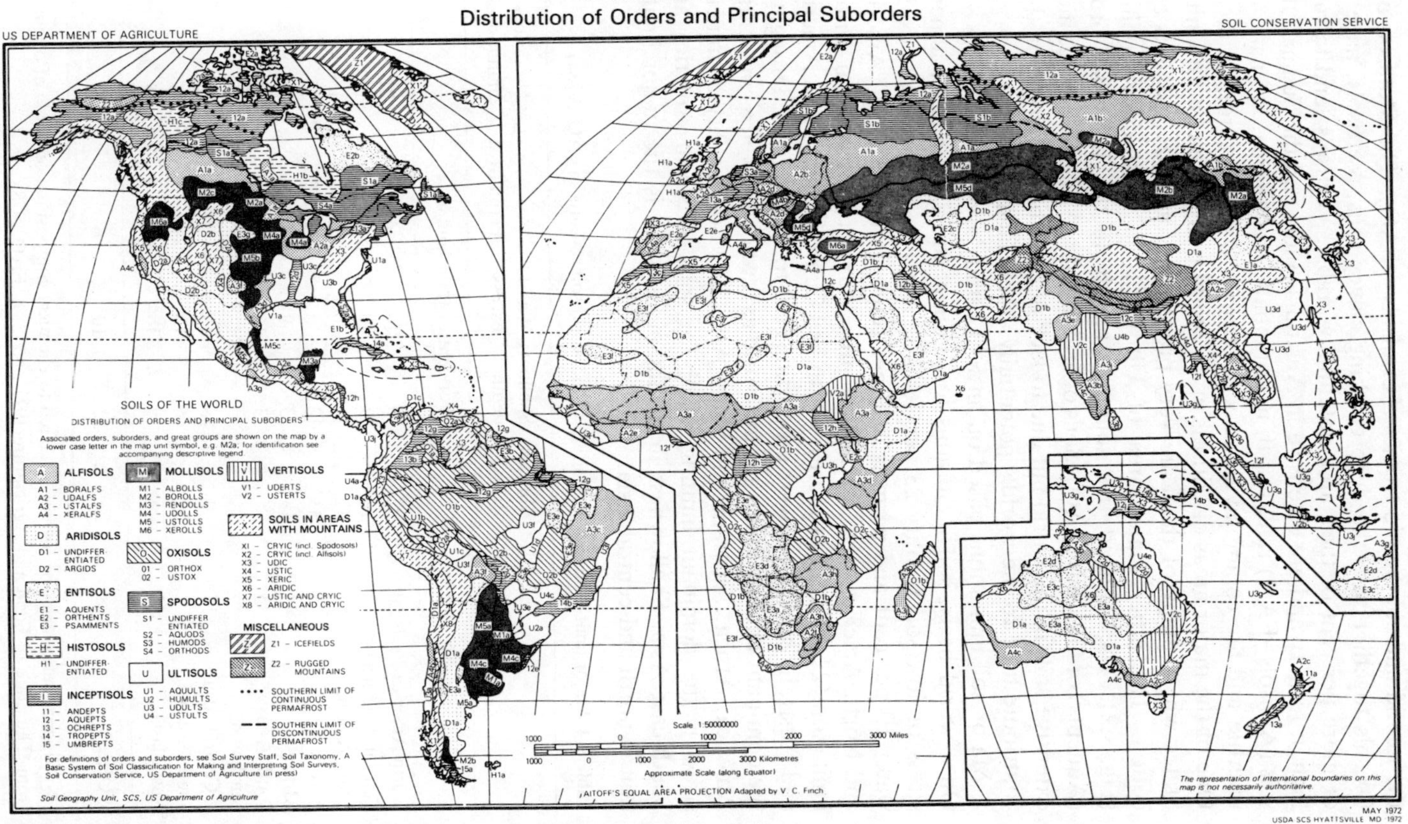

Fig. 8.2 Soils of the World. Distribution of Orders and principal Sub-orders (Soil Geography Unit, USDA–SCS Hyattsville Maryland 1972)

the humid and sub-humid tropics (Ferralsols – FAO). The Mollisols and Oxisols together with the deep cracking clayey soils, Vertisols, have 56 or more per cent of their area potentially arable. Vertisols tend to rank high in fertility and are extensive in Australia, India and the Sudan. Ultisols (Acrisols–FAO) rank fourth in potentially arable area, are naturally infertile and are commonly in association with Oxisols in tropical areas. One of the major areas of Ultisols in the world is in south-eastern United States where many of them are very productive for maize, peanuts and cotton. The other major area of Ultisols is in south-eastern China (Fig. 8.2). The wetland rice soils are mainly in the Inceptisol order (Aquepts) or Gleysols (FAO). Soils of arid regions, Aridisols, are the most extensive soils in the world but due to lack of potential for irrigation they contribute little to the world's potentially arable land.

Major soil limitations based on an FAO study are given in Table 8.4. On a world basis, soil limitations for drought are 28 per cent, mineral stress or low soil fertility 23 per cent, shallowness 22 per cent, excess water 10 per cent and permafrost 6 per cent. Europe, Central America and North America have the highest proportion of soils with no serious limitation (22 per cent or more). North and Central Asia, South America and Australia have the lowest proportions (10 to 18 per cent). Drought is the dominant limiting factor for the world as a whole as well as in Central America, Africa, Australia and South Asia. Low soil fertility or mineral stress is the dominant limiting factor for soils in South America and South-east Asia. Shallowness in mountainous areas is the major limitation in north and central Asia.

Table 8.4 Major limitations for agriculture of the world soil resources (FAO 1978a)

	Drought	Mineral stress*	Shallow depth	Water excess	Permafrost	No serious limitations
	Per cent of total land area					
North America	20	22	10	10	16	22
Central America	32	16	17	10	–	25
South America	17	47	11	10	–	15
Europe	8	33	12	8	3	36
Africa	44	18	13	9	–	16
South Asia	43	5	23	11	–	18
North and Central Asia	17	9	38	13	13	10
South-east Asia	2	59	6	19	–	14
Australasia	55	6	8	16	–	15
WORLD	28	23	22	10	6	11

* Nutritional deficiencies or toxicities related to chemical composition or mode of origin.
From 'The State of Food and Agriculture 1977', pp. 3–4, FAO 1978a

The FAO is conducting a global land evaluation study so that more and better information on global soil resources will be forthcoming in the future (Dudal 1978). This study is designed to provide more precise information for planning future agricultural development. The methodology to assess the agricultural potential of the world's land resources is based on six principles as given in Chapter 6.

These principles are among those formulated in 'A Framework for Land Evaluation' (FAO 1976) developed over the past years through international cooperation. A critical need in the future will be more and better soil surveys to make possible a matching of effective management systems and soil types to achieve high and efficient production.

8.3 SOIL RESOURCES FOR FOOD PRODUCTION

Historically, increased land area for food production has accompanied population growth. This is happening now and is expected to continue. During the period 1955 to 1977 the arable land in the world increased 9 per cent and world population increased 40 per cent. World cereal production from 1950 to 1975 increased 98 per cent, with 22 per cent due to increased land area and 66 per cent due to increased yields (Economic Research Service 1976, 1977). Developing countries have been relying more on increased land area and developed countries have been relying more on increased yields to increase food production. In North America during the period 1950 to 1975 land area in cereals decreased 10 per cent and yields per hectare increased 100 per cent, resulting in an 81 per cent increase in total production. Increased food production in the developing countries as a whole has been about half due to increased land and half due to increased yields (Hutchinson 1969). Even though there is great potential to increase arable land area, there will be pressure to increase yields in both developing and developed countries because many areas have high population density and limited unused land and many areas have low population density and very low yields that need to be increased to increase productivity of both land and people.

Most of the unused global potential to increase yields and extend arable land is in tropical Africa and South America. There are many kinds of soils in the tropics and one of the most important distinctions is that of high base status versus low base status soils or fertile versus infertile soils. The high base status soils contain significant amounts of primary minerals that weather to release plant nutrients on a sustained basis. Many high base soils are acid but not severely acid (pH generally 5.5 or higher) and some are alkaline and calcareous. High base status soils include young soils developed from volcanic ash and alluvium scattered throughout the tropics. Areas where high base soils are extensive include those dominated by Alfisols, Aridisols and Vertisols along the northern and drier tropical areas in Africa (see Fig. 8.2). Intensive agriculture has been developed on Alfisols in north-western Europe and north-central United States. The technology of this agriculture has been successfully transferred to high base soils of the tropics in the Green Revolution, resulting in large increases in wheat and rice yields. Attempts to transfer the same soil management practices to low base status soils, however, have generally been unsuccessful although some successful plantations have been established. The greatest potential for an extension of arable land (and increased yields) in the tropics, however, exists on the highly weathered, low base status soils. The area of low base soils potentially arable in the tropics is eight times the area of high base soils, excluding the alluvial soils (Hendricks 1969).

The low base status soils are mainly Oxisols (Ferralsols) and Ultisols (Acrisols) that are low in fertility due to the great amount of time for weathering and leaching. The only significant primary mineral that has not weathered is quartz (SiO_2) and bases are contained mainly in organic matter or on exchange sites. The soils are very acid with aluminium as the dominant exchangeable cation. The soils have very little or no ability to supply plant nutrients on a sustained basis, have toxic concentrations of aluminium for many plants and are very low in available phosphorus.

The importance of the difference in base status is shown by a comparison of soils on the islands of Java and Borneo (Kalimantan). Mt Merapi on Java is the dominant volcano of the region and has been very active since 1006. The soils on Java are young and derived mainly from volcanic materials or alluvium and solifluction debris containing volcanic material that produces soils rich in mineral nutrients. More than 87 per cent of Java is cultivated and the population is 80 million. By contrast, Borneo, only 400 kilometres north with similar climate and topography, has mainly low base soils. The soils on Borneo are older and less influenced by volcanic ejecta. The small amount of volcanic effusives present have a high silica content (from rhyolites and dacites) which leads to acid low base status soils. Areas of fertile soils do occur, but throughout Borneo as a whole the clays are less than 5 per cent base saturated and the soils infertile. Although the island is four times larger than Java, it supports, at present, a population of only 3 million (Hendricks 1969).

While Oxisols and Ultisols tend to have quite similar soil fertility characteristics, the soils are very dissimilar in their physical characteristics. Oxisols tend to have a fine stable micro-structure throughout that gives soil water infiltration and retention characteristics similar to quartz sand soils. Water infiltrates rapidly and soils resist erosion and show great resistance to compaction. They are droughty, however, and the droughtiness is accentuated where aluminium toxicity in subsoils produces shallow rooting. On those soils of the inter-tropical zone which have the greatest rainfall and least likelihood of drought, the yield potential is limited by

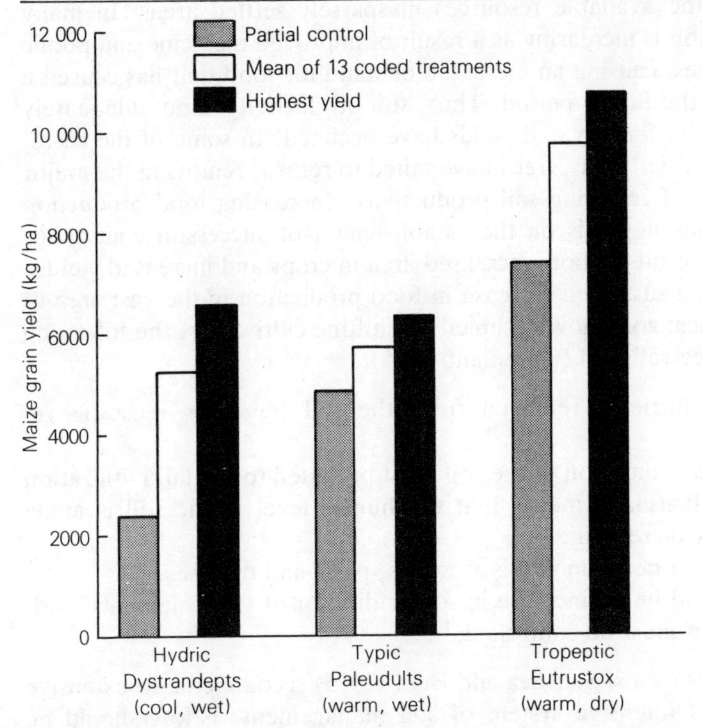

Fig. 8.3 Soils with warm and dry climate (with irrigation if needed) have greater yield potential due to increased solar radiation as compared with soils with wetter climate in the tropics (based on Benchmark Soils project, 1979)

low radiation caused by cloudy days. In comparison, on those soils of the drier regions, where rainfall is less, drought is more frequent but there is greater photosynthetic potential as a result of greater solar radiation and yields are higher. (see Fig. 8.3).

Physically, Ultisols compared to Oxisols tend to have a less stable structure. Soils are more subject to compaction and disintegration from raindrop impact and tillage, causing high water runoff and erosion on unprotected sloping land. The clay enriched subsoils (argillic horizons) also contribute to a reduced infiltration rate, but the available water capacity is better than that of the Oxisols.

About 250 million shifting cultivators use an area for cropping and fallow that greatly exceeds the present amount of world crop land (Grigg 1974). Shifting cultivators on low base soils leave the land in fallow five to 15 years for trees and shrubs to accumulate nutrients in biomass and regenerate the productivity of the soil. After a short period of one to five years of cropping the soil fertility is depleted and the land returned to fallow (Nye and Greenland 1960). It can be seen that the system makes good use of the available resources in sparsely settled areas. In many areas population is increasing as a result of improved medicine and public health measures, causing an increased demand for food that has caused a shortening of the fallow period. Thus, soil productivity is not adequately restored and soil fertility and yields have declined. In some of the cases, particularly in drier areas, trees have failed to return, removing the major natural means of restoring soil productivity. Increasing food production in these regions depends on the establishment of successful continuous cropping that results in both increased area in crops and increased yields.

To produce a sustained increase in food production in the vast area of the inter-tropical zone now occupied by shifting cultivators, the following criteria must be satisfied (Greenland 1975):

1 Chemical nutrients removed from the soil by crops must be replenished.
2 The physical condition of the soil must be suited to the land utilization type, which usually means that the humus level in the soil is made constant or increasing.
3 There should be no increase of weeds, pests and diseases.
4 There should be no increase in soil acidity, or of toxic elements, and
5 Soil erosion must be controlled.

On the low base soils, Sanchez and Buol (1975) recommend an extensive rather than an intensive system of soil management. Effort should be directed toward adapting practices to suit the soil rather than greatly modifying the soil. For example, the acid soils will need to be used with little or no lime to increase soil pH because of the general unavailability of lime and high cost. High rates of nitrogen fertilizer, so important in the Green Revolution on high base status soils, cannot be used because most nitrogen fertilizers are acid forming and this would further accentuate the already precarious acid condition with respect to aluminium toxicity and nutrient availability (Foth 1978). Additionally, the cost of the fertilizer would be prohibitive. The next step for shifting cultivators would be to develop management systems involving the application of modern scientific knowledge about plant growth, water control, use and function of organic matter, plant breeding, and so on, but with only a minimum of such products of modern industry as machinery, electric power and chemical fertilizers (Kellogg and Orvedal 1969).

Much of the work at the International Institute of Tropical Agriculture (IITA) in Nigeria and some other tropical research stations has been directed in recent years toward the development of farming systems that enable small farmers to benefit from improved varieties. Land clearing methods should be traditional because more organic matter remains in the soil and higher yields and less soil compaction occur, compared with the use of heavy machinery (Nicholaides 1979). Rudimentary tools such as sticks and hoes can be used in zero or minimum tillage systems, and when coupled with continuous plant cover, will protect the soil from erosion and degradation. Breeding programmes are needed to develop higher yielding varieties that are capable of growing in strongly acid and infertile soils in multiple cropping systems. Multiple cropping will not only help to provide continuous plant cover but will also increase variety of diets, reduce the risk of crop failure, provide some control of pests, increase total production and spread out harvesting operations. The multiple cropping system must include legumes to add nitrogen and organic matter to the soil. Kudzu is a legume that is acid tolerant and in recent experiments has shown good promise of not only adding nitrogen to the soil but also grows so abundantly that its residues can be burned and converted to ashes that replenish the soil with bases and slightly reduce soil acidity. In some cases lime may be available locally and used. Some phosphatic fertilizer will be needed to replace phosphorus removed by cropping and to increase available phosphorus to permit utilization of higher yielding varieties. The phosphate can be low grade rock phosphate that will weather in the acid soils and be nearly as effective as the more expensive superphosphate (Sanchez *et al.* 1979). In some cases micro-nutrients may be needed in small quantities. Gravity irrigation will be possible in some cases. These practices should be supplied as a package to take advantage of interactions that cause the yield from use of the total package of practices to be greater than the sum of yields increases of each practice applied alone. Preliminary results show that low-input systems can be highly productive and profitable for pasture production to produce beef on low base soils in Colombia (Sanchez *et al.* 1979).

In a similar way research knowledge can be used to increase food production in semi-arid regions. The major problem on these soils is not low soil fertility but limited water availability, because most of the people living in these regions do not irrigate or irrigation is not feasible. The emphasis is to make more efficient use of the precipitation that is available. Systems to increase water-use efficiency have been worked out at ICRISAT at Hyderabad, India for both Alfisols and Vertisols (Krantz 1979). One system includes use of bullocks to make ridges or bunds in the dry season that are effective in increasing infiltration, and reducing water runoff and erosion. Some crops like sorghum, cowpeas and maize are deep planted (5 to 7 centimetres) and fertilized a week or so before the rainy season is expected. If precipitation is slight, it will be insufficient to moisten the deep planted seeds and cause germination. Germination will be delayed until sufficient water is stored in the soil to support the seedlings following germination without danger of death from dehydration (Krantz and Kampen 1979). Water is utilized more efficiently by crops and the early planting and germination lengthens the growing season. Again, basic research is used to increase production with little dependency on high-cost inputs from the outside and with little alteration of existing systems. As in all cropping systems, varieties must be developed for the local environment and practices that produce higher yields.

To achieve success in increasing yields on the great variety of soils and in different climatic conditions will require reconnaissance soil surveys to locate promising areas, and detailed soil surveys for planning operations. Agrotechnology transfer experiments are needed to successfully match the soils with effective soil management systems. Transfer experiments on three extensive soils have been conducted at widely separated sites in Brazil, Cameroon, Hawaii, Indonesia, Phillipines and Puerto Rico (Silva and Beinroth 1979). The soil family classification formed the basis of soil management considerations. Thixotropic, isothermic, Hydric Dystrandepts, for example, are acid volcanic ash soils with cool, humid climate. These soils have high phosphorus fixing capacity and need high phosphorus fertilization and lime to reduce soil acidity for some crops. The soil requires no irrigation and is adapted to relatively cool season crops (Table 8.5).

Fertilizer response is a land quality that can be measured quantitatively and experiments have shown that soils of the same family at widely separated locations have similar fertilizer response surfaces (Fig. 8.4). Different yield levels at Hawaii versus the Phillipines are due largely to temperature and radiation differences. Higher yields of maize at Hawaii, compared to the Phillipines, appear due to lower night time temperature and less night time plant respiration. Experimental sites on Hawaii also had about 12 per cent greater solar radiation per day. Similar response surfaces to fertilizers have been obtained for clayey, kaolinitic, isohyper-

Table 8.5 Properties of thixotropic, isothermic Hydric Dystrandepts to soil management considerations (USDA Benchmark Soils Project, 1978)

Taxonomic name	Soil characteristics	Management considerations
thixotropic	High surface activity of colloids	High amounts of phosphorus must be added
isothermic	Warm soil temperatures (15°–22°C)	Crops may be grown year round
Hydric	Moist humid soils	No irrigation or only supplemental irrigation required
Dystr	Low base saturation ($< 50\%$)	High fertilizer input required; lime application required in some cases
andept	Low bulk density (< 0.85 g/cm^3)	Low power requirement for tillage; good drainage and infiltration characteristics

From Benchmark Soils Project, 1978.

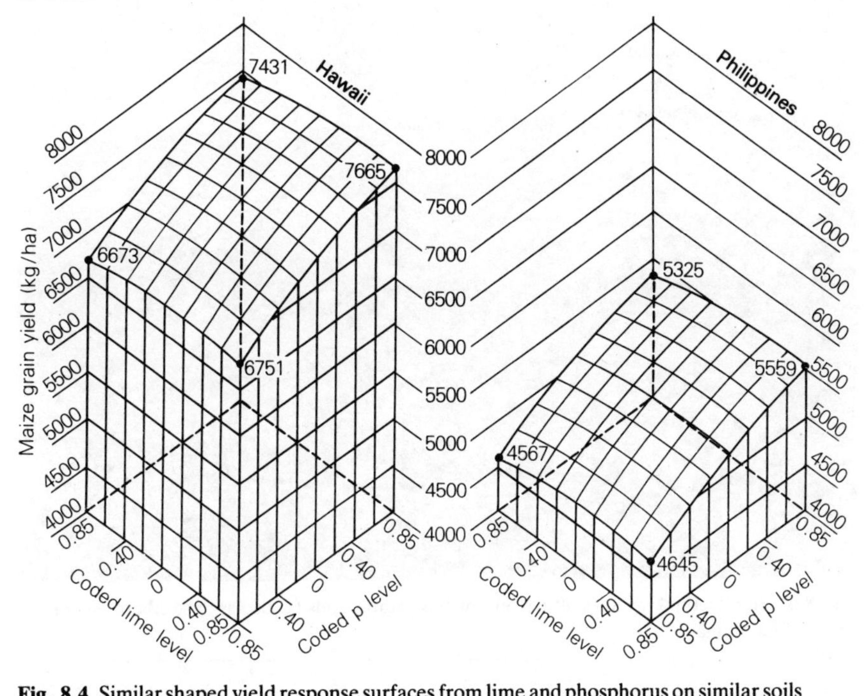

Fig. 8.4 Similar shaped yield response surfaces from lime and phosphorus on similar soils (Hydric Dystrandepts) at widely separated locations but different climate. Maximum yields (7665 and 5559 kg/ha) on both soils occurred with lowest lime and highest phosphorus treatments (Based on Benchmark Soils Project, 1978).

thermic Tropeptic Eutrustox (warm, dry Oxisols with over 50% base saturation) and clayey, kaolinitic, isohyperthermic Typic Paleudults (warm, humid Ultisols with very low base status).

Programmes to improve traditional farming systems that depend on a minimum of investment and alteration of existing living patterns and avoid dependency on large inputs of fossil fuel can be developed and successfully transferred. Where successful, the increased yields might provide extra food that could be sold. Gradually, additional investment as a knapsack sprayer could be added to the programmes for better pest control with, hopefully, additional increases in productivity and extra products for markets. An accompanying industrial development would be needed to absorb the agricultural surplus and provide goods and services for agriculture. Eventually, if incentives are sufficient and there is a favourable social and political environment, mechanization might proceed. There is the eventual possibility of evolution into a sophisticated agriculture like that on Oxisols in southern Brazil or on Ultisols as in south-eastern United States (Fig. 8.5). The period of evolution can be

Fig. 8.5 High technology agriculture on low base status soils (Oxisols) in southern Brazil, north of Campinas.

shorter than that of Europe or the United States because of the better research base and greater potential for population control today.

If governments place sufficient emphasis on developing a suitable economic, political and social environment, there is no question that the soil resources can be used to greatly expand food production in the world. Farmers must have the incentives and people must have money to buy food. Industrial development must parallel agricultural development. The use of research into methods not dependent on great inputs of fossil fuel provides the opportunity for evolution towards a much more advanced agriculture in developing areas than now exists and, perhaps, in time, other sources of fuel will become available. In the future perhaps production of biomass for energy production will be another crop for farmers. There are as many obstacles as they were in Europe 1000 years ago or in the United States 300 years ago. In parts of south-east Asia, current high population density presents a special problem in the light of limited emigration opportunities and ineffective population control. Man has shown great ingenuity and has overcome obstacles in the past, and there is no reason to suspect this will change much in the next century. Hopefully, as the third major population increase levels off, there will be a levelling of demand for food production throughout the world and with it a stable population and potential for meaningful life, including sufficient food with good health.

The use of modern scientific knowledge without reliance on fossil fuel to set in motion a more productive agriculture for large areas of the tropics seems realistic. There are, however, many difficulties. Predictions for the future are pessimistic as well as optimistic. One omen is that yields in the developed countries are levelling off (Jensen 1978). This is not all that serious in light of the fact that their population growth is slowing and excellent possibilities exist to support stabilized populations. It is not realistic to expect that food from the developed countries will feed the developing countries on a permanent basis because most food has been and will continue to be produced in regions where it is consumed. The developed countries also have significant land for future development, if needed.

Large areas of cropland are losing their productivity due to erosion, salinization, desertification, and rapid urban population growth is expanding on to agricultural land (Brown 1978). The recent southward spread of the Sahara attracted world attention through the drought in the Sahelian zone of Africa. Overgrazing, overploughing and deforestation are causing expansion of the world's major deserts. Over half of the irrigated lands have been developed since 1950, but already, 10 per cent are waterlogged and about 10 per cent have reduced productivity due to salinization. The result is about 4 per cent reduction in productivity of the world's irrigated land. Erosion preferentially removes the finer particles from soils, lowering productivity and increasing future production costs. About 2 per cent of current cropland will be lost to urban use by the year 2000. As has been described in Chapter 7, urban expansion is commonly on to good agricultural land because many cities are located on nearly level sites and fertile soils. As an example, Canada is a major grain exporter today, but concern has been raised about food and self-sufficiency by the year 2000. Population growth of 20–45 per cent is expected, cities are expanding on to prime agricultural lands and soil salinization from saline seep is reducing soil productivity (Bentley 1978). The amount of land per person for cereal grain production is expected to decline from 0.184 ha in 1975 to 0.128 ha in 2000 (Brown 1978).

If one assumes that all future development must be based on use of non-renewable resources, it is easy to calculate that there can be no development for the developing countries and the picture is indeed bleak. In part it is doubtful if the environment could tolerate the burning of fossil fuels and use of chemicals required to have western agriculture adopted everywhere. Many uncertainties exist in the future and there are some obstacles that need to be recognized as inhibitions to development. Some obstacles will be likely offset by unforeseen developments and new technology and scientific findings. More salt-tolerant plants for irrigated regions, greater or more widespread symbiotic nitrogen fixation by food crops and development of new cultivars with cloning techniques are only three of the possibilities of the future. For that time there can be hope if:

1 population control programmes become effective, and
2 incentives are provided to farmers to make use of scientifically sound practices that minimize use of non-renewable resources and farmers respond because the social and political climate is favourable.

Then, there can be an agricultural evolution in the developing countries and eventually another long period of relatively stable world population and food production.

BIBLIOGRAPHY

Aandahl, A. R. (1948) 'The characterization of slope positions and their influence on the total nitrogen content of a few virgin soils of Western Iowa', *Soil Science Society of America Proceedings*, **13**, 449–54.

Aandahl, A. R. (1958) 'Soil survey interpretation – theory and purpose', *Soil Science Society of America Proceedings*, **22**, 152–4.

Adams, W. A. and Raza, M. A. (1978) 'The significance of truncation in the evolution of slope soils in mid-Wales', *Journal of Soil Science*, **29**, 243–57.

Adams, W. A., Stewart, V. I. and Thornton, D. J. (1971) 'The construction and drainage of sportsfields for winter games in Britain', in Hornung, M. (ed.), *Soil Drainage*, Welsh Soils Discussion Group Report No. 12, 85–95.

Agricultural Land Service (1966) *Agricultural Land Classification*, Ministry of Agriculture, Fisheries and Food Technical Report No. 11.

Ahnert, F. (1970) 'A comparison of theoretical slope models with slopes in the field', *Zeitschrift für Geomorphologie, Supplementband*, **9**, 88–101

Al-Abbas, A. H., Swain, P. H. and Baumgardner, M. F. (1972) 'Relating organic matter and clay content to the multispectral radiance of soils', *Soil Science*, **114** (6), 477–85.

Allemeier, K. A. (1973) 'Application of pedological soil surveys to highway engineering in Michigan', *Geoderma*, **10**, 87–98

Anderson, K. E. and Furley, P. A. (1975) 'An assessment of the relationship between surface properties of chalk soils and slope form using principal components analysis', *Journal of Soil Science*, **26**, 130–43.

Applied Geochemistry Research Group (1978) *The Wolfson Geochemical Atlas of England and Wales*, Clarendon Press, Oxford.

Arkley, R. J. (1976) 'Statistical methods in soil classification and research', *Advances in Agronomy*, **28**, 37–70.

Arnett, R. R. and Conacher, A. J. (1973) 'Drainage basin expansion and the nine unit landsurface model', *Australian Geographer*, **12**, 237–49.

Arnold, P. W. and Close, B. M. (1961) 'Release of non-exchangeable potassium from some British soils cropped in the glasshouse', *Journal Agricultural Science, Cambridge*, **57**, 295–304.

Aubert, G. (1965) 'La classification pédologique utilisée en France', *Pédologie*, Gand, numéro spécial 3, 25–56.

Aubert, G. and Duchaufour, Ph. (1956) 'Projet de classification des sols', *Transactions 6th International Congress of Soil Science*, Paris, France, E, 597–604.

Avery, B. W. (1955) *Soils of the Glastonbury district of Somerset*, HMSO, London.

Avery, B. W. (1956) 'A classification of British soils', *Transactions 6th International Congress of Soil Science*, Paris, France, E, 279–85.

Avery, B. W. (1962) 'Classification of soils', *Aslib Proceedings*, **14**, 234–8.

Avery, B. W. (1965) 'Soil classification in Britain', *Pédologie*, Gand, numéro special 3, 75–90

Avery. B. W. (1968) 'General soil classification: Hierarchial and co-ordinate systems', *Transactions 9th International Congress of Soil Science*, Adelaide, Australia, **4**, 169–75.

Avery, B. W. (1970) 'Soil surveying and soil variability', in E. M. Bridges (ed), *Soil Heterogeneity and Podzolization*, Welsh Soils Discussion Group Report No. 11.

Avery, B. W. (1973) 'Soil classification in the Soil Survey of England and Wales', *Journal of Soil Science*, **24**, 324–38.

Avery, B. W. (1980) *System of Soil Classification for England and Wales (Higher Categories)*, Technical Monograph 14, Soil Survey of Great Britain, Harpenden.

Avery, B. W. and Bullock, P. (1969) 'The soils of Broadbalk – Morphology and Classification of Broadbalk Soils', *Rothamsted Experimental Station Report for 1968*, Harpenden.

Avery, B. W., Findlay, D. C. and Mackney, D. (1975) *Soil Map of England and Wales, 1 : 1 000 000*, Ordnance Survey, Southampton.

Baldwin, M., Kellogg, C. E. and Thorp, J. (1938) 'Soil classification', in *Soils and Men, Yearbook of Agriculture*, US Department of Agriculture, US Government Printing Office, Washington, 979–1001.

Ball, D. F. (1960) *The Soil and Land use of the District Around Rhyl and Denbigh*, Memoir of the Soil Survey of England and Wales, Harpenden.

Ball, D. F. and Williams, W. M. (1968) 'Variability of soil chemical properties in two uncultivated brown earths', *Journal of Soil Science*, **19**, 379–91.

Ballantyne, A. K. (1963) 'Recent accumulation of salts in the soils of south-eastern Saskatchewan', *Canadian Journal of Soil Science*, **43**, 52–8.

Barker, G. and Webley, D. (1978) 'Causewayed camps and early Neolithic economies in central southern England', *Proceedings of the Prehistoric Society*, **44**, 161–86.

Barshad, I. (1958) 'Factors affecting soil formation', *Clays and Clay Minerals, Sixth National Conference*, pp. 110–32.

Bartelli, L. J. *et al.* (eds.) (1966) *Soil Surveys and Land Use Planning*, Soil Science Society America and American Society of Agronomy, Madison, Wisconsin.

Bashir, H. E. S. E., Bushara, Y. I. M., Mohammed, Y. Y., Hassan, E. M. A., Ibrahim, K. A. K. and Mohammed, A. R. A. A. (1978) *Remote sensing in the Sudan* International Development Research Centre, Ottawa, IDRC-TS9e, 36 pp.

Basinki, J. J. (1959) 'The Russian approach to soil classification and its recent development', *Journal of Soil Science*, **10**, 14–26.

Bauer, K. W. (1973) 'The use of soils data in regional planning', *Geoderma*, **10**, 1–26.

Beckett, P. H. T. (1978) 'The rate of soil survey in Britain', *Journal of Soil Science*, **29**, 95–101.

Beckett, P. H. T. and Bie, S. W. (1975) 'Reconnaissance for soil survey 1. Pre-survey estimates of the density of soil boundaries necessary to produce pure mapping units', *Journal of Soil Science*, **26**, 144–54.

Beckett, P. H. T. and Webster, R. (1971) 'Soil variability: a review', *Soils and Fertilizers*, **34**, 7–15.

Beek, K. J. (1978) *Land Evaluation for Agricultural Development*, Publication No. 23, International Institute for Land Reclamation and Improvement, Wageningen, The Netherlands.

Bentley, C. F. (1978) Canada's agricultural land resources and the world food problem in *Trans 11th Int. Congr Soil Sci*, **2**, 1–26, Edmonton.

Bibby, J. S. and Mackney, D. (1969) *Land Use Capability Classification*, Technical Monograph No. 1, Soil Survey, Harpenden.

Bidwell, D. W. and Hole, F. D. (1965) 'Man as a factor of soil formation', *Soil Science*, **99**, 65–72.

Bie, S. W. and Beckett, P. H. T. (1971) 'Quality control in soil survey. Introduction: I. The choice of mapping unit', *Journal of Soil Science*, **22** (1), 32–49.

Birch, H. P. and Friend, M. T. (1956) 'The organic matter and nitrogen status of East African Soils', *Journal of Soil Science*, **7**, 156–67.

Birkeland, P. (1974) *Pedology, Weathering and Geomorphological Research*, Oxford University Press, Oxford.

Blume, H.-P. (1968) 'Die pedogenetische Deutung einer Catena durch die Untersuchung der Bodendynamik', *Transactions of the Ninth International Congress of Soil Science*, Adelaide, **4**, 441–9.

Blume, H.-P. and Schlichting, E. (1965) 'The relationships between historical and experimental pedology', in *Experimental Pedology*, Hallsworth, E. G. and Crawford, D. V. (eds.), 340–53, Butterworths, London.

Boast, C. W. (1973) 'Modeling the movement of chemicals by soil water', *Soil Science*, **115**, 224–30.

Borchardt, G. A., Hole, F. D. and Jackson, M. L. (1968) 'Genesis of layer silicates in representative soils in a glacial landscape of southeastern Wisconsin', *Proceedings of the Soil Science Society of America*, **32**, 399–403.

Bown, C. J. and Heslop, R. E. F. (1979) *The Soils of the Country round Stranraer and Wigtown*, Memoir of the Soil Survey of Great Britain.

Boyd, D. A. and Dermott, W. (1964) 'Fertilizer experiments on main-crop potatoes 1955–1961', *Journal of Agricultural Science, Cambridge*, **63**, 249–63.

Bradley, R. (1978) *The Prehistoric Settlement of Britain*, Routledge and Kegan Paul, London.

Bradley, R. I., Rudeforth, C. C. and Wilkins, C. (1978) 'Distribution of some chemical elements in the soils of north west Pembrokeshire', *Journal of Soil Science*, **29** (2), 258–70.

Bridges, E. M. (1966) *The Soils and Land Use of the District North of Derby*, Memoir of the Soil Survey of England and Wales, Harpenden.

Bridges, E. M. (1978a) *World Soils*, 2nd edition, Cambridge University Press.

Bridges, E. M. (1978b) 'Interaction of soil and mankind in Britain', *Journal of Soil Science*, **29**, 125–39.

Bridges, E. M. (1978c) 'Soil, the vital skin of the earth', *Geography*, **63**, pt. 4, 354–61.

Brinkman, R. and Smyth, A. J. (eds.) (1973) *Land Evaluation for Rural Purposes*, Publication No. 17, International Institute for Land Reclamation and Improvement, Wageningen, The Netherlands.

Brown, G. (1974) 'The agricultural significance of clays', in Mackney, D. (ed.), *Soil Type and Land Capability*, 27–42, Technical Monograph No. 4, Soil Survey, Harpenden.

Brown, L. R. (1975) 'The world food prospect', *Science*, **190**, 1053–59.

Brown, L. R. (1978) *The Worldwide Loss of Cropland*, Worldwatch Institute, Washington, DC.

Bryan, W. H. and Teakle, L. J. H. (1949) 'Pedogenic inertia – a concept in soil science', *Nature*, **164**, 969.

BSI (1957) *Site Investigations*, BS Code of practice CP 2001, British Standards Institution.

Bullock, P. (1974) 'The use of micromorphology in the new system of soil classification for England and Wales', in Rutherford, G. K. (ed.), *Soil Microscopy*, Limestone Press. Kingston, Ontario, 607–31.

Bunting, B. T. (1965) *The Geography of Soil*, Hutchinson, London.

Buol, S. W., Hole, F. D. and McCracken, R. J. (1980) *Soil Genesis and Classification*, Iowa State University Press, Ames, Iowa.

Buringh, P. (1978) 'Food production potential of the world', in Shina, R. (ed), *The World Food Problem: Concensus and Conflict*, Pergamon, Oxford, 477–85.

Burnham, C. P. and Dermott, W. (1964) 'Preliminary studies on the agricultural significance of soil series in the West Midlands', *Welsh Soils Discussion Group Report*, No. 5, 17–43.

Burns, I. G. (1975) 'An equation to predict the leaching of surface-applied nitrate', *Journal of Agricultural Science, Cambridge*, **85**, 443–454.

Burns, I. G. (1976) 'Equations to predict the leaching of nitrate uniformly incorporated to a known depth or uniformly distributed throughout a soil profile', *Journal of Agricultural Science, Cambridge*, **86**, 305–13.

Burns, I. G. (1977) 'Nitrate movement in soil and its agricultural significance', *Outlook on Agriculture*, **9**, 144–8.

Butler, B. E. (1959) 'Periodic phenomena in landscapes as a basis for soil studies', *CSIRO Australia, Soil Publication*, 14.

Butzer, K. W. (1964) *Environment and Archaeology*, Methuen, London.

Butzer, K. W. (1974) 'Accelerated soil erosion: a problem of man-land relationships', in Manners, I. R. and Mikezell, M. W. (eds.), *Perspectives on Environment*, Association of American Geographers, Washington DC, 57–77.

Caldwell, T. H. and Jones, J. L. O. (1965) 'Preliminary studies using P_{32} on availability of phosphate in problem soils in the Eastern Region', in *Soil Phosphorus*, Technical Bulletin No. 13, MAFF.

Campbell, J. B. (1979) 'Spatial variability of soils', *Annals of the Association of American Geographers*, **69**, 544–56.

Canada Land Inventory (1965) *Soil Capability Classification for Agriculture*, Report No. 2, Department of the Environment, Ottawa.

Canada Soil Survey Committee (1978) *The Canadian System of Soil Classification*, Canada Department of Agriculture, Publication 1646, Ottawa, Ontario.

Cannell, R. Q., Davies, D. B., Mackney, D. and Pidgeon, J. D. (1979) 'The suitability of soils for sequential drilling of combine-harvested crops in Britain, a provisional classification', 1–23, in Jarvis, M. G. and Mackney, D. (eds.) *Soil Survey Applications*, Technical Monograph No. 13, Soil Survey, Harpenden.

Carbon, B. A. and Galbraith, K. A. (1975) 'Simulation of the water balance for plants growing on coarse-textured soils', *Australian Journal of Soil Research*, **13**, 21–31.

Carmean, W. H. (1967) 'Soil Survey refinements for predicting Black Oak site quality in Southeastern Ohio', *Soil Science Society of American Proceedings*, **31**, 805–10.

Carroll, D. M. and Bascomb, D. L. (1967) *Notes on the Soils of Lesotho (1 : 250 000 Map)*, Technical Bulletin No. 1, Land Research Division, Department of Overseas Survey, Tolworth.

Carter, V. G. and Dale, T. (1974) *Topsoil and Civilization*, rev. edn, University of Oklahoma Press.

Casagrande, A. (1947) 'Classification and identification of soils', *Proceedings of the American Society of Civil Engineers*, **73**, 783–810.

Catt, J. A. (1979) 'Soils and Quaternary geology in Britain', *Journal of Soil Science*, **30**, 607–42.

Chorley, R. J. (1966) 'The application of statistical methods to geomorphology', in *Essays in Geomorphology*, Dury, G. H. (ed.), 275–387, Heinemann, London.

Cipra, J., Unger, B. and Bidwell, O. W. (1969) 'A computer program to "key out" world soil orders', *Soil Science*, **108**, (3), 153–9.

Cipra, J. E., Bidwell, O. W., Whitney, D. A. and Feyerherm, A. M. (1972) 'Variations with distance in selected fertility measurements of pedons of a western Kansas Ustoll', *Soil Science Society of American Proceedings*, **36**, 111–5.

Clarke, G. R. (1940) *Soil Survey of England and Wales: Field Handbook*, University Press, Oxford.

Clarke, G. R. (1951) 'The evaluation of soils and the definition of quality classes from studies of the physical properties of the soil profile in the field', *Journal of Soil Science*, **2**, 50–60.

Cline, M. G. (1949) 'Basic principles of soil classification', *Soil Science*, **67**, 81–91.

Cline, M. G. (1963) 'Logic of the new system of classification', *Soil Science*, **95**, 17–22.

Coffey, C. N. (1912) *A Study of the Soils of the United States*, Bulletin Bureau of Soils, US Department of Agriculture, No. 85.

Coleman, A. (1976) 'Is planning really necessary?', *Geographical Journal*, **142**, 411–37.

Coles, J. M. (1976) 'Forest farmers: some archaeological, historical and experimental evidence relating to the prehistory of Europe', in de. Laet, S. J. (ed.), *Acculturation and Continuity in Atlantic Europe*, De Tempel, Brugge, 59–66.

Conacher, A. J. (1975) 'Throughflow as a mechanism responsible for excessive soil salinization in non-irrigated, previously arable lands in the Western Australia wheatbelt: a field study', *Catena*, **2**, 31–68.

Conacher, A. J. and Dalrymple, J B. (1977) 'The nine-unit land-surface model: an approach to pedogeomorphic research', *Geoderma*, **18**, 1–154.

Conry, H. J. (1971) 'Irish plaggen soils – their distribution, origin and properties', *Journal of Soil Science*, **22**, 401–16.

Cooke, G W. (1965) 'The response of crops to phosphate fertilizers in relation to soluble phosphorous in soils', in *Soil Phosphorus* Technical Bulletin No. 13, MAFF, London.

Cooke, G. W. (1967) *The Control of Soil Fertility*, Crosby Lockwood, London.

Cooke, G. W. (1972) *Fertilizing for Maximum Yields*, Crosby Lockwood, London.

Corbett, W. M. (1972) 'Norfolk: County Map', in *Soil Survey Annual Report*, Harpenden.

Coulter, J. K. (1974) 'Soil types and soil fertility', in *Soil Type and Land Capability*, Mackney, D. (ed.), Technical Monograph No. 4, Soil Survey, Harpenden.

CPCS (Commission de Pédologie et de Cartographie des Sols) (1967) *Classification des sols*, Ecole Nationale Supérieure Agronomique, Grignon (mimeographed).

Crocker, R. L. (1952) 'Soil genesis and the pedogenic factors', *Quarterly Review of Biology*, **27**, 139–68.

Crocker, R. L. and Major J. (1955) Soil development in relation to vegetation and surface age at Glacier Bay, Alaska. *Jour. Ecol*, **43**, 427–48.

Crowther, E. M. (1953) 'The sceptical soil chemist', *Journal of Soil Science*, **4**, 107–22.

Curtis, C. D. (1976a) 'Chemistry of rock weathering: fundamental reactions and controls', in *Geomorphology and Climate*, Derbyshire, E. (ed.), 25–57, Wiley, London.

Curtis, C. D. (1976b) 'Stability of minerals in surface weathering reactions: a general thermochemical approach', *Earth Surface Processes*, **1**, 63–70.

Dan, J. and Yaalon, D. H. (1968) 'Pedomorphic forms and pedomorphic surfaces', *Transactions of the Ninth International Congress of Soil Science*, Adelaide, **4**, 577–84.

Daniels, R. B., Gamble, E. E. and Cady, J. G. (1971) 'The relationship between geomorphology and soil morphology and genesis', *Advances in Agronomy*, **23**, 51–88.

Darby, H. C. (1948) 'The economic geography of England, AD 1000–1250', in Darby, H. C. (ed.), *An Historical Geography of England Before 1800*, Cambridge University Press, Cambridge, 165–229.

Davidson, D. A. (1980) *Soils and Land Use Planning*, Longman, London.

Davies, B. E. (1976) 'Mercury content of soils in western Britain with special reference to contamination from base metal mining', *Geoderma*, **16**, 183–92.

Davies, B. E. (1978) 'Plant-available lead and other metals in British garden soils', *Science of the Total Environment*, **9**, 243–62.

Davies, B. E. and Roberts, L. J. (1978) 'The distribution of heavy metal contaminated soils in northeast Clwyd, Wales', *Water, Air and Soil Pollution*, **9**, 507–18.

Davidson, D. A. (1980) 'Erosion in Greece during the first and second millennia BC ', in Cullingford, R. A., Davidson, D. A. and Lewin, J. (eds.), *Timescales in Geomorphology*, Wiley, Chichester, 143–58.

de Bakker, H. (1970) 'Purposes of soil classification', *Geoderma*, **4**, 195–208.

de Bakker, H. (1979) *Major Soils and Soil Regions in the Netherlands*, Dr W. Junk B. V. Publishers, The Hague.

de Bakker, H. and Schelling, J. (1966) *Systeem van bodem classificatie voor Nederland*, de hogere niveaus, Pudoc, Wageningen, The Netherlands.

Deevey, E. S. Jr. (1960) 'The human population', *Scientific American*, **203**, 195–204

de Gruijter, J. J. (1977) *Numerical Classification of Soils and its Application in Survey*,

Agricultural Research Reports, No. 885, Centre for Agricultural Publishing and Documentation, Wageningen. (Also Soil Survey Papers, No. 12, Netherlands Soil Survey Institute, Wageningen.)

Dermott, W., Roberts, E. and Wilkinson, B. (1965) 'The use of soil maps in advisory work,' *NAAS Quarterly Review* No. 69, 16–22.

de Wit, C. T. and van Keulen, H. (1972) *Simulation of Transport Processes in Soils*, Centre for Agricultural Publishing and Documentation, Wageningen, The Netherlands.

Dimbleby, G. W. (1961) 'Soil pollen analysis', *Journal of Soil Science*, **12**, 1–11.

Dimbleby, G W. (1962) *The Development of British Heathlands and Their Soils*, Oxford Forestry Memoir No. 23.

Dimbleby, G. W. (1976) 'Climate, soil and man', *Philosophical Transactions of the Royal Society, London* B, **275**, 197–208.

Duchaufour, Ph. (1960) *Précis de pédologie*, Masson, Paris.

Duchaufour, Ph. (1965) *Précis de pédologie*, 2nd edn. Masson, Paris.

Duchaufour, Ph. (1970) *Précis de pédologie*, 3rd edn. Masson, Paris.

Duchaufour, Ph. (1976) *Atlas écologique des sols du monde*, Masson, Paris. Translated as '*Soils of the World*', by Mehuys, G. R., de Kimpe, C. R. and Martel, Y. A., Masson, New York, 1978.

Duchaufour, Ph. (1977) *Pedologie. I Pedogenése et classification*, Masson, Paris.

Dudal, R. (1968) 'Definitions of soil units for the soil map of the world', *World Soil Resources Report* No. 33, Rome, FAO.

Dudal R. (1978) 'Land resources for agricultural development', *Plenary Session Papers, 11th International Congress of Soil Science*, Edmonton, Vol. 2, 314–40.

Dumanski, J. (ed.) (1978) '*The Canada Soil information system* (*CanSIS*). Manual for describing soils in the field', Agriculture Canada, Land Resource Research Institute, Central Experimental Farm, Ottawa.

Dumanski, J. (ed.) (1978) 'Resource information systems – Foreword', *Proceedings of the 11th International Congress of Soil Science*, Edmonton, **3**, 130–1.

Dumanski, J., Kloosterman, B. and Brandon, S. E. (1975) 'Concepts, objectives and structure of the Canada soil information system', *Canadian Journal of Soil Science*, **55**, 181–7.

Dumbleton, M. J. and West, G. (1971) *Preliminary Sources of Information for Site Investigation in Britain*, Transport and Road Research Laboratory Report LR 403.

East, W. G. (1948) 'England in the eighteenth century', in Darby, H. C. (ed.), *An Historical Geography of England before 1800*, Cambridge University Press, Cambridge, 465–528.

Economic Research Service (1976) *Twenty-six Years of World Cereal Statistics*, United States Department of Agriculture, Washington, DC.

Economic Research Service (1977) *Indexes of World Agriculture and Food Production*, United States Department of Agriculture, Washington, DC.

Edelman, C. H. (1950) *Soils of the Netherlands*, North-Holland, Amsterdam.

Ehwald, E. (1968) 'Some new approaches to soil classification in the German Democratic Republic', *Soviet Soil Science*, **10**, 1329–36.

Ehwald, E., Lieberoth, I. and Schwanecke, W. (1966) *Zur Systematik der Böden der Deutschen Demokratischen Republik, besonders in Hinblick auf die Bodenkartierung*, Sitzungsberichte Deutsche Akademie der Landwirtschaftswissenschaften, Berlin 15 (18).

Evans, J. G. (1975) *The Environment of Early Man in the British Isles*, Paul Elek, London.

Fallou, F. A. (1862) *Pedologie oder allgemeine und besondere Bodenkunde*, Dresden, Germany.

FAO (1974) *Soil Map of the World, 1 : 5 000 000*, Vol. 1, Legend, UNESCO, Paris.

FAO (1974–8) *Soil Map of the World*, UNESCO, Paris.

FAO (1975) *Land Evaluation in Europe*, FAO Soils Bulletin No. 29, Rome.

FAO (1976) *A Framework for Land Evaluation*, FAO Soils Bulletin No. 32, Rome.

FAO (1977) *A Framework for Land Evaluation*, Soils Bulletin No. 34, Rome.

FAO (1978a) *The State of Food and Agriculture 1977*, Food and Agriculture Organization of the United Nations, Rome.

FAO (1978b) *Report on the Agro-ecological Zones Project. Vol. I. Methodology and Results for Africa*, World Soil Resources Report 48, FAO, Rome.

FAO (1979) *Report on the Agro-ecological Zones Project. Vol. 2. Results for Southwest Asia*, World Soil Resources Report 48/2, FAO, Rome.

Farquharson, F. A., Mackney, D., Newson, M D. and Thomasson, A. J. (1978) *Estimation of Run-off Potential of River Catchments from Soil Surveys*, Soil Survey Special Survey No. 11, Harpenden.

Feachem, R. (1973) 'Ancient agriculture in the highlands of Britain', *Proceedings Prehistoric Society*, **39**, 332–53.

Finch, A. and Hotson, J. (1974) 'CAMAP and GRID CAMAP', *Inter-University Research Councils, Research and Development Notes* No. 13.

Finch, T. F. (1971) *Soils of County Clare*, Soil Survey Bulletin No. 23, National Soil Survey of Ireland, Dublin.

Fleming, A. (1971) 'Bronze age agriculture on the marginal lands of northeast Yorkshire', *Agricultural Historical Review*, **19**, 1–24.

FitzPatrick, E. A. (1971) *Pedology*, Oliver and Boyd, Edinburgh.

Foth, H. D. (1978) *Fundamentals of Soil Science*, John Wiley, New York, 346–8.

Furley, P. A. (1968) 'Soil formation and slope development. 2. The relationship between soil formation and gradient in the Oxford area', *Zeitschrift für Geomorphologie* (NF), **12**, 25–42.

Furley, P. A (1971) Relationship between slope form and soil properties over chalk parent materials, in *Slopes: Form and Process* (ed. D. Brunsden) *Spec. Publ* **No. 3** *Trans. Inst. Brit. Geogr.*, 141–163.

Gilmour, J. S. L. (1937) 'A taxonomic problem', *Nature, London*, **139**, 1040–2.

Gilmour, J. S. L. (1951) 'Taxonomy', in MacLeod, A. M. and Cobley, L. S. (eds.), *Contemporary Botanical Thought*, Oliver and Boyd, Edinburgh, 27–45.

Glazovskaya, M. A. (1968) 'Geochemical landscapes and geochemical soil sequences', *Transactions of the Ninth International Congress of Soil Science*, Adelaide, **4**, 303–12.

Goddard, T. M., Runge, E. C. A. and Ray, B. W. (1973) 'The relationship between rainfall frequency and amount to the formation and profile distribution of clay particles', *Soil Science Society of America Proceedings*, **37**, 299–303.

Greene, H. (1963) 'Soils of Western Germany', *Soils and Fertilizers*, **26**, 1–3.

Greenland, D. J. (1975) 'Bringing the green revolution to the shifting cultivator', *Science*, **190**, 841–4.

Grigal, D. F. and Arneman, H. F. (1969) 'Numerical classification of some forested Minnesota soils', *Soil Science Society of America Proceedings*, **33**, 433–8.

Grigg, D. B. (1974) *The Agricultural Systems of the World – An Evolutionary Approach*, Cambridge, London.

Grimes, W. F. (1945) 'Early man and the soils of Anglesey', *Antiquity*, **19**, 169–74.

Haggett, P. (1965) *Locational Analysis in Human Geography*, Arnold.

Hall, B. R. and Folland, C. J. (1964) *Soils of the Southwest Lancashire Coastal Plain*, Memoir of the Soil Survey of England and Wales, Harpenden.

Hall, D. G. M., Reeve, M. J., Thomasson, A. J. and Wright, V. F. (1977) *Water Retention, Porosity and Density of Field Soils*, Soil Survey Technical Monograph No. 9, Harpenden.

Hardan, A. (1970) 'Dating of soil salinity in the Mesopotamian plain', paper presented at the Symposium on the Age of Parent Materials and Soils, Amsterdam, The Netherlands, 10–15 August 1970

Harradine, F. and Jenny, H. (1958) 'Influence of parent material and climate on texture and nitrogen and carbon content of virgin Californian soils', *Soil Science*, **85**, 235–43.

Harrod, M. F. (1975) 'Field behaviour of light soils', in *Soil Physical Conditions and Crop Production*, Technical Bulletin No. 29, 22–51, MAFF, Lonon.

Harrod, T. R. (1979) 'Soil suitability for grassland', in Jarvis, M. G. and Mackney, D. (eds.), *Soil Survey Applications*, Technical Monograph No. 13, Soil Survey, Harpenden.

Hendricks, S. B. (1969) 'Food from the land', in *Resources and Man*, National Academy of Sciences – National Research Council, W. H. Freeman, San Francisco.

Hill, A. R. (1976) 'The effects of man-induced erosion and sedimentation on the soils of a portion of the Oak Ridges moraine', *Canadian Geographer*, **20**, 384–404.

Hodgson, J. M. (ed.) (1976) *Soil Survey Field Handbook*, Technical Monograph 5, Soil Survey of Great Britain, Harpenden.

Hodgson, J. M. (1978) *Soil Sampling and Soil Description*, Oxford University Press.

Hoffer, R. M., Anuta, P. E. and Phillips, T. L. (1972) 'ADP, Multiband and multiemulsion digitized photos', *Photogrammetric Engineering*, **38**, 989–1001.

Holowaychuk, N., Gersper, P. L., and Wilding, L. P. (1969) 'Strontium-90 content of soils near Cape Thompson, Alaska', *Soil Science*, **107**, 137–44.

Hooper, L. J. (1974) 'Land use capability in farm planning', in Mackney, D. (ed.), *Soil Type and Land Capability*, 135–49. Technical Monograph No. 4, Soil Survey, Harpenden.

Hooper, L. J. and Crampton, C. B. (1964) 'The use of potential horticultural soils in Glamorgan', *Welsh Soils Discussion Group Report* No. 5, 83–98.

Hoskins, W. G. (1955) *The Making of the English Landscape*, Hodder and Stoughton, London.

Huggett, R. J. (1973) *Soil landscape systems: theory and field evidence*, unpublished Ph. D. thesis, University of London.

Hugget, R. J. (1975) 'Soil landscape systems: a model of soil genesis', *Geoderma*, **13**, 1–22.

Huggett, R. J. (1976a) 'Lateral translocation of soil plasma through a small valley basin in the Northaw Great Wood, Hertfordshire', *Earth Surface Processes*, **1**, 99–109.

Huggett, R. J. (1976b) 'Conceptual models in pedogenesis: a discussion', *Geoderma*, **16**, 261–2.

Huggett, R. J. (1980) *Systems Analysis in Geography*, Oxford University Press, Oxford.

Hutchinson, Sir J. (1969) *Population and Food Supply*, Cambridge, London.

Hutton, F. Z. and Rice, C. E. (1977) *Soil Survey of Onondaga County, New York*, USDA Soil Conservation Service, Washington DC.

Hutton, J T. (1968) 'The redistribution of the more soluble chemical elements associated with soils as indicated by analysis of rainwater, soils and plants', *Transactions of the Ninth International Congress of Soil Science*, Adelaide, **4**, 313–21.

Imeson, A. C. and Jungerius, P. D. (1974) 'Landscape stability in the Luxembourg Ardennes as exemplified by hydrological and (micro)pedological investigations of a catena in an experimental watershed', *Catena*, **1**, 273–95.

Jacks, G. V. (1954) *Soil*, Nelson, London.

Jackson, M. L., Gillette, D. A., Danielson, E. F., Blifford, I. H., Bryson, R. A. and Syers, J. K. (1973) 'Global dustfall during the Quaternary as related to environments', *Soil Science*, **116**, 135–45.

Jacobsen, T. and Adams, R. M. (1958) 'Salt and silt in ancient Mesopotamian agriculture', *Science*, **128**, 1251–8.

Jarvis, M. G. (1973) *Soils of the Wantage and Abingdon district*, Memoir Soil Survey of England and Wales, Harpenden.

Jansen and Arnold, R. W. (1976) 'Defining ranges of soil characteristics', *Soil Science Society of America Proceedings*, **40**, 89–92.

Jenny, H. (1941) *Factors of Soil Formation. A System of Quantitative Pedology*, McGraw-Hill, New York.

Jenny, H. (1958) 'The role of the plant factor in pedogenic functions', *Ecology*, **39**, 5–16.

Jenny, H. (1961) 'Derivation of state factor equations of soils and ecosystems', *Soil Science Society of America Proceedings*, **25**, 385–8.

Jenny, H. (1965) 'Tessera and pedon', *Soil Survey Horizons*, **6**, 8–9.

Jenny, H., Bingham, F. and Padilla-Saravina, B. (1948) 'Nitrogen and organic contents of equatorial soils of Columbia, South America', *Soil Science*, **66**, 173–86.

Jenny, H. and Raychandhuti, S. P. (1960) *Effect of Climate and Cultivation on Nitrogen and Organic Matter Reserves in Indian Soils*, ICAR, New Delhi.

Jenny, H. S., Salem, A. E. and Wallis, J. R. (1968) 'Interplay of organic matter and soil fertility with state factors and soil properties', in *Organic Matter and Soil Fertility*, *Pontif. Acad. Sci. Scr.*, **32**, 5–44.

Jensen, N. F. (1978) 'Limits to growth in world food production', *Science*, **201**, 317–20.

Jewitt, T. N. (1955) 'Gezira soil', *Bulletin of the Ministry of Agriculture of the Sudan*, No. 12.

Johnson, J. A. (1979) 'Image and reality: initial assessments of soil fertility in New Zealand', *Australian Geographer,* **14**, 160–6.

Jones, M. J. (1973) 'The organic matter content of the savanna soils of West Africa', *Journal of Soil Science*, **24**, 42–53.

Jones, R. J. A. (1979) 'Soils of the western Midlands grouped according to ease of cultivation', in Jarvis, M. G. and Mackney, D. (eds.), *Soil Survey Applications*, Technical Monograph No. 13, Soil Survey, Harpenden, 24–42.

Jones, R. L. (1976) 'The activities of Mesolithic man: further palaeobotanical evidence from north-east Yorkshire', in Davidson, D. A. and Shackley, M. L. (eds.), *Geoarcheology*, Duckworth, London, 355–66.

Kane, T. T. (1979) '*World Population Data Sheet*, Population Reference Bureau, Washington, DC.

Keef, P. A. M., Wymer, J. J. and Dimbleby, G. W. (1965) 'A Mesolithic site on Iping Common, Sussex, England', *Proceedings of the Prehistoric Society*, **31**, 85–92.

Kellogg, C. E. (1949) 'Introduction to special issue on soil classification', *Soil Science*, **67**, 77–80.

Kellogg, C. E. (1962) 'Soil surveys for use', *Transactions of Joint Meeting of Commissions IV and V, International Society of Soil Science*, Soil Bureau, Lower Hutt, New Zealand, 529–36.

Kellogg, C. E. (1963) 'Why a new system of soil classification?', *Soil Science*, **96**, 1–5.

Kellogg, C. E. (1974) 'Soil genesis classification and cartography 1924–74', *Geoderma*, **12**, 347–62.

Kellogg, C. E. and Orvedal, A. C. (1969) 'Potentially arable soils of the world and critical measures for their use', in Brady, N. C. (ed.), *Advances in Agronomy*, Academic Press, New York, 109–70.

Kirkby, M. J. (1971) 'Hillslope process-response models based on the continuity equation', in *Slopes: Forms and Process*, in Brunsden, D. (ed.), *Institute of British Geographers Special Publication No. 3*, 15–30.

Kirkby, M. J. (1977) 'Soil development models as a component of slope models', *Earth Surface Processes*, **2**, 203–30.

Kleiss, H. J. (1970) 'Hillslope sedimentation and soil formation in northeastern Iowa', *Soil Science Society of America Proceedings*, **34**, 287–90.

Kline, J. R. (1973) 'Mathematical simulation of soil–plant relationships and soil genesis', *Soil Science*, **115**, 240–9.

Klingebiel, A. A. and Montgomery, P. H. (1961) *Land-Capability Classifications*, Agriculture Handbook No. 210, US Dept. Agriculture Soil Conservation Service, Washington, DC.

Knox, E. G. (1965) 'Soil individuals and soil classification', *Soil Science Society of America Proceedings*, **29**, 79–84.

Kohnke, H., Stuff, R. G. and Miller, P. A. (1968) 'Quantitative relations between climate and soil formation', *Z. Pflanzernähr. Düng. Bodenkd*, **119**, 24–33.

Koinov, V. Y., Boneva, K. K. and Djokova, M. Y. (1972) 'Certain geochemical features of eluvial landscapes on the most important soil-forming rocks in south Bulgaria', *C. R. Acad. Agric. G. Dimitrov, Sofia*, **5**, 23–27.

Krantz, B. A. (1979) *Small Watershed Development for Increased Food Production*, International Crops Research Institute for the Semi-Arid Tropics, Hyderabad.

Krantz B. A and Kampen, J. (1979) 'Crop production systems in semi-arid tropical zones', in Thorne, D. W. and Thorne, M. D. (eds.), *Soil, Water and Crop Production*, AVI, Westport, Connecticut, 278–98.

Kristoff, S. J. and Zachary, A. L. (1974) 'Mapping soil features from multispectral scanner data', *Photogrammetric Engineering*, **40**, 1427–34.

Kubiena, W. L. (1953) *The Soils of Europe*, Murby, London.

Kubiena, W. L. (1958) 'The classification of soils', *Journal of Soil Science*, **9**, 9–19.

Kwaad, F. J. P. M. and Müchler, H. J. (1975) 'A micromorphological study of some soils and slope deposits in the Luxembourg Ardennes near Wiltz in view of the latest phases of landscape development', *Bulletin de la Societé des Naturalistes Luxembourgeois*, **80**, 8–58.

Kwaad, F. J. P. M. and Müchler, H. J. (1977) 'The evolution of soils and slope deposits in the Luxembourg Ardennes near Wiltz', *Geoderma*, **17**, 1–37.

Låg, J. (1968) 'Some principles in the study of the influence of soil-forming factors and the capacity of the soil for plant production', *Acta Agriculturae Scandinavica*, **18**, 95–96.

Låg, J. (1974) 'The influence of soil conditions on the distribution of plant communities', *Acta Agriculterae Scandinavica*, **24**, 13–16.

Leeper, G. W. (1956) 'The classification of soils', *Journal of Soil Science*, **7**, 59–64.

Lieberoth, I. (1969) 'Mapping of soils in the German Democratic Republic', *Soviet Soil Science*, **10**, 1350–63. (Oct. 1968) (translated June 1969).

Limbrey, S. (1975) *Soil Science and Archaeology*, Academic Press, London.

Linton, R. and White, C. (1973) 'Production of multi-coloured soil maps by automated methods', Paper presented to Symposium E, *Computers in Soil Science*, British Society of Soil Science, Wye.

Low, D. (1976) *A comparison of the physiographic and parametric approaches to forest capability assessment in the upper Don basin*, unpublished M.Sc. Thesis, University of Aberdeen.

Lowenstein, P. L. and Howarth, R. J. (1972) 'Automated colour-mapping of three-component systems and its application to regional geochemical reconnaissance', *Geochemical Exploration*, 297–304.

Macaulay Institute for Soil Research (1975–6) *Annual Report for 1975–76*, Aberdeen.

Mackney, D. (1969) 'The agronomic significance of soil mapping units', in *The Soil Ecosystem*, Sheals, J. G. (ed.), 55–62, Systematics Association Publication No. 8, Harpenden.

Mackney, D. (1974) 'Soil survey in agriculture', in *Soil Type and Land Capability* Mackney, D. (ed.) Technical Monograph No. 4, Soil Survey, Harpenden.

Mackney, D. and Burnham, C. P. (1964) *The Soils of the West Midlands*, Bulletin 2, Soil Survey of Great Britain, Harpenden.

Mackney, D. and Burnham, C. P. (1966) *The soils of the Church Stretton district of Shropshire*, Memoir of the Soil Survey of England and Wales, Harpenden.

McCormack, D. E., Moore, A. W. and Dumanski, J. (1978) 'A review of soil information systems in Canada, the United States and Australia', *Proceedings of the 11th International Congress of Soil Science*, Edmonton, **3**, 143–60.

McCormack, R. J. (1972) *Land Capability Classification for Forestry*, The Canada Land Inventory Report No. 4, Ottawa.

McKeague, J. A., Day, J. H. and Shields, J. A. (1971) 'Evaluating relationships among soil properties by computer analysis', *Canadian Journal of Soil Science*, **51**, 105–11.

McNeil, Mary (1964) 'Lateritic soils', *Scientific American*, **211**, No. 5, 97–102.

Major, J. (1951) 'A functional, factorial approach to ecology', *Ecology*, **32**, 392–412.

Mansell, R. S., Selim, H. M. and Fiskell, J. G. A. (1977) 'Simulated transformations and transport of phosphorus in soil', *Soil Science*, **124**, 102–9.

Marbut, C. F. (1922) 'Soil classification', *American Association of Soil Survey Workers*, 2nd Annual Report, Bulletin, **3**, 24–32.

Marbut, C. F. (1935) 'Soils of the United States', in *Atlas of American Agriculture*, US Department of Agriculture, part 3, advance sheets No. 8.

Margerison, T. (1978) 'How the map could be put on the computer', *New Scientist*, 716–18.

Mathews, H. L., Cunningham, R. L., Cipra, J. E. and West, T. R. (1973) 'Application of multispectral remote sensing to Soil Survey research in Southeastern Pennsylvania', *Soil Science Society of America Proceedings*, **37**, 88–93.

Mauldin, W. P. (1978) 'World population situation: problems and prospects', in Shina, R. (ed.), *The World Food Problem: Concensus and Conflict*, Pergamon, Oxford, 395–404.

Mayer, Jean (1976) 'The dimensions of human hunger', *Scientific American*, **235**, No. 3, 40–9.

Mellars, P. (1975) 'Ungulate populations, economic patterns, and the Mesolithic landscape', in *The Effect of Man on the Landscape: the Highland Zone*, Council for British Archaeology Report No. 11, 49–56.

Michigan Department of State Highways (1970) *Field Manual of Soil Engineering* (fifth edn.), Lansing, Michigan.

Milne, G. (1935a) 'Some suggested units of classification and mapping particularly for East African soils', *Soil Research, Berlin*, **4**, 183–98.

Milne, G. (1935b) 'Composite units for the mapping of complex soil associations', *Transactions of the Third International Congress of Soil Science*, **1**, 345–7.

Moore, T. R. (1979) 'Land use and erosion in the Machakos Hills', *Annals of the Association of American Geographers*, **69**, 419–31.

Moormann, F. R. and Van Wambeke, A. (1978) 'The soils of the lowland rainy tropical climates: their inherent limitations for food production and related climate restraints', Plenary Papers, Vol. 2, *11th International Congress of Soil Science*, Edmonton, 272–91.

Morison, C. G. T., Hoyle, A. C. and Hope-Smith, J. F. (1948) 'Tropical soil-vegetation catenas and mosaics. A study in the south-western part of the Anglo-Egyptian Sudan', *Journal of Ecology*, **36**, 1–84.

Mückenhausen, E. (1962) *Enstehung, Eigenschaften und Systematik der Böden der Bundersrepublik*, Deutschland, DLG, Verlag, Frankfurt (Main).

Mückenhausen, E. (1965) 'The soil classification system of the Federal Republic of Germany', Pédologie, Gand, numéro spécial 3, 57–89.

Mückenhausen, E. (1975) *Die Bodenkunde und ihre geologischen, geomorphologischen, mineralogischen und petrologischen Grundlagen*, DLG, Verlag, Frankfurt (Main).

Mückenhausen, E. (1977) *Entstehung, Eigenschaften und Systematik der Böden der Bundesrepublik Deutschland (2 Auflage)*, DLG, Verlag, Frankfurt (Main).

Muir, A. (1961) 'Soil survey in Britain', in *Soil Survey of Great Britain, Report No. 13* (1960), Agricultural Research Council.

Muir, J. W. (1969) 'A natural system of soil classification', *Journal of Soil Science*, **20**, 153–66.

Mulcahy, M. J. and Humphries, A. W. (1967) 'Soil classification, soil surveys and land use', *Soils and Fertilizers*, **30**, 1–8.

Nicholaides, J. III (1979) 'Crop production systems on acid soils in humid tropical America', in Thorne, D. W. and Thorne, M. D. (eds.), *Soil, Water and Crop Production*, AVI, Westport, Connecticut, 243–277.

Nichols, J. D. (1975) 'Characteristics of computerized soil maps', *Soil Science Society of America Proceedings*, **39**, 917–31.

Nikiforoff, C. C. (1959) 'Reappraisal of the soil', *Science*, **129**, 186–96.

Nortcliff, S. (1978) 'Soil variability and reconnaissance soil mapping: a statistical study in Norfolk', *Journal of Soil Science*, **29**, 404–18.

Northcote, K. H. (1971) *A Factual Key for the Recognition of Australian Soils*, 3rd edn, Rellim Technical Publications, Glenside, South Australia.

Northcote, K. H., with Beckmann, G. G., Bettenay, E., Churchward, H. M., van Dijk, D. C., Dimmock, G. M., Hubble, G. D., Isbell, R. F., McArthur, W. M., Murtha, G. G., Nicholls, K. D., Paton, T. R., Thompson, C. H., Webb, A. A. and Wright, M. J. (1960–8) *Atlas of Australian Soils Sheet 1–10, with explanatory booklets* (CSIRO and Melbourne Univ. Press, Melbourne).

Northcote, K. H., Hubble, G. D., Isbell, R. F., Thomson, C. H. and Bettenay, E. (1975) *A Description of Australian Soils*, CSIRO, Australia.

Noy-Meir, I. (1974) 'Multivariate analysis of semi-arid vegetation in south-eastern Australia. II. Vegetation catenae and environmental gradients', *Australian Journal of Botany*, **22**, 115–40.

Nye, P. H. and Greenland, D. J. (1960) *The Soil under Shifting Cultivation*, Commonwealth Agricultural Bureau, Farnham Royal, Bucks.

Olson, G. W. (1973) *Soil Survey Interpretation for Engineering Purposes*, Food and Agriculture Organization of the United Nations, Rome.

Osmond, D. A., Swarbrick, T., Thompson, C. R. and Wallace, T. (1949) *A Survey of the Soils and Fruit in the Vale of Evesham, 1926–1934*, Ministry of Agriculture Bulletin 116, MAFF, London.

Page, G. (1970) 'Quantitative site assessment: some practical applications in British Forestry', *Forestry*, **43**, 45–56.

Palmer, R. C. and Jarvis, M. G. (1979) 'Land for winter playing fields, golf course fairways and parks', in Jarvis, M. G. and Mackney, D. (eds.), *Soil Survey Applications*, Soil Survey Technical Monograph No. 13.

Pennington, W. (1975) 'The effect of Neolithic man on the environment in north west England; the use of absolute pollen diagrams', in *The Effect of Man on the Landscape: the Highland zone*, Council for British Archaeology Report No. 11, 74–86.

Perring, F. (1958) 'A theoretical approach to a study of chalk grassland', *Journal of Ecology*, **46**, 665–79.

Pimentel, D. *et al.* (1976) 'Land degradation: effects on food and energy resources', *Science*, **194**, 149–55.

President's Science Advisory Committee, The White House (1967) 'Water and land', in *The World Food Problem*: a *Report of the President's Science Advisory Committee*, US Government Printing Office, Washington, DC.

Protz, R., Presant, E. W. and Arnold, R. W. (1968) 'Establishment of the modal profile and measurement of variability within a soil landform unit', *Canadian Journal of Soil Science*, **48**, 7–19.

Proudfoot, B. (1971) 'Man's occupance of the soil', in Buchanan, R. H., Jones, E. and McCourt, D. (eds.), *Man and his Habitat*, Routledge and Kegan Paul, London, 8–33.

Pyatt, D. G. (1977) 'Guide to site types in forests of north and south Wales', *Forestry Commission Forest Record* No. 69, 2nd edition.

Ragg, J. M. (1977) 'The recording and organisation of soil field data for computer areal mapping', *Geoderma*, **19**, 81–9.

Ragg, J. M. and Clayden, B. (1973) *The Classification of Some British Soils According to the Comprehensive System of the United States*, Technical Monograph 3, Soil Survey of Great Britain, Harpenden.

Rayner, J. H. (1966) 'Classification of soils by numerical methods', *Journal of Soil Science*, **17** (1), 79–92.

Rayner, J. H. (1969) 'The numerical approach to soil systematics', in Sheals, J. G. (ed.), *The Soil Ecosystem*, the Systematics Association Publication No. 8, the Systematics Association, London, 31–9.

Richthofen, F. F. von (1886) *Führer für Forschungsreisende*, Berlin.

Riezebos, P. A. and Slotboom, R. T. (1978) 'Pollen analysis of the Husberbaach peat (Luxembourg): its significance for the study of subrecent geomorphological events', *Boreas*, **7**, 75–82.

Robinson, G. W. (1943) 'Mapping the soil of Britain', *Discovery*, **4**, 118–21.

Robson, J. D. and Thomasson, A. J. (1977) *Soil Water Regimes*, Technical Monograph No. 11, Soil Survey, Harpenden.

Rode, A. A. (1961) *The Soil-forming Process and Soil Evolution*, Israel Programme for Scientific Translations, Jerusalem.

Rothamsted Experimental Station (1966) 'Potato experiments', *Report Rothamsted Experimental Station for 1965*, 50–1, Harpenden.

Rothamsted Experimental Station (1968) 'Special surveys', *Report Rothamsted Experimental Station for 1967*, 312–13, Harpenden.

Rozov, N. N. and Ivanova, E. N. (1967) 'Classification of the soils of the USSR', *Soviet Soil Science*, **2**, 147–56.

Rudeforth, C. C. (1975) 'Storing and processing data for soil and land use capability surveys', *Journal of Soil Science* **26** (2), 155–68.

Rudeforth, C. C. (1977) *A quantitative approach to soil survey in Wales*, Ph.D. Thesis, University of London.

Rudeforth, C. C. (1978) 'Soil probability maps from air-photographs: a quality and cost comparison with conventional soil maps in Wales', *Transactions of the 11th International Soil Science Congress Commission Papers*, **1**, 185–6, Edmonton.

Rudeforth, C. C. and Thomasson, A. J. (1970) *Hydrological Properties of Soils in the River Dee Catchment*, Special Survey No. 4, Soil Survey, Harpenden.

Rudeforth, C. C. and Webster, R. (1973) 'Indexing and display of soil survey data by means of feature-cards and Boolean maps', *Geoderma*, **9**, 229–48.

Ruhe, R. V. (1960) 'Elements of the soil landscape', *Transactions of the Seventh International Congress of Soil Science, Madison*, **4**, 165–70.

Ruhe, R. V. and Walker, P. H. (1968) 'Hillslope models and soil formation. I. Open systems', *Transactions of the Ninth International Congress of Soil Science*, Adelaide, **4**, 551–60.

Runge, E. C. A. (1973) 'Soil development sequences and energy models', *Soil Science*, **115**, 183–93.

Salisbury, E. J. (1925) 'Note on the edaphic succession in some dune soils with special reference to the time factor', *Journal of Ecology*, **13**, 322–8.

Salter, P. J. and Williams, J. B. (1965) 'The influence of texture on the moisture characteristics of soils. II. Available-water capacity and moisture release characteristics', *Journal of Soil Science*, **16**, 310–17.

Sanchez, P. *et al.* (1979) 'Low input soil management and conservation technology for Oxisol-Ultisol regions of tropical America', *Agronomy Abstracts*, **48**.

Sanchez, P. A. and Buol, S. W. (1975) 'Soils of the tropics and the world food crisis', *Science*, **188**, 598–603.

Sasscer, D. S., Jordan, C. F. and Kline, J. R. (1971) 'Mathematical model of tritiated and stable water movement in an old-field system', in *Radionuclides in Ecosystems*, Nelson, D. J. (ed.), 915–23. Proceedings of the Third National Symposium on Radioecology, CONF-710501-P1, US Atomic Energy Commission.

Schelling, J. and Bie, S. W. (1978) 'Soil information systems – considerations for the future', *Proceedings of the 11th International Congress of Soil Science, Edmonton*, **3**, 208–13.

Scrivner, C. L., Baker, J. C. and Brees, D. R. (1973) 'Combined daily climatic data and dilute solution chemistry in studies of soil profile formation', *Soil Science*, **115**, 213–23.

Segalen, P., Fauck, R., Lamouroux, M., Perraud, A., Quantin, P. and Roederer, P. (1978) 'Pour une nouvelle classification objective des sols', *Transactions 11th Congress International Society of Soil Science*, Edmonton, Canada, Vol. 1 (abstracts), 69–70.

Shreve, R. L. (1966) 'Statistical law of stream numbers', *Journal of Geology*, **74**, 17–37.

Sibirtsiev, N. M. (1901) 'Soil science', in *Selected Works*, vol. 1, Israel Programme for Scientific Translations, Jerusalem, 1966.

Silva, J. A. and Beinroth F. H. (1979) 'Development of the transfer model and soil taxonomic interpretations on a network of three soil families', *Benchmark Soils Project Progress Report No. 2*, College of Agriculture, University of Hawaii and Puerto Rico.

Simmons, I. G. (1969a) 'Evidence for vegetation changes associated with Mesolithic man in Britain', in Ucko, P. J. and Dimbleby, G. W. (eds.), *The Domestication and Exploitation of Plants and Animals*, Duckworth, London, 110–19.

Simmons, I. G. (1969b) 'Environment and early man on Dartmoor, Devon, England', *Proceedings Prehistoric Society*, **35**, 203–19.

Simonett, D. S. (1960) 'Soil genesis on basalt in North Queensland', *Transactions of the Seventh International Congress of Soil Science*, Madison, **4**, 238–43.

Simonson, R. W. (1959) 'Outline of a generalized theory of soil genesis', *Soil Science Society of America Proceedings*, **23**, 152–6.

Simonson, R. W. (ed.) (1974) *Non-agricultural Applications of Soil Survey*, Elsevier, Amsterdam.

Sinclair, J. (1793) *The Statistical Account of Scotland*, Vol. **9**.

Singer, A. (1966) 'The mineralogy of the clay fraction from basaltic soils in the Galilee, Israel', *Journal of Soil Science*, **17**, 136–47.

Smeck, N. E. and Runge, E. C. A. (1971) 'Phosphorus availability and redistribution in relation to profile development in an Illinois landscape segment'. *Soil Science Society of America Proceedings*, **35**, 952–9.

Smith, A. G. (1970) 'The influence of Mesolithic and Neolithic man on British vegetation; a discussion', in Walker, D. and West, R. G. (eds.), *Studies in the Vegetational History of the British Isles*, Cambridge University Press, Cambridge, 81–96.

Smith, O. L. (1976) 'Nitrogen, phosphorus and potassium utilization in the plant-soil system: an analytical model', *Soil Science Society of America Proceedings*, **40**, 704–14.

Sneath, P. H. A. and Sokal, R. R. (1973) *Numerical Taxonomy*, W. H. Freeman, San Francisco.

Snedecor, G. W. and Cochran, W. G. (1967) *Statistical Methods*, Iowa State University Press, Ames.

Soil Survey Staff (1951) *Soil Survey Manual*, US Department of Agriculture, Agricultural Handbook 18, Washington DC.

Soil Survey Staff (1960) *Soil Classification: a Comprehensive System: 7th Approximation*, US Department of Agriculture, Washington, DC.

Soil Survey Staff (1971) *Guide for Interpreting Engineering Uses of Soils*, USDA Soil Conservation Service, Washington DC.

Soil Survey Staff (1975) *Soil Taxonomy – a Basic System of Soil Classification for Making and Interpreting Soil Surveys*, US Department of Agriculture, Agricultural Handbook 436, Washington, DC.

Spaargaren, O. C. (1979) *Weathering and Soil Formation in a Limestone Area near*

Pastena (Fr., Italy), Publicaties van het Fysisch Geografischen Bodemkundig Laboratorium van de Universiteit van Amsterdam, No. 30.

Speirs, R. B. (1976) 'The use of soil maps in the east of Scotland', *Welsh Soils Discussion Group Report*, **17**, 120–33.

Stace, H. C. T., Hubble, G. D., Brewer, R., Northcote, K. H., Sleeman, J. R., Mulcahy, M. J. and Hallsworth, E. G. (1968) *A Handbook of Australian Soils*, Rellim Technical Publishers, Glenside, South Australia.

Stephens, C. G. (1947) 'Functional synthesis in pedogenesis', *Transactions of the Royal Society of Australia*, **71**, 168–81.

Stephens, C. G. (1962) *A manual of Australian Soils*, 3rd edn, CSIRO, Australia, Melbourne.

Steur, G. G. L. *et al.* (1961) 'Methods of soil surveying in use at the Netherlands Soil Survey Institute', *Boor and Spade*, **XI**, 59–77.

Stewart, D. C. (1956) 'Fire as the first great force employed by man', in Thomas W. L. (ed.), *Man's Role in Changing the Face of the Earth*, University of Chicago Press, Chicago, 115–33.

Stobbs, A. R. (1970) 'Soil Survey procedures for development purposes', in Cox, I. H. (ed.), *New Possibilities and Techniques for Land Use and Related Surveys*, Occasional papers No. 9, 41–64, Geographical Publications Ltd, Berkhamsted, England.

Story, R., Galloway, R. W., McAlpine, J. R., Aldrick, J. M. and Williams, M. A. J. (1976) *Lands of the Alligator Rivers Area Northern Territory*, Land Research Series 38, CSIRO, Australia.

Strutt, N. (1970) *Modern Farming and the Soil*, Ministry of Agriculture, Fisheries and Food, HMSO, London.

Sturdy, R. G. (1971) *Soils in Essex 1. Sheet TQ59* (Harold Hill), Soil Survey Record No. 7, Harpenden.

Tanada, T. (1951) 'Certain properties of inorganic colloidal fractions of Hawaiian soils', *Journal of Soil Science*, **2**, 83–96.

Tavernier, R. and Marechal, R. (1962) 'Soil survey and soil classification in Belgium', *Transactions International Congress of Soil Science*, New Zealand, 298–307.

Tavernier, R. and Smith, G. D. (1957) 'The concept of Braunerde (brown forest soil) in Europe and the United States', *Advances in Agronomy*, **9**, 217–89.

Taylor, J. A. (1960) 'Methods of soil study', *Geography*, **45**, 52–67.

Taylor, N. H. and Cox, J. E. (1956) 'The soil pattern of New Zealand', *New Zealand Institute of Agricultural Science Proceedings*, 28–44.

Taylor, N. H. and Pohlen, I. J. (1968) 'Classification of New Zealand soils', in *Soils of New Zealand, Part 1*, New Zealand Soil Bureau Bulletin **26**(1), 15–46.

Thaer, A. D. (1853) *Grundsätze der rationellen Landwirtschaft*.

Thomas, R. W. and Huggett, R. J. (1980) *Modelling in Geography: A Mathematical Approach*, Harper and Row, London.

Thomasson, A. J. (1971) *Soils of the Melton Mowbray district*, Memoir of the Soil Survey of England and Wales, Harpenden.

Thomasson, A. J. (ed.) (1975) *Soils and Field drainage*, Technical Monograph No. 7, Soil Survey, Harpenden.

Thornton, I. and Webb, J. S. (1979) 'Geochemistry and health in the United Kingdom', *Philosophical Transactions of the Royal Society*, London, B288, 151–68.

Thorp, J. and Smith, G. D. (1949) 'Higher categories of soil classification: order, suborder and great soil groups', *Soil Science*, **67**, 117–26.

Tiurin, I. V. (1965) 'The system of soil classification in the USSR', *Pédologie*, Gand, numéro spécial 3, 7–24.

Troeh, F. R. (1964) 'Landform parameters correlated to soil drainage', *Proceedings of the Soil Science Society of America*, **28**, 808–12.

Trudgill, S. T. (1977) *Soil and Vegetation Systems*, Oxford University Press, Oxford.

Turner, J. (1970) 'Post-neolithic disturbance of British vegetation', in Walker, D. and West, R. G. (eds.), *Studies in the Vegetational History of the British Isles*, Cambridge University Press, Cambridge, 97–116.

Turner, J. (1975) 'The evidence for land use by prehistoric farming communities: the use of three-dimensional pollen diagrams', in *The Effect of Man on the Landscape: the Highland Zone*, Council for British Archaeology Report No. 11, 86–95.

Turner, J. (1979) 'The environment of northeast England during Roman times as shown by pollen analysis', *Journal of Archaeological Science*, **6**, 285–90.

Turrill, W. B. (1952) 'Some taxonomic aims, methods and principles', *Nature, London*, **169**, 388–93.

USDA (1971) *Estimates of Areas of Potentially Arable Land*, Soil Geography Unit, USDA, Mimeo.

USDD (1968) *Unified Soil Classification System for Roads, Airfields, Embankments and Foundations*, United States Department of Defense MI L-STD 619A.

Valentine, K. W. G. and Dalrymple, J. B. (1976) 'The identification of a buried paleosol developed in place at Pitstone, Buckinghamshire', *Journal of Soil Science*, **27**, 541–53.

Veldkamp, W. J. (1979) *Land Evaluation of Valleys in a Tropical Rain Area – A Case Study*, State Agricultural University, Wageningen, the Netherlands.

Vink, A. P. A. (1963) *Planning of Soil Surveys in Land Development*, Publication 10, International Institute for Land reclamation and Improvement, 55 pp., Wageningen, The Netherlands.

Vink, A. P. A. (1975) *Land Use in Advancing Agriculture*, Springer-Verlag, Berlin.

Vita-Finzi, C. (1969) *The Mediterranean Valleys*, Cambridge University Press, London.

Vreeken, W. J. (1973) 'Soil variability in small loess watersheds: clay and organic matter content', *Catena*, **1**, 181–96.

Vreeken, W. J. (1975a) 'Variability of depth to carbonates in fingertip loess watersheds in Iowa', *Catena*, **2**, 321–36.

Vreeken, W. J. (1975b) 'Principal kinds of chronosequences and their significance in soil history', *Journal of Soil Science*, **26**, 378–94.

Walker, D. (1966) 'The Late Quaternary history of the Cumberland lowland', *Philosophical Transactions Royal Society, London*, B **251**, 1–210.

Walker, P. H. (1966) 'Postglacial environments in relation to landscape and soils on the Cary drift, Iowa', *Iowa State University Experimental Station, Research Bulletin*, **549**, 838–875.

Walker, P. H. and Ruhe, R. V. (1968a) 'Hillslope models and soil formation. 2. Closed systems', *Transactions of the Ninth International Congress of Soil Science*, Adelaide, **4**, 561–8.

Walker, P. H., Hall, G. F. and Protz, R. (1968b) 'Soils trends and variability across selected landscapes in Iowa', *Soil Science Society of America Proceedings*, **32**, 97–101.

Walker, P. H., Hall, G. F. and Protz, R. (1968c) 'Relation between landform parameters and soil properties', *Soil Science Society of America Proceedings*, **32**, 101–4.

Warren, A. and Cowie, J. (1976) 'The use of soil maps in education, research and planning', in *Soil Survey Interpretation and Use, Welsh Soils Discussion Group Report* No. 17, Davidson, D. A. (ed.), 1–14.

Webster, R. (1968) 'Fundamental objections to the 7th approximation', *Journal of Soil Science*, **19**, 354–66.

Webster, R. (1975) 'Sampling, classification and quality control', in Bie, S. W. (ed.), *Soil Information Systems*, Proceedings of the meeting of the International Society of Soil Science Working Group on soil information systems, Centre for Agricultural Publishing and Documentation, Wageningen, the Netherlands, 65–72.

Webster, R. (1977a) 'Canonical correlation in pedology: How useful?', *Journal of Soil Science*, **28** (1), 196–221.

Webster, R. (1977b) 'Quantitative and numerical methods in soil classification and survey', Clarendon Press, Oxford.

Webster, R. and Beckett, P. H. T. (1968) 'Quality and usefulness of soil maps', *Nature, London*, **219**, 680–2.

Webster, R. and Cuanalo, de la C, H. E. (1975) 'Soil transect correlograms of north Oxfordshire and their interpretation', *Journal of Soil Science*, **26** (2), 176–94.

Webster, R., Lessells, C. M. and Hodgson, J. M. (1976) 'DECODE – A computer program for translating coded soil profile descriptions into text', *Journal of Soil Science*, **27** (2), 218–26.

Webster, R. and Wong, I. F. T. (1969) 'A numerical procedure for testing soil boundaries interpreted from air photographs', *Photogrammetria*, **24**, 59–72.

Weismiller, R. A., Persinger, I. D. and Montgomery, O. L. (1977) 'Soil inventory from digital analysis of satellite scanner and topographic data', *Soil Science Society of America Proceedings*, **41**, 1166–70.

Westerveld, G. J. W. and Van Den Hurk, J. A. (1973) 'Application of soil and interpretive maps to non-agricultural land use in the Netherlands', *Geoderma*, **10**, 47–66.

Weston, S. (1978) *Soil Survey Contracts and Quality Control*, Oxford University Press, England.

Whitfield, W. A. D. and Furley, P. A. (1971) 'The relationship between soil patterns and slope form in the Ettrick Association, south-east Scotland', in *Slopes: Form and Process*, Brunsden, D. (ed.), *Institute of British Geographers, Special Publication No. 3*, 165–7.

Wilde, S. A. (1953) 'Soil science and semantics', *Journal of Soil Science*, **4**, 1–4.

Wilkinson, B. (1968) 'Land capability – has it a place in agriculture', *Agriculture*, **75**, 343–7.

Wilkinson, B. (1974) 'Quantitative basis for land capability interpretation', *Land Capability Classification*, Technical Bulletin No. 30, MAFF, HMSO, London.

Wischmeier, W. H. and Mannering, J. V. (1969) 'Relation of soil properties to erodibility', *Soil Science Society of America Proceedings*, **33**, 131–7.

Wooldridge, S. W. (1948) 'The Anglo-Saxon settlement', Darby, H. C. (ed.), *An Historical Geography of England before 1800*, Cambridge University Press, Cambridge, 88–132.

Yaalon, D. H. (1963) 'On the origin and accumulation of salts in ground-water and in soils of Israel', *Bulletin of the Research Council of Israel*, 11G, 105

Yaalon, D. H. (1960) 'Some implications of fundamental concepts of pedology in soil classification', *Transactions of the Seventh International Congress of Soil Science*, Madison, **4**, 119–23.

Yaalon, D. H. (1965) 'Downward movement and distribution of anions in soil profiles with limited wetting', in *Experimental Pedology*, Hallsworth, E. G. and Crawford, D. V. (eds.), 157–64, Butterworths, London.

Yaalon, D. H. (1975) 'Conceptual models in pedogenesis: can soil-forming functions be solved?', *Geoderma*, **14**, 189–205.

Yaalon, D. H. (1971) 'Soil-forming processes in time and space', in *Paleopedology*, Yaalon, D. H. (ed.), 29–39, International Society of Soil Science and Israel Universities Press, Jerusalem.

Yaalon, D. H. (1971) 'Criteria for the recognition and classification of paleosols', in *Paleopedology: Origin, Nature and Dating of Paleosols*, International Society of Soil Science and Israel University Press.

Yaalon, D. H., Brenner, I. and Koyumdjisky, H. (1974) 'Weathering and mobility sequence of minor elements on a basaltic pedomorphic surface', *Geoderma*, **12**, 233–44.

Yaalon, D. H. and Ganor, E. (1973) 'The influence of dust on soils during the Quaternary', *Soil Science*, **116**, 146–55.

Yaalon, D. H. and Yaron, B. (1966) 'Framework for man-made soil changes – an outline of meta-pedogenesis', *Soil Science*, **102**, 272–7.
Yahia, H. M. (1971) *Soil and Soil Conditions in Sediments of the Ramadi Province (Iraq)*, Publicaties van het Fysisch Geografisch en Bodemkundig Laboratorium van de Universiteit van Amsterdam, No. 18.
Young, A. (1973) 'Soil survey procedures in land development planning', *Geographical Journal*, **139**, 53–64.
Young, A. (1976) *Tropical Soils and Soil Survey*, Cambridge University Press, Cambridge.
Young, A. (1979) 'Mapping the world's soils', *New Scientist*, 252–4.
Young, A. and Goldsmith, P. F. (1977) 'Soil survey and land evaluation in developing countries: a case study in Malawi', *Geographical Journal*, **143**, 407–31.

INDEX